SOLAR RECEIVERS FOR THERMAL POWER GENERATION

SOLAR RECEIVERS FOR THERMAL POWER GENERATION

Fundamentals and Advanced Concepts

Amos Madhlopa

Chancellor College, University of Malawi, Zomba, Malawi

ELSEVIER

ACADEMIC PRESS

An imprint of Elsevier

Academic Press is an imprint of Elsevier
125 London Wall, London EC2Y 5AS, United Kingdom
525 B Street, Suite 1650, San Diego, CA 92101, United States
50 Hampshire Street, 5th Floor, Cambridge, MA 02139, United States
The Boulevard, Langford Lane, Kidlington, Oxford OX5 1GB, United Kingdom

Notices

Knowledge and best practice in this field are constantly changing. As new research and experience broaden our understanding, changes in research methods, professional practices, or medical treatment may become necessary.

Practitioners and researchers must always rely on their own experience and knowledge in evaluating and using any information, methods, compounds, or experiments described herein. In using such information or methods they should be mindful of their own safety and the safety of others, including parties for whom they have a professional responsibility.

To the fullest extent of the law, neither the Publisher nor the authors, contributors, or editors, assume any liability for any injury and/or damage to persons or property as a matter of products liability, negligence or otherwise, or from any use or operation of any methods, products, instructions, or ideas contained in the material herein.

ISBN: 978-0-323-85271-5

For information on all Academic Press publications visit our website at https://www.elsevier.com/books-and-journals

Publisher: Charlotte Cockle
Acquisitions Editor: Edward Payne
Editorial Project Manager: Andrae Akeh
Production Project Manager: Prem Kumar Kaliamoorthi
Cover Designer: Christian Bilbow

Typeset by TNQ Technologies

Contents

Preface

Energy plays a vital role in the socio-economic development of any nation, and it can be produced by converting primary or secondary energy sources. For primary sources, energy is directly produced from the actual resource. Secondary energy sources are derived from primary sources. The conversion of primary energy to thermal power requires technologies that come in the form of heat engines.

At present, thermal power plants are predominantly driven by fossil resources. Nevertheless, it has been established that the burning of fossil fuels is contributing to climate change of the earth through the emission of carbon dioxide (a major greenhouse gas) into the atmosphere. Consequently, the international community has initiated various interventions, including the transformation of policy and regulatory instruments, to promote environmental protection. Some of these interventions include the United Nations Framework Convention on Climate Change (UNFCCC) and the Kyoto Protocol. For instance, at the 21st Conference of the Parties to the UNFCCC held in Paris in December 2015, delegates agreed to limit the global temperature rise below 2 K (2°C) above the preindustrial levels. Achievement of this goal will require significant reduction in greenhouse gas (GHG) emissions from different sources, including thermal power plants. Moreover, fossil fuels occur in finite quantities and are therefore depletable. In contrast, renewable energy sources are replenished through natural mechanisms and have generally low levels of GHG emissions. In this vein, solar radiation is one of the most important renewable energy sources, that can be converted to solar thermal power by using a suitable heat engine.

Solar thermal power generation is the use of solar thermal energy to generate electricity. The basic feature of a solar thermal power system is to capture heat from solar radiation. In this system, solar radiation is concentrated on a small receiving area to produce medium- to high-temperature heat, and such a thermal power system is commonly known as a concentrating solar power (CSP) system. It comprises three essential components to function: a concentrator, a receiver, and an engine cycle. The concentrator collects and focuses solar radiation on the receiver that converts concentrated solar radiation to heat. The generated heat is evacuated from the receiver to other components of the CSP system by using a heat transfer medium (often a fluid) and then it is used to drive an engine cycle, such as the Rankine or Brayton cycle in the power block, to produce electricity.

Thus, the receiver is a linchpin between the concentrator and the power block, and it significantly influences the performance of the whole CSP system. In this book, the terms "receiver" and "solar receiver" are used interchangeably.

CSP plants are classified into four main types: parabolic trough concentrator, linear Fresnel reflector, solar tower, and parabolic dish concentrator. The parabolic trough concentrator and linear Fresnel reflector focus solar radiation on a line receiver, while the solar tower and parabolic dish concentrators focus solar radiation on a point receiver. Generally, line focus receivers attain lower temperatures than point focus receivers. At present, the solar-to-electric efficiency of CSP systems is in the range of 8%–35%, which is relatively low. The thermodynamic efficiency of the engine cycle and the performance of whole CSP system increase with the receiver outlet temperature. Thus, the evolution of receivers has enormous potential to elevate the performance of CSP power plants.

Many studies have been conducted to improve the performance of solar receivers. Nevertheless, collated information on these advances is scarce in the form of a book. So, the aim of this book is to provide the reader with fundamental and advanced concepts of receivers. It is the first book to pool together the scattered pieces of knowledge on the various major aspects of receivers, and this is done with adequate illustrations and coherent treatment. Broad concepts of the CSP technology are introduced in Chapter 1 to provide the context of receivers while Chapter 2 focuses on solar radiation as the major source of heat supplied to the receiver. Different types of receivers have been developed through research over the past decades. Thus, classification of the receivers is necessary, and the main criteria for classifying them are presented in Chapter 3. The performance of different categories of receivers is affected by optical and thermal properties of materials. For instance, it is desirable that transparent cover surfaces should have high transmittance of the solar spectrum and high reflectance of the infrared radiation, while the solar absorber element of the receiver is supposed to have high absorptance for the solar spectrum and low emittance for the infrared band to optimize solar collection. Nevertheless, naturally occurring materials with these combinations of optical properties are scarce. Consequently, this has led to the development of new optical materials and concepts as shown in Chapter 4. Similarly, the characteristics of heat transfer media affect the receiver's thermal efficiency. So, Chapter 5 is dedicated to concepts of conventional and advanced heat transfer media. Part of the heat generated by the receiver can be stored for use when insolation is low or not available at all, thereby raising the capacity factor of the CSP system. For high system performance, storage media and systems must possess required characteristics (see Chapter 6). Thermodynamic parameters, such as the energy and exergy of receivers, also influence the overall performance of a CSP system. In view of this, the first major step in the CSP system design is thermodynamic analysis, and the thermodynamic parameters of receivers are presented in Chapter 7. Convective heat transfer is the most important mode of heat transportation from the receiver to other CSP components, and it involves flow of fluids in a receiver. New heat transfer media, such as nanofluids and suspended solid particles, have been developed to improve the rate of heat transportation and, therefore, the thermal efficiency of receivers. However, some of the innovative heat transfer media are presenting new hydrodynamic challenges to the scientific research community. Consequently, Chapter 8 has been included to provide the reader with basic and advanced concepts of receiver hydrodynamics. A high temperature of the heat transfer medium, as it exits the receiver, is desirable. Theoretically, some designs of receivers can achieve temperatures of up to 1773 K, which is suitable for driving advanced thermodynamic cycles with high thermal efficiency, but most of the CSP plants in the world are currently operating at temperatures below 900 K due to the limitation of material characteristics at elevated temperatures. Therefore, the current research direction is to develop materials and concepts that can support operation of receivers at higher temperatures (see Chapter 9). The performance of receivers is influenced by multiple factors, and the impetus to advance receivers is resulting in complex configurations and processes that require suitable computational tools for designing this CSP component. In view of this, featured in Chapter 10 are methods of modeling the optical and thermal performances of different receivers. Finally, results obtained from the process of modeling are often used in constructing prototype systems. However, mathematical models and the process of manufacturing (fabrication) have certain degrees of accuracy. It is therefore imperative to test any new receiver design by using standard methods for testing solar collecting systems to establish its actual performance. In this vein, the last chapter of this book (Chapter 11) contains selected standard test methods and progress in the development of receivers. This chapter includes case studies, and challenges and opportunities around the world.

Acknowledgments

Peer reviewing is an important component of academic publishing. In this vein, I am very grateful to reviewers for their useful suggestions. My sincere thanks to Sellina (my wife) and our children for their love, care, and moral support.

Chapter 1

Introduction to concentrating solar power

1.1 Introduction

Concentrating solar power (CSP) is one of the most promising renewable energy technologies for electricity production. It comprises the solar field (concentrator and solar receiver) and power block. The solar receiver is a linchpin between the concentrator and the power block, and it has significant influence on the performance of the entire CSP system. In view of this, a lot of research attention has been given to the development of solar receivers. In this chapter, broad concepts about CSP technology are introduced. Section 1.1.1 provides a general sustainability context of electricity generation. It is argued that production of electricity from renewable energy is a promising sustainable pathway for development, and solar radiation has enormous potential for sustainable electricity generation. Major technologies for solar-to-electric conversion are photovoltaic (Section 1.1.2) and CSP (Section 1.1.3).

In a CSP system, the concentration of solar radiation is achieved through the utilization of a concentrator (Section 1.2). Concentrated solar radiation is incident on a solar receiver, which transforms it to heat. Therefore, a solar receiver is one of the most critical components of a CSP system, and it is introduced in Section 1.3. It is also a well-known fact that solar radiation is an intermittent resource, which results in relatively low capacity factors of solar technologies. So, thermal storage and backup heating are required to enhance the dispatchability of electricity. Particularly, thermal storage is an outstanding advantage of CSP over other generation technologies that are based on intermittent renewable energy resources. Therefore, the concepts of thermal storage and backup heating are presented in Section 1.4. Heat produced by the receiver is transferred to the power block to generate electricity. In this vein, the concepts of the power block are examined in Section 1.5. Energy efficiency is an important metric for assessing the performance of electricity generation technologies. So, Section 1.6 deals with system efficiency. Finally, the various types of the CSP technology are presented in the last section of this chapter (Section 1.7).

Solar Receivers for Thermal Power Generation. https://doi.org/10.1016/B978-0-323-85271-5.00003-3

1.1.1 Sustainable production of electricity

Electricity can be generated from different primary energy sources by using a variety of energy conversion technologies. So, there are different competing technologies in the electricity market (Kim et al., 2019), and technologies with lower costs are most preferred from an affordability standpoint. At present, most of the electricity is globally generated by thermal power plants, and these plants are predominantly driven by fossil fuels (coal, oil, and natural gas), as shown in Fig. 1.1 (International Energy Agency, 2020). Overdependence on fossil fuels is contributing to environmental degradation, and climate change is the major environmental challenge of the 21st century. It has been established that anthropogenic activities are generating greenhouse gases (GHGs) that account for most of the rise in ambient air temperature (Saikku et al., 2008). The burning of fossil fuels is significantly contributing to climate change through the emission of carbon dioxide (major GHG) and other substances (IPCC, 2007). Consequently, the goal of augmenting access to electricity is

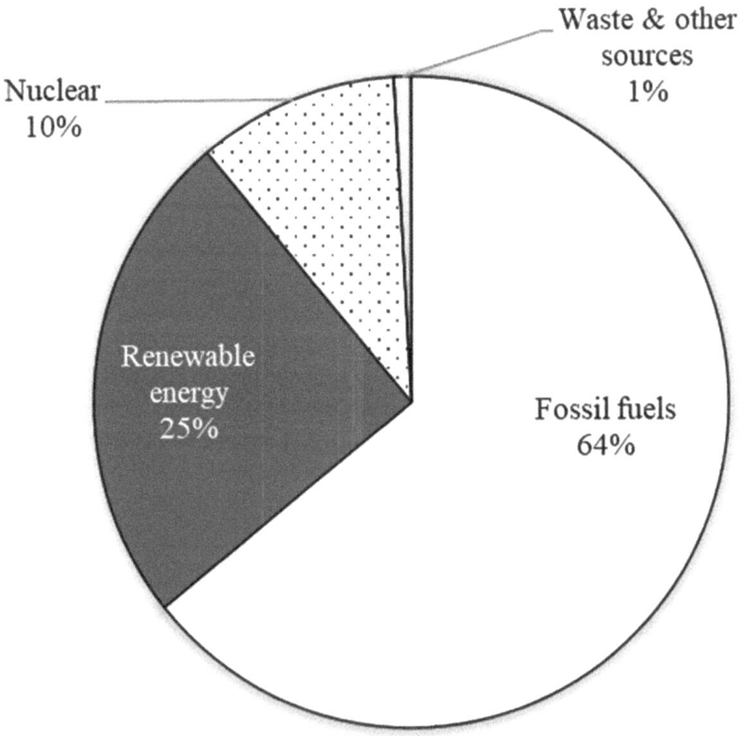

FIGURE 1.1 Global production of electricity by primary energy. *Drawn by author using data from International Energy Agency. (2020).*

also linked to climate change policies (van Vuuren et al., 2012) that aim at achieving a balance between anthropogenic emissions by sources and removals by sinks of GHGs by the second half of the 21st century at the global level, including decarbonization of the electricity sector. In this connection, it is argued that carbon-neutral electricity can compete against the largest carbon emitter (coal power) if its generation costs fall below approximately 0.08USD/kWh (International Renewable Energy Agency, 2013). Moreover, fossil fuels are depletable, which poses insecurity to the energy system. It has also been reported that the dominance of a single energy source and system is ultimately unsustainable even if it may be perfect at one time (Li, 2005). In contrast, renewable energy has low intensity of carbon emissions and replenishes itself by natural mechanisms. Consequently, it is perceived as a more sustainable option for developing the electricity sector.

Among the available renewable energy resources in the world, solar radiation has the greatest potential for electricity generation (Pillot et al., 2019), and the major technologies for converting solar energy to electricity are photovoltaic (PV) and CSP.

1.1.2 Photovoltaic technology

A photovoltaic (PV) technology directly converts solar radiation to electricity by using the principle of photoelectric effect (Khan & Arsalan, 2016). The basic building block of the PV system is a solar cell (Fig. 1.2), which is made up of p-type and n-type semiconductor materials (Tyagi et al., 2013). Incident solar radiation comprises discrete energy packets called photons, and each photon contains energy that depends on the frequency of the radiation. The n-type layer absorbs part of the incident photons, resulting in the creation of excess electrons that flow through the external circuit to the p-type electrode. Thus, conventional current flows from the p-type electrode through the external circuit to the n-type electrode. A group of cells are interconnected to form a module, and modules are assembled to produce an array. The efficiency of PV modules can vary between 10% and 30% (Aberoumand et al., 2018).

To mitigate the intermittency of the solar resource, PV systems can be integrated with storage batteries. Thus, many microgrid projects based on PV farms have lately been developed, but only a limited number of them possess large-scale energy storage systems because of the high cost of batteries (Cen et al., 2018). Due to these constraints, it is perceived that increasing onsite consumption of PV-generated electricity will become important to preserve the stability of the grid when the global PV market exceeds 76 GW (Akbari et al., 2019). Nonetheless, the demand profile of electricity may not match that of the fluctuating supply from PV power plants without storage. In contrast, the development of CSP with thermal storage has already attained large-scale commercial production of electricity.

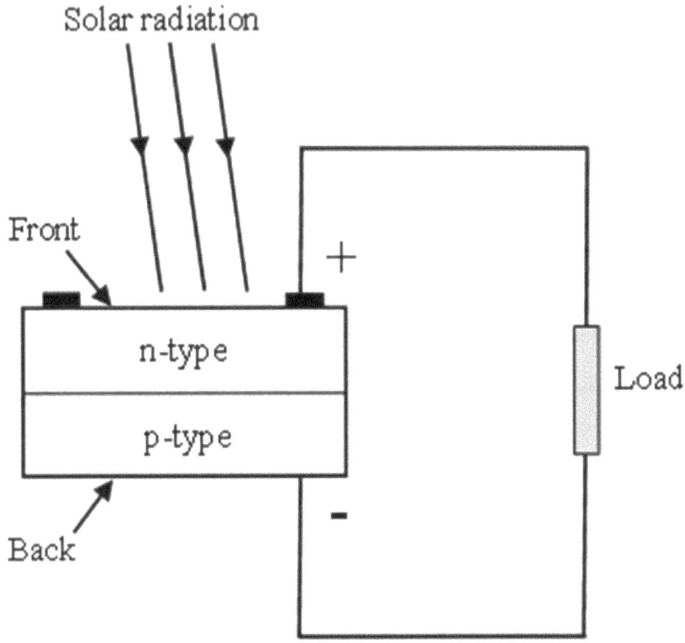

FIGURE 1.2 Schematic representation of a solar cell with an external electric load.

1.1.3 Concentrating solar power

Solar radiation can be converted to solar thermal power by using a suitable technology, and the use of solar thermal energy to generate electricity is known as solar thermal power generation. The concentrating solar power (CSP) technolgy generates electricity by utilizing heat from solar radiation (Dowling et al., 2017). A basic CSP system comprises a solar field and a power block, and the solar field is made up of a concentrator and a receiver (Fig. 1.3). The receiver converts concentrated solar radiation to heat. Part of this heat is transferred to the engine, while a fraction of it may be stored and utilized later during the night (or when the level of insolation is low). It is also possible to integrate a CSP power plant with a backup heater in order to enhance the capacity factor of the plant (see Section 1.4). Research is still going on to improve the performance of the CSP system.

Technology is developed in stages from conceptualization of the idea to commercialization, and the degree of maturity increases with stage number. Technology maturity is a good indicator of the closeness of a technology to commercial application (Selman & Chen, 2012). In this context, CSP is a

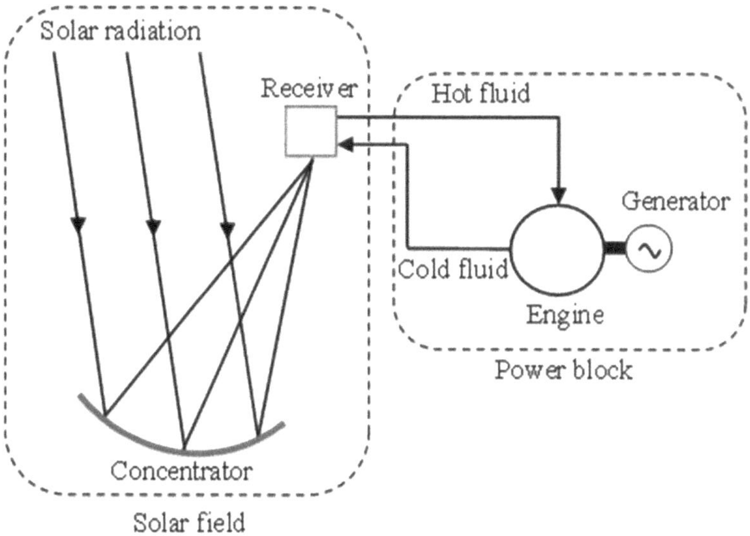

FIGURE 1.3 Components of a basic concentrating solar power system.

matured technology, with commercial power plants operating in different parts of the world. The weighted average levelized cost of CSP-based electricity dropped from 0.346 USD/kWh in 2010 to 0.182 USD/kWh in 2019 and predicted to decline further to 0.075 USD/kWh for projects commissioned in 2021 (International Renewable Energy Agency, 2020b). These data show that CSP is currently uncompetitive with coal (0.080 USD/kWh), but it is expected to attain parity with this fossil fuel in future. Nevertheless, CSP is competitive with other fossil-fuel technologies such as gas turbines and diesel generators at present (Labordena et al., 2017). In this vein, the global cumulative capacity of the CSP technology has been increasing over the past decade, reaching a level of 6289 MW in 2019 as shown in Fig. 1.4 that is based on data from (International Renewable Energy Agency, 2020a).

1.2 Concentrator

A concentrator is an optical device that focuses solar radiation from the sun onto a small area of the receiver. It is made up of a set of reflecting mirrors (or lenses). The plane opening of the concentrator through which solar radiation passes is referred to as an aperture. In this vein, the aperture of a point concentrator is characterized by the diameter of the opening, while the aperture of a linear concentrator is characterized by the width of the opening.

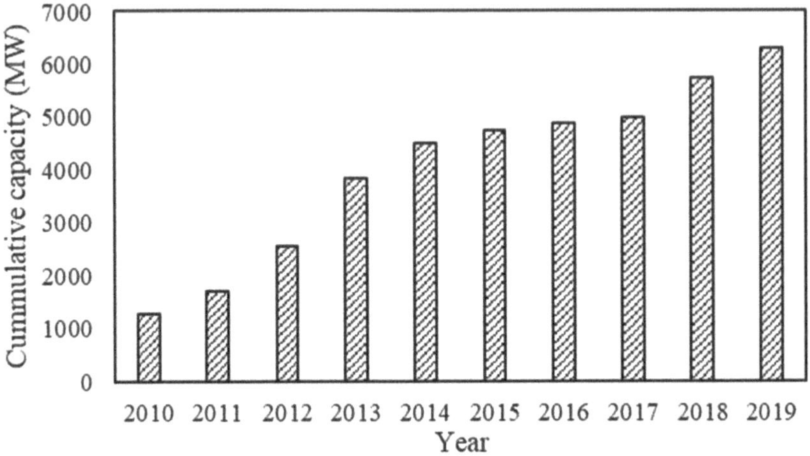

FIGURE 1.4 World cumulative capacity of concentrating solar power. *Drawn by author using data from International Renewable Energy Agency. (2020a).*

1.2.1 Classification of concentrators

Concentrators are classified into imaging and nonimaging categories. In imaging systems, parallel rays from the sun are focused on a line or point. Linear concentrators such as the parabolic trough and linear Fresnel reflector (LFR) are imaging concentrators that focus solar radiation on a line. Linear imaging concentrators with parabolic cross section are exploited in applications requiring intermediate concentration ratios and temperatures. Circular concentrators such as the parabolic dish and solar tower (ST) are imaging concentrators that focus solar radiation on a point, and these systems are utilized in applications that require high concentration ratios and temperatures.

On the other hand, nonimaging concentrators do not form an image of the sun, and they have a low concentration ratio. Consequently, they are not suitable for use in industrial applications that require a high range of temperature such as driving engine cycles to generate electricity. These concentrators can operate seasonally or annually with negligible need of tracking. In addition, nonimaging concentrators can reflect to the receiver all of the incident solar radiation on the aperture across ranges of incidence angles within broad limits (Duffie & Beckman, 2013). The acceptance angle of the concentrator is defined by these limits. Even the diffuse radiation within these angles contributes to the input solar energy to the collector because all the radiation incident within the acceptance angle is reflected to the receiver. A compound parabolic concentrator is an example of nonimaging concentrators. This system can harness both beam and diffuse components of solar radiation, and it can be operated without sun tracking.

1.2.2 Concentration ratio

The concentration ratio of a concentrator (C_r) can be defined in two ways (Ameer & Shahad, 2017). The first definition is based on the geometry of the concentrator, and it is therefore called the geometric or area concentration ratio, while the second definition is based on the radiation intensity, and it is known as the intensity or flux concentration. The area concentration ($C_{r,area}$) is the ratio of the effective area of the aperture to the surface area of the absorber:

$$C_{r,area} = \frac{A_{ap}}{A_{sr}} \tag{1.1}$$

Where A_{ap} is the effective area of the aperture and A_{rec} is the surface area of the solar receiver.

The upper limit of $C_{r,area}$ depends on the number of dimensions of concentration. A two-dimensional (linear) concentrator such as a parabolic trough focuses radiation on a line, while a three-dimensional concentrator such as a parabolic dish focuses radiation on a point. There is an upper limit of the concentration ratio, and this limit is derived from the second law of thermodynamics applied to the radiative heat exchange between the sun and the receiver (Rabl, 1976). Suppose a circular concentrator with aperture area (A_{ap}) focuses radiation from the sun of radius (r) onto a receiver area (A_{rec}) as shown in Fig. 1.5. The concentrator is located at distance R from the sun that subtends an angle $2\theta_s$ on the concentrator.

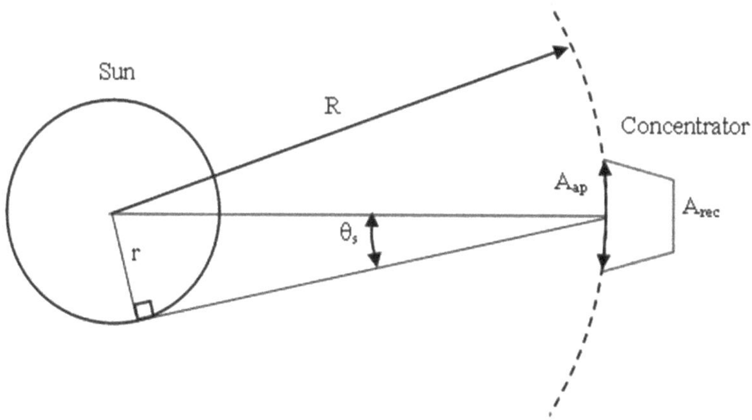

FIGURE 1.5 Schematic representation of sun at a temperature T_s and distance R from a concentrator with aperture area A_{ap} and receiver area A_{rec}. *Adapted from Rabl (1976).*

Assuming the concentrator is perfect, then the aperture intercepts a proportion of the radiation emitted by the sun. For approximate analysis, the sun can be assumed to be a blackbody at a temperature $T = T_s$:

$$Q_{s-rec} = A_{ap}\frac{r^2}{R^2}\sigma_{sb}T_s^4 \tag{1.2}$$

A perfect receiver (blackbody) radiates energy equal to $A_{rec}\sigma_{sb}T^4$, and a proportion of this ($E_{rec\text{-}s}$) reaches the sun.

$$Q_{rec-s} = A_{rec}\,\sigma_{sb}\,T_{rec}^4 E_{rec-s} \tag{1.3}$$

When T_{rec} and T_s are not changing, then, based on the second law of thermodynamics, $Q_{s\text{-}rec} = Q_{rec\text{-}s}$.

Consequently, Eq. (1.4) can be derived from Eqs. (1.2) and (1.3):

$$\frac{A_{ap}}{A_{rec}} = \frac{r^2}{R^2}E_{rec-s} \tag{1.4}$$

Nevertheless, the maximum value of $E_{rec\text{-}s}$ is one and so the maximum value of the concentration ratio for circular concentrators:

$$\left(\frac{A_{ap}}{A_{rec}}\right)_{circular,max} = \frac{r^2}{R^2} = \frac{1}{\sin^2\theta_s} \tag{1.5}$$

The maximum concentration ratio for linear concentrators can be derived using a similar approach to obtain:

$$\left(\frac{A_{ap}}{A_{sr}}\right)_{linear,max} = \frac{1}{\sin\,\theta_s} \tag{1.6}$$

The sun subtends a maximum angle of about $2\theta_s = 0.533$ degrees on the earth. Therefore, the half-angle $\theta_s = 0.27$ degrees and:

$$\left(\frac{A_{ap}}{A_{rec}}\right)_{circular,max} = 45000$$

$$\left(\frac{A_{ap}}{A_{rec}}\right)_{linear,max} = 212$$

The flux concentration ratio ($C_{r,flux}$) is the ratio of the intensity of radiation on the aperture to that on the receiver.

$$C_{r,flux} = \frac{G_{rec}}{G_{ap}} \tag{1.7}$$

Where G_{ap} is the intensity of radiation on the aperture (Wm^{-2}), and G_{rec} is the intensity of radiation on the receiver (Wm^{-2}).

Both and G_{rec} need to be measured in order to determine the flux concentration ratio empirically. Measurement of G_{ap} is easy but not G_{rec} because of the high operating temperatures of solar receivers. Consequently, the area concentration ratio is commonly utilized in design and performance analyses of CSP systems. Values of the concentration ratio vary from one for a flat-plate collector to a few thousands for the parabolic dish concentrator (PDC). Usually, high values of C_r are achieved by increasing the area of the aperture. As a result of this, a large land surface is required for CSP technology. For instance, the aperture area is $A_{ap} = 2{,}600{,}000\ m^2$ for the Ivanpah Solar Electric Generating System with a capacity of 392 MW in the United States of America (see Section 1.7.3).

Example 1.1

A solar tower has a concentration ratio of 1000, and the effective beam radiation on the aperture is 600 Wm^{-2} at a certain time of the day. Calculate the flux of solar radiation incident on the receiver.

Solution

From Eq. (1.7),

$$\frac{G_{rec}}{G_{ap}} = C_{r,flux}$$

This implies that, $G_{rec} = C_{r,flux}G_{ap} = 1000*600\ Wm^{-2} = 600\ kWm^{-2}$

1.3 Solar receiver

A solar receiver is a device that converts concentrated solar radiation to heat, which is transferred to a heat transfer fluid that flows through the receiver (Fig. 1.6). Part of the incident concentrated solar radiation is reflected into the environment. For receivers with a transparent cover on the aperture, the intercepted radiation undergoes reflection, absorption, and transmission on the cover surface, thereby contributing to optical losses. Thus, the performance of a receiver is influenced by optical and thermal factors.

1.3.1 Energy balance of solar receiver

Concentrated solar radiation is incident on the receiver that absorbs a fraction of the radiation, while the rest of it is lost through optical processes. It is necessary for the concentrator and receiver to have low optical losses in order to achieve high efficiencies (see Section 1.6). The absorbed flux of solar

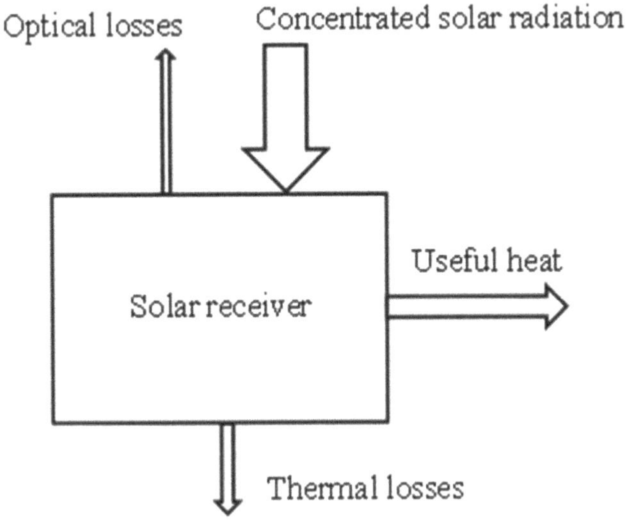

FIGURE 1.6 Schematic representation of energy balance in a solar receiver.

radiation is transformed into heat, and a fraction of this thermal energy is lost through conductive, convective, and radiative modes of heat transfer. Thermal losses are often estimated from the coefficient of heat loss and temperature gradients (Duffie & Beckman, 2013). Temperature gradients in the receiver can be accounted for by the collector heat removal factor, and it can be assumed that losses are independent of the intensity of the incident radiation. This assumption may not be quite correct if a transparent cover absorbs a significant proportion of the concentrated solar radiation. Consequently, cover materials need to have high transmittance but low absorbance characteristics. In addition, cover materials need to withstand high temperatures that prevail in receivers. Due to the high operating temperatures of receivers, most of the heat is lost to the ambient environment via the mode of radiative heat transfer, and the coefficient of heat loss is dependent on the operating temperature. The difference between input solar power and losses yields useful heat that is available to drive the engine cycle. Useful heat is transported by the working fluid to directly or indirectly drive the engine cycle. In the direct mode of heat transfer, the working fluid is used as a heat transfer medium from the receiver to the power block (Fig. 1.7A). In the indirect mode, the solar receiver heats up a heat transfer fluid that in turn heats up the temperature of the working fluid via a heat exchanger (Fig. 1.7B).

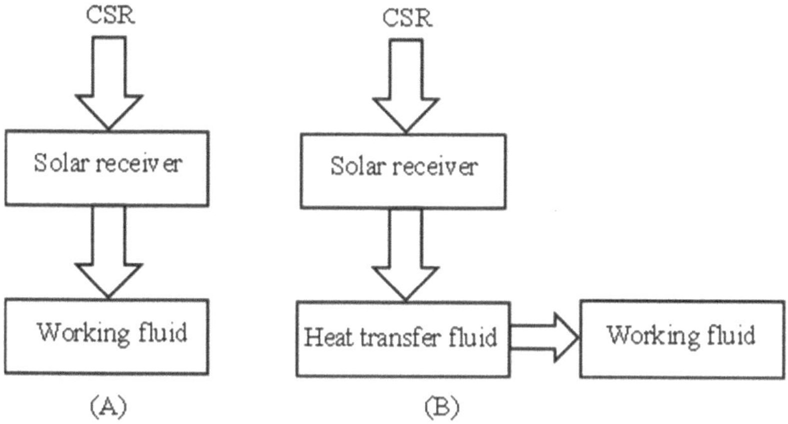

FIGURE 1.7 Mode of heat transfer between solar receiver and working fluid: (A) direct heat transfer and (B) indirect heat transfer. *CSR*, concentrated solar radiation.

1.3.2 Heat classification and operation of solar receivers

Heat can be classified into low-temperature ($T < 373$ K), medium-temperature ($373 \leq T < 673$ K) and high-temperature ($T \geq 673$ K) grades (Forman et al., 2016). Solar receivers can raise the temperature of a working fluid up to 393–1773 K (Zhang et al., 2013). Consequently, receivers produce heat at medium to high temperatures. Nevertheless, the radiative and convective heat losses of a receiver increase with operating temperature (Ho, 2017). Another drawback of high operating temperatures is the limitation of material properties. It is a challenge to find suitable materials for construction of receivers and other components that are exposed to high temperatures. For example, synthetic oil such as Therminol VP-1 (which can operate up to 673 K) is used as a heat transfer fluid in most parabolic trough concentrator (PTC) power plants at present. Molten salt is commonly used as a heat transfer fluid as well as a storage medium. However, this medium can only be applied in the temperature range of 533–894 K (Zhang et al., 2013), which limits the operating temperature to 894 K even though the CSP technology is capable of achieving higher temperatures.

The heat transfer medium influences the receiver efficiency, type, and performance of a thermodynamic cycle, and kind of thermal storage to utilize (Benoit et al., 2016). Conventionally, heat transfer can be achieved through direct steam generation, compressed gases, molten salts, and thermal oil (see Chapter 5). Advanced concepts of heat transfer media include super critical steam and carbon dioxide, metallic liquids, nanofluids, and suspended solid particles. For example, it is possible to increase the temperature of the particle medium above 1273 K (Ho, 2017). Production of dispatchable electricity is also necessary, and, in this vein, thermal storage and backup heating are promising thermal processes of enhancing the reliability of CSP systems (see Section 1.4).

1.4 Enhancement of capacity factor

Power plants do not operate all the time in a year. There are periods when a plant is idle for various reasons, thereby generating less electricity than the nominal annual production level. In this regard, capacity factor is a measure of the proportion of time that a power plant produces electricity at full power (Izquierdo et al., 2010). This parameter varies with location and time because of the following factors: a) planned and unplanned outages of units of power plants for maintenance and repair, b) daily electricity load patterns, c) operating costs, d) efficiency of power plants, and e) climatic conditions for renewable resources. Coal power plants produce electricity for 24 h without daily start-up and shutdown (Kim et al., 2019), and so their capacity factors are relatively high. A capacity factor of 0.60 (or higher) shows that the power plant is operated at off-peak loading conditions (De Oliveira-De Jesus, 2019). For intermittent renewable energy technologies, capacity factors vary between 0.20 and 0.40 (Green et al., 2015). Consequently, these technologies require an energy storage or backup heating to augment their capacity factors, and the CSP system is no exception.

1.4.1 Thermal storage

It is possible to integrate a CSP system with a thermal storage component in order to increase its capacity factor (Fig. 1.8). Common techniques of thermal storage are sensible, latent, and thermochemical heat with different configurations of receivers and storage components (see Chapter 6). With storage, the capacity factor of CSP can exceed 50% (Green et al., 2015), which shows that CSP has great potential to supply a baseload. Therefore, many recent CSP power plants are equipped with storage components.

A total of 56 out of 14 power plants operating by mid-2020 were equipped with thermal storage. The largest power plant (250 MW) equipped with thermal storage is the Solana PTC Generating Station (Solana) in the United States of America. This plant utilizes molten salt as a heat transfer fluid and a thermal storage medium with a storage capacity of 6 h. The Atacama-1 ST power plant (110 MW), under construction in Chile, exhibits the largest storage capacity (17.5 h). Nonetheless, the intensity of solar radiation is sometimes so diminished that the temperature of the storage medium may rise by an insignificant margin to provide the required heat for the engine, thereby limiting the continuity of electricity production. Consequently, there may be a need for backup heating in order to achieve the required capacity factor of a plant.

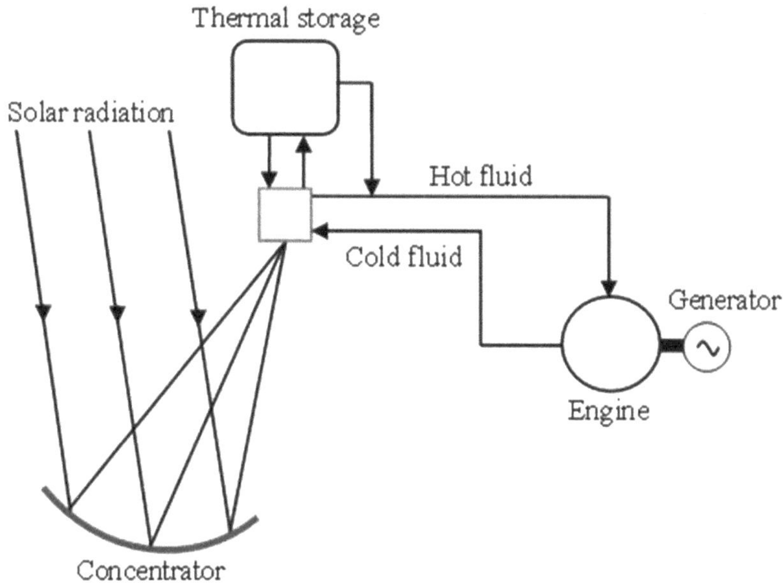

FIGURE 1.8 Schematic representation of a concentrating solar power system with thermal storage.

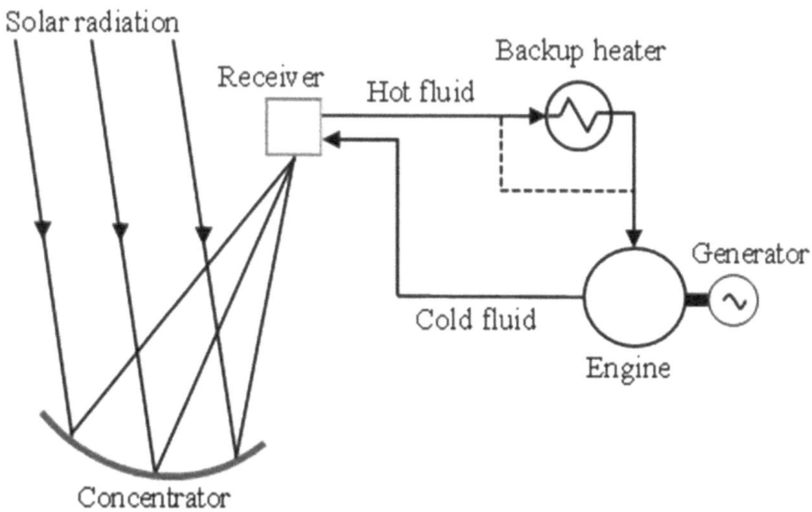

FIGURE 1.9 Schematic representation of a basic concentrating solar power system with the backup heater.

1.4.2 Backup heating

In a CSP system with backup heating, the receiver and other sources provide the required heat to drive the engine cycle (Fig. 1.9). In this case, the hot fluid from the receiver is further heated by a backup heater to achieve the stipulated inlet temperature of the engine. Depending on weather conditions, the fluid can by-pass the backup heater when its exit temperature from the receiver is comparable to the inlet temperature of the engine, thereby saving fuel (in case of fuel-based backup heating).

Worldwide, some operating CSP power plants are equipped with backup heaters. By mid-2020, 13 out of 14 operating power plants (12.5%) were equipped with backup heaters, which is significantly lower than the proportion of power plants equipped with storage components (53.8%). Of the 13 plants with backup heating, 10 are fueled by natural gas, and the Solana Generating Station (located in Arizona, United States of America) is the largest operating plant (250 MW) with a backup heater (fueled by natural gas). Clearly, the exploitation of fossil-fueled backup heaters still contributes to environmental pollution. In this context, biofuel has enormous potential to sustainably back up CSP systems. For instance, the thermodynamic analysis of a hybrid solar-biogas open cycle gas turbine (OCGT) power plant with recuperation has shown the benefits of fuel saving and increase in the availability of thermal output (Cameretti et al., 2015). Biogas-driven microgas turbines are in the most advanced stage of development, but this technology has not yet reached the commercial level of technology maturity (Okoroigwe & Madhlopa, 2019). Moreover, there is a limited number of power plants with biomass backup heating. By mid-2020, the Borges Termosolar CSP power plant (25 MW) was the largest operating CSP plant with a biomass backup heater.

1.5 Power block

Main components of a basic power block are a heat engine and an electric generator. A heat engine is a closed-cycle device that extracts heat from a hot reservoir, performs useful work, and rejects heat to a cold reservoir (heat sink). The engine transforms heat to mechanical work. In combustion engines, the required thermal energy is generated by burning fuels, and there are two types of these engines: internal combustion (IC) and external combustion (EC). The combustion of fuel takes place with an oxidizer in a combustor that forms part of the flow circuit of the working fluid of an IC engine. Some examples of IC engine cycles are the Brayton (gas turbine), Otto, and Diesel cycles. The working fluid of an EC is heated by burning fuel in an external combustion unit, and the required heat is transferred to the working fluid via a heat exchanger. Examples of EC cycles include the Rankine, Stirling, and Kalina cycles. Fundamentals of engine cycles are discussed elsewhere in books on thermodynamics but the Rankine, Bryton, Stirling, and Kalina cycles are briefly presented in this book because of their potential of exploitation in CSP systems. The Carnot cycle is also included as a reference point. This cycle gives the maximum thermal efficiency when operated under the same heat source and sink as any other cycle.

Engine cycles are affected by thermodynamic properties, and the commonest of these properties are temperature, pressure, and specific volume (or density) (ASHRAE, 2017). Temperature is the degree of hotness of a system, while pressure is the amount of force per unit area. Specific volume is a measure of the volume per unit mass. Other thermodynamic properties can be derived from these fundamental properties. Entropy (s) and enthalpy (i) are also important factors in thermodynamic systems. Entropy is a measure of the molecular disorder of a system. This parameter increases with the extent of mixing (an unmixed system exhibits diminished levels of entropy). Enthalpy includes internal energy (e) and flow work (Pv). Internal energy is produced by the movement of molecules and/or intermolecular forces, and it is affected by the work and heat flow of the system. Work done by the system decreases internal energy, while work done on the system raises the level of internal energy. Similarly, heat flow out of the system reduces internal energy while heat flow into the system elevates internal energy. In a solar receiver, heat is added to the working (or heat transfer) fluid, resulting in the rise of its internal energy. The hot working fluid does work, and it may also lose some of the heat as it expands through a turbine. Consequently, the internal energy of the working fluid at the turbine outlet is lower than the inlet internal energy at the turbine inlet. Flow work is the amount of energy transferred into the system boundary because of a pumping process that takes place outside the system, causing a fluid to flow into the system. The relation between internal energy and flow work is given by:

$$i = e + Pv \tag{1.8}$$

Where P is pressure and v is specific volume of the system.

In a heat engine, the engine–fuel interaction passes through different stages from one thermodynamic state to another and back to its initial state, thereby constituting an engine cycle. In this connection, the working fluid is an important component of the cycle, and it changes from one state to another within the cycle (such as sub-cooled liquid, liquid-vapor, or superheated vapor states) as shown in Fig. 1.10. The fluid attains a critical state at critical pressure and temperature. For example, the critical temperature and pressure of water are, respectively, 647.3 K and 22.1×10^6 Pa (Dunham & Iverson, 2014). It is not possible to liquefy the vapor of a fluid above its critical temperature, regardless of the magnitude of the pressure.

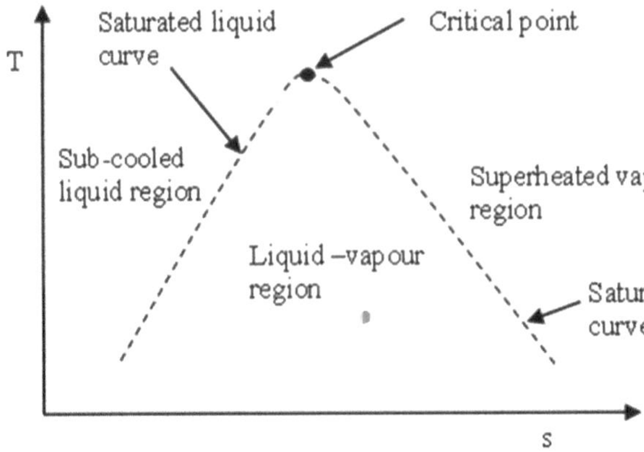

FIGURE 1.10 Temperature—entropy (T—s) diagram of a generalized fluid.

1.5.1 Carnot cycle

A Carnot cycle is an ideal engine cycle in which heat is extracted from a hot source at a constant temperature (T_h) and rejected to a cold sink at a constant temperature (T_c) as shown in Fig. 1.11A. In this cycle, there are two reversible isothermal processes at $T = T_h$ and $T = T_c$ and two reversible isentropic processes (Fig. 1.11B). For condensable vapor, the two isothermal processes can be driven through isobaric heating and cooling, within the region in which the vapor is wet. The temperature of the working fluid does not change in an isothermal process. An isentropic process proceeds at constant entropy of the working fluid, while the pressure of the working fluid remains constant in an isobaric process.

The efficiency of a Carnot engine (η_{car}) can be given by:

$$\eta_{car} = \frac{W}{Q_h} = \frac{T_h - T_c}{T_h} \tag{1.9}$$

Where W is work, T_h is the temperature of a hot reservoir, T_c is the temperature of a cold reservoir and $T_h > T_c$.

The efficiency of the Carnot engine is the maximum efficiency of converting the extracted heat into work by an engine that operates in thermodynamic contact with a cold reservoir. Consequently, the thermal efficiency of any other engine is less than that of the Carnot engine operating between the same reservoirs. It should also be noted that the efficiency of the Carnot engine draws near to unity as T_h approaches infinity. In practical heat engines, T_h does not exceed the blackbody temperature at which the sun emits radiation ($T_s = 6000$ K).

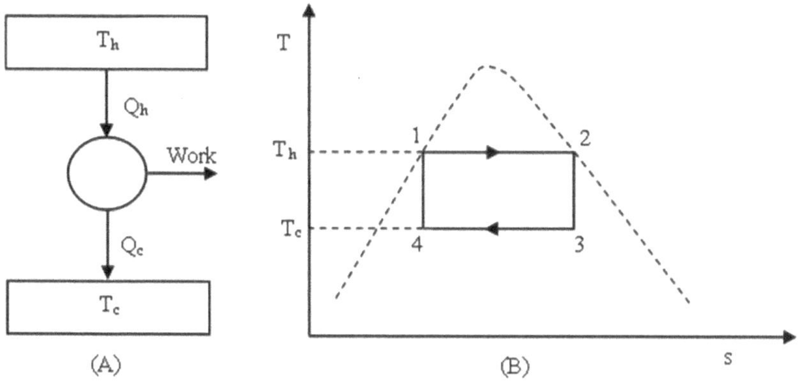

FIGURE 1.11 (A) Energy flow and (B) temperature—entropy (T—s) diagram of a Carnot cycle.

Example 1.2

A Carnot engine is operating between hot (at 800 K) and cold (at 300 K) reservoirs. Calculate the efficiency of the engine a) at these temperatures of the hot and cold reservoirs, b) when the temperature of the hot reservoir is increased by 100 K, and c) when the temperature of the cold reservoir is reduced by the same margin (100 K). Which is more effective, increasing the temperature of the hot reservoir or decreasing the temperature of the heat sink by the same margin?

Solution

From Eq. (1.9),

$$\eta_{car} = \frac{T_h - T_c}{T_h}$$

$$(a)\eta_{car} = \frac{800 - 300}{800} = 0.625$$

$$(b)\eta_{car} = \frac{900 - 300}{900} = 0.667$$

$$(c)\eta_{car} = \frac{800 - 200}{800} = 0.750$$

Increasing the temperature of the hot reservoir is less effective than reducing the temperature of the cold reservoir by the same margin.

1.5.2 Rankine cycle

A basic Rankine cycle comprises a heat source, a turbine, a condenser, and a feed pump (Fig. 1.12), and the cycle utilizes water as a working fluid. Steam and gas turbines belong to a group of machines called turbomachines. In a turbomachine, mechanical energy in the form of shaft work is transferred to or from a continuously flowing fluid by the dynamic action of rotating turbine blades that are arranged in rows. Depending on whether energy is transferred to or from the fluid, turbomachines fall into two categories: a) power producing and b) power absorbing. Power producing turbomachines extract energy, via a rotor system, from a moving fluid with high kinetic energy to produce shaft work and a fluid with low kinetic energy. Steam, gas, and wind turbines are examples of turbomachines. In contrast, power-absorbing turbomachines transfer energy from shaft work to a low-energy fluid, through an impeller, to produce a moving fluid with high contentment of kinetic energy. Examples of power absorbing turbomachines include compressors, fans, propellers, and pumps.

At the beginning of a Rankine cycle (stage 1–2), water is pumped into the heat source unit where it is heated to generate steam as shown in Fig. 1.12. In this mode of heat transfer, the amount of thermal energy added to the engine cycle is equal to the useful heat from the receiver. In hybrid CSP systems, the steam may be reheated in a boiler before expanding through the turbine to reduce shaft work. This mechanical energy is then used to run the electric generator. The Rankine cycle comprises: a) isentropic water compression (stage 1–2), b) isobaric addition of heat to produce steam in the receiver (stage 2–3), c) isentropic expansion of the steam (stage 3–4), and d) isobaric steam condensation (stage 4-1). In a standard Rankine cycle, output work is produced during the isentropic expansion of the steam (stage 3–4), while the feed pump provides input work during the isentropic compression of the working fluid. In a CSP system, the receiver adds heat to the working fluid to produce steam

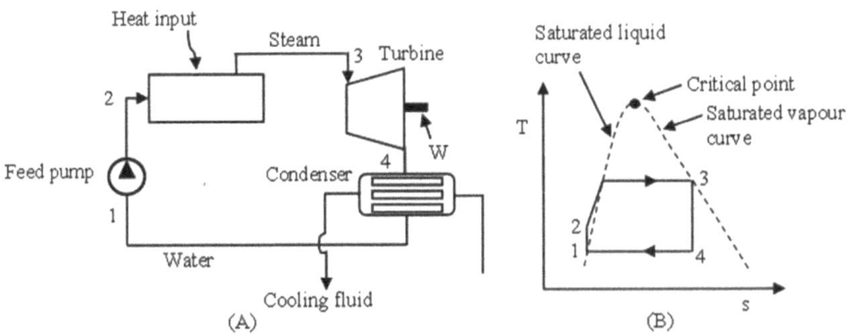

FIGURE 1.12 (A) Energy flow and (B) temperature-entropy (T-s) diagram of a standard Rankine cycle. *W*, shaft work.

(stage 2−3), which drives the steam turbine in stage 3−4. Some heat is rejected to the environment by the condenser in stage 4-1. In a typical Rankine cycle, more than 50% of the input heat is rejected as waste heat, which implies that the thermal efficiency of this cycle is lower than 50%.

Thermal efficiency is a critical parameter of the performance of a power plant, and it is defined as the ratio of net output work to the total thermal energy added to the cycle. There is output work (W_{34}) during the isentropic expansion of the steam in a standard Rankine cycle. The feed pump supplies the input work (W_{12}) during the isentropic compression of the water. Heat (Q_{23}) is added to the working fluid by the receiver between points 2 and 3, while some heat (Q_{41}) is rejected to the environment by the working fluid in the condenser. The various energy inputs and outputs can be calculated as follow:

$$Q_{41} = \dot{m}_v(i_4 - i_1) = \dot{m}_v C_v (T_4 - T_1) \tag{1.10}$$

$$Q_u = Q_{23} = \dot{m}_v(i_3 - i_2) = \dot{m}_v C_v (T_3 - T_2) \tag{1.11}$$

$$W_{12} = \dot{m}_v(i_2 - i_1) = \dot{m}_v C_v (T_2 - T_1) \tag{1.12}$$

$$W_{34} = \dot{m}_v(i_3 - i_4) = \dot{m}_v C_v (T_3 - T_4) \tag{1.13}$$

Where C_v is the specific heat capacity of the working fluid, $\dot{m}_v$ is the rate of vapor mass flow, i is the enthalpy and T is the temperature of the working fluid at the specified points.

Eq. (1.13) shows that shaft work (W_{34}) increases with the turbine inlet temperature (T_3). High values of T_3 are attainable through reheating and/or superheating the steam from the solar heater. In this regard, superheating can enable the power plant to operate in the supercritical phase of the working fluid, which elevates the thermodynamic efficiency. The thermal efficiency (η_{ra}) of the Rankine cycle is computed as follows:

$$W_{ra} = W_{34} - W_{12} = \dot{m}_v(i_1 - i_2 + i_3 - i_4) = \dot{m}_v C_v (T_1 - T_2 + T_3 - T_4) \tag{1.14}$$

$$\eta_{ra} = \frac{W_{ra}}{Q_u} = \frac{(i_1 - i_2 + i_3 - i_4)}{(i_3 - i_2)} = \frac{(T_1 - T_2 + T_3 - T_4)}{(T_3 - T_2)} \tag{1.15}$$

Where W_{ra} is the net output work.

One of the good strategies for promoting sustainable energy production is the recovery of waste heat from thermal power plants. Nevertheless, most of the available waste heat is of low grade ($T < 373$ K), and the critical temperature of water is 647.04 K (Güngören Madenoğlu, 2017). So, the conventional Rankine cycle (with water as the working fluid) cannot be driven by a wide range of waste heat sources. This observation has led to the development of the concept of the organic Rankine cycle (ORC) in which the working fluid is an organic substance (with lower critical temperature than water). Many organic substances are available for exploitation in ORCs by utilizing low-grade heat sources (Xu & Yu, 2014). By mid-2020, there were three operational ORC-based CSP plants around the world. The largest of these plants is the Aalborg CSP-Brønderslev project (with a capacity of 16.6 MW) located in Denmark. This power plant is driven by a PTC and started operating in 2016. Rankine steam cycles are well-established, and researchers have extensive experience with this category of heat engines acquired from the development of conventional thermal power plants driven by fossil and nuclear sources of energy. Consequently, principles of steam cycle power plants have provided a strong basis for CSP development, and all the commercial power plants operating on a global scale, by mid-2020, used (with slight changes) a subcritical steam Rankine cycle. Theoretically, a wet-cooled steam cycle can achieve an efficiency of up to 50% at a temperature and a pressure of 1023 K and 30 MPa, respectively (Dunham & Iverson, 2014), but the material factor limits the maximum temperature to 900 K based on the current state of the art. Through research and development, future designs of steam turbines may operate in the supercritical region.

One major drawback of the steam Rankine cycle is the high proportion (~50%) of the low-grade heat that is lost through the condenser (Stein & Buck, 2017). By the end of 2017, the steam cycles were able to achieve efficiencies of up to 42% (Petrakopoulou et al., 2017). On the other hand, gas cycles have higher thermodynamic efficiency and are therefore promising for CSP development.

1.5.3 Gas turbine

A standard gas turbine is an engine in which a gas (working fluid) is compressed, heated, and then allowed to expand through a set of turbine blades where part of the heat is converted to rotational energy. The gas turbine cycle (Brayton cycle) was first developed by George Brayton in 1873, and based on the fluid flow path, gas turbines are classified into open cycle gas turbine (OCGT), closed cycle gas turbine (CLCGT), and semi-closed cycle gas turbine (SCGT) categories. Major components of a basic cycle gas turbine are a compressor, a combustor (heater), and a turbine. In addition to these components, the CLCGT has a cooler, while the semi-closed cycle gas turbine combines the principles of the OCGT and CLCGT systems.

In an open cycle gas turbine, the working fluid (air) is mixed with fuel (such as natural gas) in the combustor where the mixture is burnt. The hot mixture expands through a turbine. The exhaust from the turbine in an OCGT system is rejected to the ambient environment. The combustion of fuel in the working fluid enables achievement of higher turbine inlet temperatures, and this is one of major advantages of the OCGT system. In this vein, the thermodynamic performance of an open cycle gas turbine is superior to that of the closed cycle. However, this OCGT system can only use selected fuels (such as gaseous and liquid) because the exhaust from the combustor directly drives the turbine, and so, it has to meet the inlet requirements of the turbine. Combustible fuels for gas turbines normally include natural gas, process gas, low-Btu coal gas, and vaporized fuel oil gas (Boyce, 2002). Among these fuels, natural gas is the most preferred conventional fuel for running gas turbines. On the other hand, the working fluid of a closed cycle gas turbine is indirectly heated by a fuel-based heater (such as coal, oil, nuclear) and the exhaust from the turbine is recirculated in the system. So, there is flexibility in the selection of fuels for CLCGT systems. Some main similarities between the closed cycle and the Rankine cycle are: the combustion gases do not themselves pass through the turbine and b) the exhaust from the turbine is recirculated. A lot of research attention has been given to the solarization of the OCGT and CLCGT, but there is limited similar work on the semi-closed cycle (Madhlopa, 2018b). Two main factors that influence the performance of gas turbines are: component efficiencies and turbine working temperature. The implication of this is that the efficiency and outlet temperature of a receiver have significant influence on the performance of a solar gas turbine.

Compression, combustion, and expansion of gases take place in separate components of a conventional gas turbine. These components can be assembled together to form a gas turbine system in different ways that give a wide range of possible configurations (Cohen et al., 1996). There is no limit to the number of each component, and so multiple compressors, combustors (and/or solar receivers), turbines, and shafts (spools) can be used in a given system. The system configuration affects the optimal thermal efficiency and the change of efficiency with output power and of output torque with rotational speed. It is worth noting that a given configuration may be suitable for a specific application such as driving: a) an electric generator under transient load at constant speed and b) a ship's propeller where the power varies as the cube of the speed. Consequently, it is necessary to establish a configuration that nearly matches the application at hand, including the suitability of a configuration for solarization. For example, the use of multiple compressors enables recompression of the working fluid to obtain a high compression ratio that raises the efficiency of the system. In this case, the working fluid is cooled after each stage of compression. Sometimes, it is desirable to reheat the working fluid in order to realize high turbine inlet temperatures. This is achievable through the exploitation of multiple heaters. Details about possible configurations of solar gas turbines are reported elsewhere (Madhlopa, 2018a).

A solar gas turbine is a device in which concentrated solar radiation is utilized to heat up the working fluid that expands through the turbine section. The working fluids of both the open and closed cycle gas turbines can be directly heated by a solar receiver. The gas turbine cycle has two isentropic and two isobaric processes: a) isentropic gas compression (stage 1−2), b) isobaric heat addition to the gas (stage 2−3), c) isentropic expansion of the gas (stage 3−4), and d) isobaric cooling of the gas (stage 4-1) as shown in Fig. 1.13. The closed-cycle solar gas turbine has an additional stage for

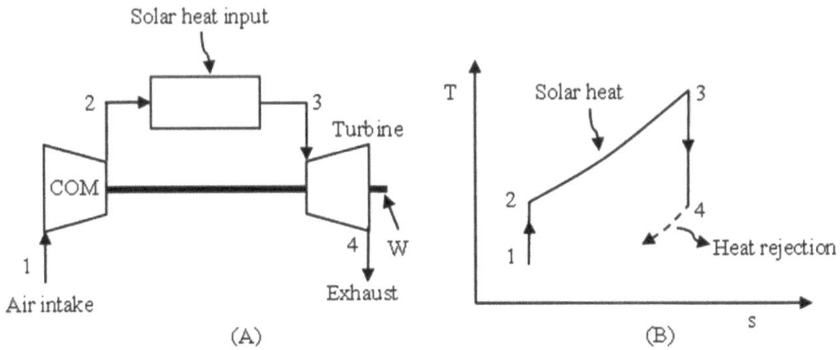

FIGURE 1.13 Schematic representation of an open cycle solar gas turbine, showing: (A) energy flow and (B) temperature—entropy (T—s) diagram. *COM*, compressor and *W*, shaft work.

cooling the working fluid downstream the turbine Fig. 1.14. The waste heat from a gas turbine is of high grade, and so it can be recovered to drive another engine cycle. Consequently, the gas turbine can be combined with a bottoming cycle, often a Rankine cycle or Organic Rankine cycle to obtain a combined cycle. A combined cycle is broadly defined as a system of one or more gas turbines with heat recovery from the exhaust for driving steam turbines (Fig. 1.15). It allows generation of electricity from the gas and steam turbines for the same amount of input heat to the topping cycle, and the net work for the combined cycle is higher than that of individual cycles. Consequently, thermal efficiencies of advanced combined cycle gas turbines for sole production of electricity fall in the range of 50%—58%, which is superior to that of the best

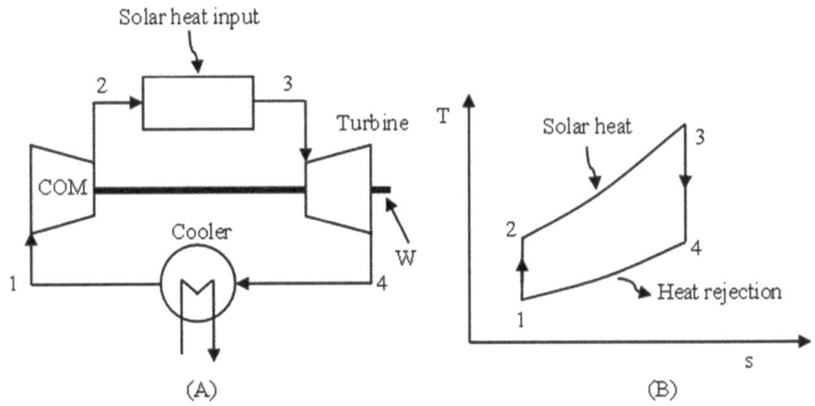

FIGURE 1.14 Close-cycle solar gas turbine, showing: (A) energy flow and (B) temperature—entropy (T—s) diagram. *COM*, compressor and *W*, shaft work.

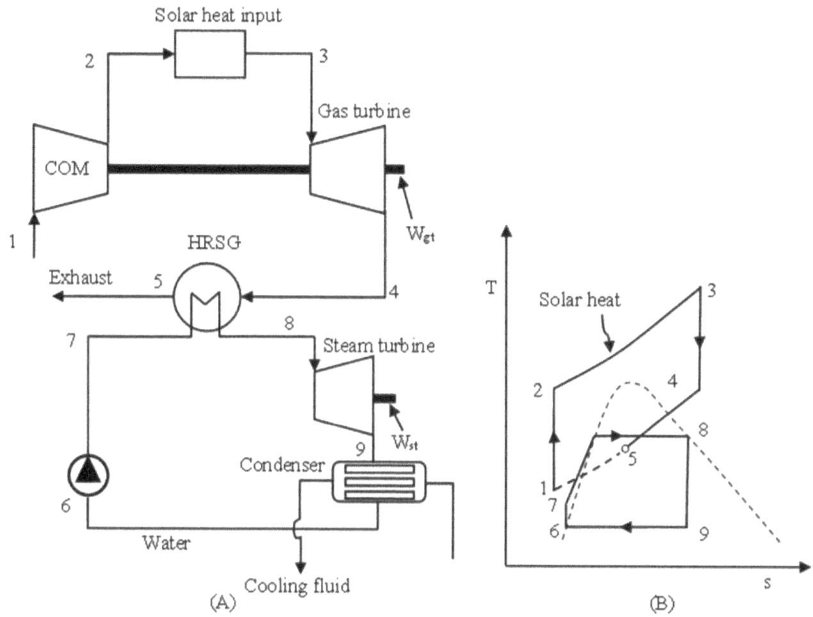

FIGURE 1.15 A combined cycle gas turbine with a heat recovery steam generator (HRSG): (A) energy flow and (B) temperature–entropy (T–s) diagram. *COM*, compressor, *Wgt*, gas turbine shaft work, and *Wst*, steam turbine shaft work.

oil and coal-fired power plants (43%–45%) (Poullikkas, 2005). It is possible to include a backup heater between the solar heater and turbine, and several other possible configurations of the solar gas turbine technology are possible (Madhlopa, 2018a).

The thermal efficiency of a gas turbine cycle can be computed by considering the energy balance of the components. Compressors consume power (W_{12}) to compact the working fluid (wf). This power is supplied by the turbine (W_{34}). The difference between the work done by the compressor and the turbine work yields net work from the gas turbine cycle (W_{gt}).

$$W_{12} = \dot{m}_{wf}(i_2 - i_1) = \dot{m}_{wf}C_{p,wf}(T_2 - T_1) \tag{1.16}$$

$$W_{34} = \dot{m}_{wf}(i_4 - i_3) = \dot{m}_{wf}C_{p,wf}(T_3 - T_4) \tag{1.17}$$

$$W_{gt} = \dot{m}_{wf}(i_1 - i_2 + i_3 - i_4) = \dot{m}_{wf}C_{p,wf}(T_1 - T_2 + T_3 - T_4) \tag{1.18}$$

Where $C_{p,wf}$ is specific heat capacity the working fluid at constant pressure, and $\dot{m}_{wf}$ is the flow rate of the working fluid.

Eq. (1.17) shows that output work (W_{34}) increases with the turbine inlet temperature (T_3). As expected, higher values of T_3 are required to elevate the thermodynamic efficiency of the cycle. The heat input to the working fluid (Q_{23}) is supplied by the solar receiver in a solar-only system, and so the thermal efficiency of the gas turbine cycle (η_{gt}) can be calculated as follows:

$$Q_u = Q_{23} = \dot{m}_{wf}(i_3 - i_2) = \dot{m}_{wf}C_{p,wf}(T_3 - T_2) \tag{1.19}$$

$$\eta_{gt} = \frac{W_{gt}}{Q_u} = \frac{(i_1 - i_2 + i_3 - i_4)}{(i_3 - i_2)} = \frac{(T_1 - T_2 + T_3 - T_4)}{(T_3 - T_2)} \tag{1.20}$$

Air is the working fluid of a conventional gas turbine cycle, but other gases, such as carbon dioxide (CO_2), helium (He), and hydrogen (H_2) are alternative candidate fluids for the closed-cycle gas turbine. In 1950, Sulzer patented the first work on the CO_2 closed cycle gas turbine (Olumayegun et al., 2016). Since then, the CO_2 Brayton cycle has evolved by a significant margin. There are three options for operating the CO_2 Brayton cycle in the gaseous phase: a) conventional (subcritical) cycle (sn-CO_2), b) transcritical noncondensing cycle (tn$-CO_2$), and c) supercritical cycle (s-CO2) (Garg et al., 2013). In the sn-CO_2 cycle, the upper side pressure falls below the critical pressure of carbon dioxide. On the other hand, the upper side pressure of the tn$-CO_2$ cycle falls in the supercritical range, with the lower side pressure in the subcritical region. As for the s-CO_2 cycle, the lower side pressure lies in the supercritical regime. A comparison of the performances of these three CO_2 Brayton cycles has shown that the s-CO_2 cycle possesses the greatest potential for integration with concentrated solar power technology. The s-CO_2 cycle is an emerging technology and some of its advantages are: a) high cycle efficiency because of decrease in required power for compression arising from an abrupt change in the properties of carbon dioxide, b) smaller components (particularly, turbomachinery), c) fewer components, and d) simpler engine architecture.

Many studies have investigated the feasibility of solarizing a gas turbine cycle. It has been found that the PDC and ST technologies have the most favorable thermodynamic characteristics to drive this engine cycle (Madhlopa & Okoroigwe, 2017). However, the maturity level of the PDC is very low as shown in Section 1.7.4. The first operational hybrid solar gas turbine power plant was built at Kibbutz Samar in southern Israel by AORA Solar and has been operating since 2009 (Merchant, 2009). This power plant generates 100 kW_e and 170 kW_{th}, and it is driven by an ST that heats up compressed air to about 1273 K before injecting it into a gas turbine. An important milestone in the development of solar gas turbines was also achieved through the SOLUGAS project. This is the first demonstration project at megawatt scale where compressed air is preheated by a 3.2 MW_{th} receiver up to 1073 K and injected to the combustion chamber of the 4.6 MW Mercury 50 gas turbine. The aim of this project was to demonstrate the performance of an operational solar hybrid gas turbine system at a precommercial scale (Korzynietz et al., 2016). Other demonstration projects have been initiated, and the ST was used to drive the gas turbine in all these projects except for the Optimized

TABLE 1.1 Projects on solar gas turbines.

Project name	Country	Capacity (MW)	Year start	Reference
Thémis Solar Power	France	2.5	1979	Drouot and Hillairet (1984)
SOLGATE	Spain	0.23	2001	SOLGATE (2005)
SOLHYCO	Spain	0.25	2006	SOLHYCO (2011)
PEGASE	France	1.4	2006	Grange et al. (2014)
AORA Solar	Israel	0.27	2009	Merchant (2009)
SOLUGAS	Spain	4.6	2012	Quero et al. (2014)
OMSoP project	Italy	0.003–0.010	2013	Lanchi et al. (2015)
Next-CSP project	Israel	4.0	2017	Next-CSP (2016)

Microturbine Solar Power (OMSoP) project in which a microgas turbine was powered by a PDC Table 1.1 (Drouot & Hillairet, 1984; Grange et al., 2014; Lanchi et al., 2015; Next, 2016; Quero et al., 2014; SOLGATE, 2005; SOLHYCO, 2011). Nevertheless, by mid-2020, there was no commercial solar gas turbine power plant in operation.

1.5.4 Stirling cycle

A basic Stirling engine comprises cylinders (with cold and hot spaces), pistons, a regenerator, a heater, and a cooler. The heater supplies heat to maintain the hot end of the regenerator at a constant temperature of $T = T_h$, while the cooler keeps the temperature of the cold end of the regenerator at a uniform temperature of $T = T_c$. In this engine, the working fluid (gas) does not leave the cylinders of the engine. Air is commonly used as the working fluid, but other gases have also been exploited in the engine. Based on how the cylinders are connected, the Stirling engine can be classified into three types: alpha, beta, and gamma configurations (Rahmati et al., 2020). The alpha engine configuration has two power pistons in two separate cylinders, a heater, a regenerator, and a cooler (Fig. 1.16A). The two pistons concurrently perform the tasks of expansion and compression of the working fluid and that of transferring the working fluid transfer between the hot and cold spaces. The beta type of the Stirling possesses only one cylinder (Fig. 1.16B). Only the power piston is responsible for densifying and expanding the working fluid, while the displacer piston transfers the working fluid between hot and cold

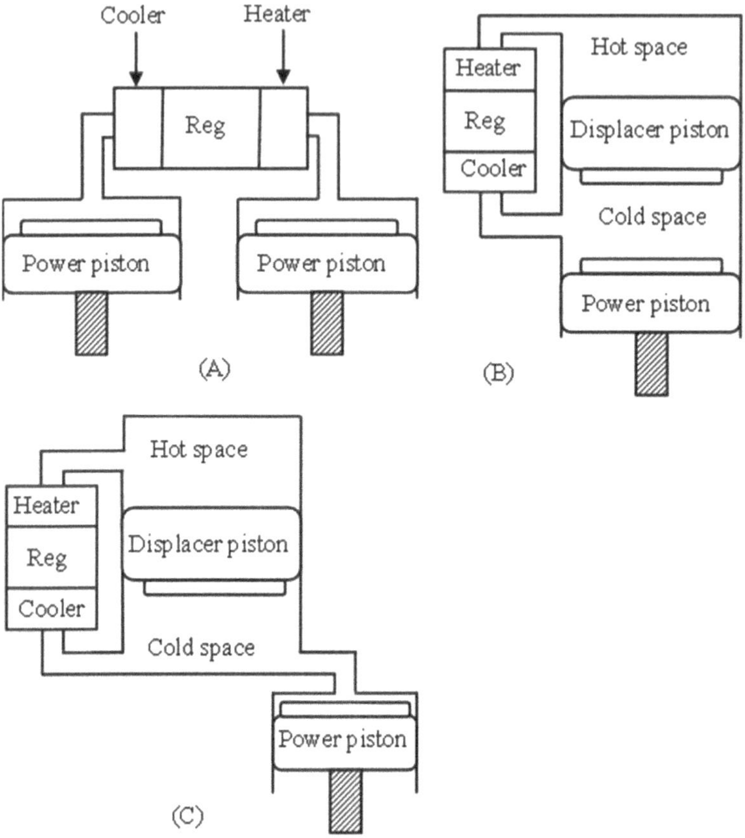

FIGURE 1.16 Schematic representation of various configurations of the Stirling engine: (A) alpha type, (B) beta type, and (C) gamma type. *Reg*, regenerator.

chambers. There is some similarity between the beta and gamma configurations of the Stirling engine (Fig. 1.16C) except for the provision of distinct cylinders for the power piston and displacer pistons, and the flow of fluid between the two cylinders.

The beta engine was the first layout invented by Dr Robert Stirling in 1816. This engine has two isothermal (compression and expansion) and two isochoric (heating and cooling) processes (Fig. 1.17).

(a) Isothermal compression (stage 1−2): The power piston moves inward to compress the gas at constant temperature $T = T_c$, thereby increasing its pressure. During this process, work is done by the power piston, and the cooler removes heat from the gas and rejects it to the environment.

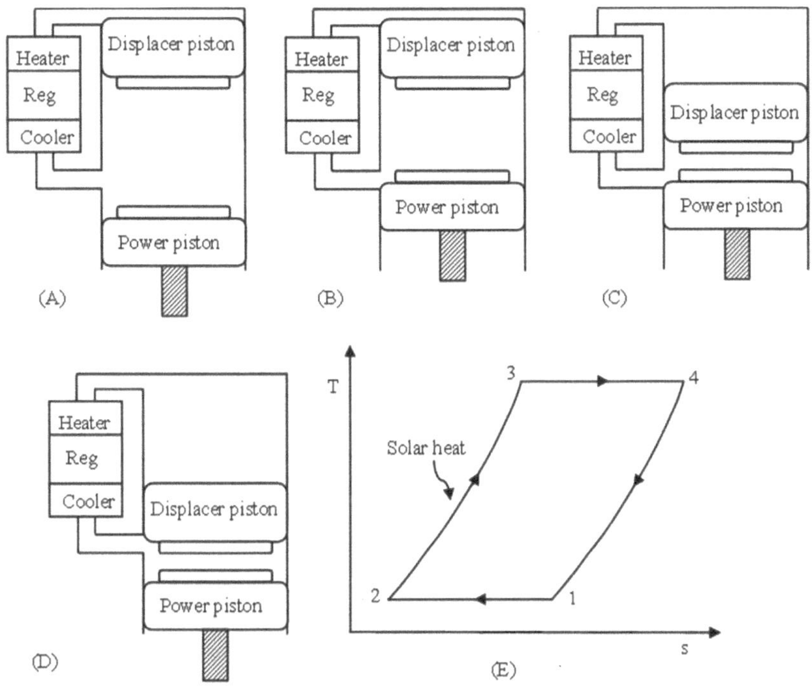

FIGURE 1.17 Schematic representation of: (A–D) Four processes of the beta-type Stirling engine and (E) temperature–entropy (T–s) diagram of the engine.

(b) Isochoric heating (stage 2–3): This process takes place when the power piston has reached its top dead center where it momentarily remains motionless, thereby maintaining a constant volume. The gas is charged with heat from the regenerator as the displacer piston translates inwards, raising the gas temperature from $T = T_c$ to $T = T_h$ and the gas pressure to a maximum level. This heat input is supplied by the receiver in a solarized Stirling cycle. Entropy is minimum at the beginning of this part of the cycle ($s = s_2$). This implies that there is a maximum amount of available energy to do work. Nevertheless, no work is done in this segment of the cycle.

(c) Isothermal expansion (stage 3–4): In this stage, the power piston is pushed outward by the hot expanding gas, thereby increasing the overall volume of the gas and reducing the gas pressure. Work is done on the piston by the expanding gas, and heat is added to the gas from the heater (which maintains the temperature of the gas).

(d) Isochoric cooling (stage 4-1): The power piston is at the bottom dead center in this stage of the cycle, where it momentarily remains still and keeps a constant volume of the gas. Heat is transferred from the gas to the regenerator as the displacer moves through its range of motion. The absorption of heat by the regenerator results in the decline of gas temperature from $T = T_h$ to $T = T_c$ and pressure to a minimum level. Work is not done in this stage.

The thermal efficiency of an ideal Stirling engine is equal to that of the Carnot cycle. However, the nonideal efficiency (η_{sti}) can be computed from:

$$\eta_{sti} = \frac{W_{sti}}{Q_u} = \eta_{car}\left\{1 + \frac{(1-\eta_{reg})mC_v(T_h - T_c)}{nRT_h \ln\left(\frac{V_{max}}{V_{min}}\right)}\right\}^{-1} \tag{1.21}$$

Where C_v is the specific heat capacity of the gas at constant volume, m is the mass of the gas, n is the number of gas moles, R is the gas constant, T_c is the temperature of the cooler, T_h is the heater temperature, V_{max} is the maximum gas volume, V_{min} is the minimum gas volume, W_s is the net output work, and η_{reg} is the regenerator effectiveness. It should be noted that the effectiveness of an ideal regenerator is $\eta_{reg} = 1.0$, which implies that Eq. (1.21) reduces to the expression of efficiency of the Carnot engine in this case.

Dish Stirling engine systems can achieve an efficiency of about 25% for the net solar energy to electrical energy conversion efficiency (Mansiri et al., 2014). However, solarization of the Stirling cycle is still at infant stage, and there were no demonstration and commercial power plants based on this cycle by mid-2020. Research is still going on to develop CSP systems that are based on this engine cycle.

1.5.5 Kalina cycle

A novel thermodynamic power cycle known as Kalina cycle was proposed by Dr. Alexander Kalina in 1984 as an alternative bottoming cycle for combined cycle power plants. Different designs of the Kalina cycle, known as Kalina cycle systems (KCS) have evolved over the time (Shan, 2020). A simple Kalina cycle comprises a boiler, separator, turbine, recuperator, absorber, condenser, and pump. The cycle can be driven by a range of heat sources, from low to high grade. A mixture of two fluids is utilized as the working fluid of the cycle. Ammonia—water (NH_3—H_2O) mixture is used as a conventional working fluid for the Kalina cycle. The ammonia—water mixture is pumped into a recuperator for preheating before it flows into the boiler. The hot mixture exits the boiler in liquid and vapor phases, and it undergoes phase separation in the separator as shown in Fig. 1.18. Ammonia is more volatile than water, and therefore, the vapor component is richer in ammonium gas. This vapor flows into the turbine that extracts work from it and then exits into the absorber (mixer). The ammonia-deficient liquid component from the separator is

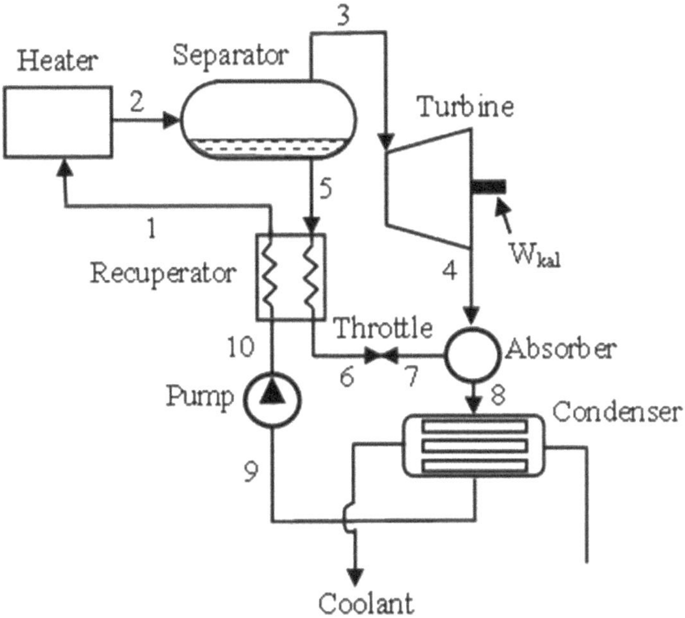

FIGURE 1.18 Schematic representation of Kalina cycle KCS 11.

channeled through the recuperator where heat is transferred from the liquid to the cold ammonia–water mixture from the pump. Then, the ammonia-deficient liquid is throttled into the absorber where it mixes with the exhaust from the turbine, and this mixture is condensed into the ammonia–water working fluid that flows into the pump to start another cycle. For a heat input of Q_{in}, the Kalina cycle efficiency can generally be computed from:

$$\eta_{kal} = \frac{W_{kal}}{Q_{in}} \tag{1.22}$$

The major difference between the Kalina Cycle and Rankine cycle (and organic Rankine cycle) is that its condensation and evaporation proceed at variable temperature (Ahmad & Karimi, 2013). This implies that boiling does not take place at one temperature but in a range of temperature for specific pressure and composition. The main advantage of using the zeotropic fluid mixture is that it can match the heat source temperature profile better, thereby yielding higher efficiency. Even though the Kalina cycle is more complex than the Rankine cycle, exploitation of the ammonia–water mixture at an elevated temperature has been demonstrated at the Canoga Park demonstration plant in the United States of America with turbine inlet temperature of 788 K and pressure of 11.0 MPa (Modi et al., 2016). This is the first Kalina cycle–based power plant (6.5 MW), which was commissioned in 1992. It is driven by nuclear waste heat. In addition, the patent by Kalina shows that, with nitridation prevention, a power plant based on ammonia–water mixtures can operate stably up to T = 1366 K and P = 68.95 MPa (Kalina, 2002).

1.6 Overall system efficiency

Four major sequential steps are involved in the production process of electricity from solar radiation: a) concentration and absorption of solar radiation (optical performance), b) conversion of absorbed solar radiation to heat by the receiver, c) transformation of useful heat to mechanical energy by the engine, and d) change of mechanical energy to electricity by a generator. Each step has a corresponding efficiency of energy conversion. Consequently, production of CSP-based electricity entails optical, receiver, engine cycle, and generator efficiencies.

1.6.1 Optical efficiency

There is a variety of designs of concentrating solar collectors, and therefore, their optical performance depends on the geometric and surface properties of the concentrator and receiver. It is, however, possible to apply a generic mathematical model in describing the optical performance of all of them. In general, the intensity of solar radiation absorbed per unit area (G_{ab}) of the unshaded aperture can be given by (Duffie & Beckman, 2013):

$$G_{ab} = G_{ap}\rho(\alpha\tau\gamma)_n K_{\alpha\tau\gamma} \tag{1.23}$$

Where G_{ap} is the effective solar radiation measured on the aperture plane (which comprises only beam radiation except for concentrators with low concentration ratio), ρ is the specular reflectance except when the reflector is diffuse (ρ is replaced by transmittance when the concentrator is a refractor) , α is the absorbance of the absorber on the solar receiver, τ is the transmittance of the transparent cover on the solar receiver ($\tau = 1.0$ for solar receivers without a transparent cover), γ is the proportion of the reflected solar radiation that is incident on the absorbing surface of the receiver (values of γ may be greater than 0.9), and $K_{\alpha\tau\gamma}$ is the incidence angle modifier (established from empirical data).

All the three factors α, τ, and γ depend on the angle of incidence (subscript n = normal incidence), and computation of the parameters α and τ may be difficult because of the uncertainty that exists in establishing the angle of incidence of solar radiation from the reflector onto the transparent cover. The optical efficiency (η_{opt}) of the solar collector can therefore be expressed as:

$$\eta_{opt} = \frac{G_{ab}}{G_{ap}} = \rho(\alpha\tau\gamma)_n K_{\alpha\tau\gamma} \tag{1.24}$$

It should be noted that the optical efficiency of the system involves parameters of both the concentrator and solar receiver.

1.6.2 Efficiency of solar receivers

There are wide variations in the design of receivers, operational temperatures, edge effects, conduction terms, and uniformity of radiation flux incident on receivers. In view of this, each receiver design needs to be analyzed as a distinct case. Nevertheless, the useful heat output (Q_u) from the receiver can be defined as:

$$Q_u = A_{ap} G_{ab} - Q_{loss} \tag{1.25}$$

Where Q_{loss} is the thermal loss of power. Eq. (1.25) is a generalized model of the energy balance of the receiver.

The useful heat can also be computed from:

$$Q_u = \dot{m}_{htf} C_{p,htf} \left(T_{htf,out} - T_{htf,in}\right) \tag{1.26}$$

Where $\dot{m}_{htf}$ is the flowrate of the heat transfer fluid, $C_{p,htf}$ is the specific heat capacity of the heat transfer fluid at constant pressure, $T_{htf,in}$ is the inlet temperature of the heat transfer fluid, and $T_{htf,out}$ is the outlet temperature of the heat transfer fluid.

$$\eta_{rec} = \frac{Q_u}{A_{ap} S_{ab}} \tag{1.27}$$

A higher receiver efficiency is desirable for optimal production of electricity. Detailed optical and thermal models are presented in Chapter 10.

Example 1.3

A solar receiver uses air as a heat transfer fluid. At a certain time of the day, the inlet and outlet temperatures of the air are 300 and 900 K, at a flow rate of 8.0 kg s^{-1}. Calculate the useful heat from the receiver, assuming the specific heat capacity of the air is constant at 1010 $Jkg^{-1}K^{-1}$ throughout the receiver.

Solution

From Eq. (1.26),

$$Q_{net} = \dot{m}_{htf} C_{htf} \left(T_{htf,out} - T_{htf,in}\right)$$
$$= 8.0 * 1010 * (900 - 300)$$

$$= 4848000 \text{ W}$$

Efficiencies of selected engine cycles, commonly exploited in the CSP technology, have been considered in Section 1.5. The engine cycle is driven by useful heat (Q_u) from the receiver. If Q_u is the only input heat, then the efficiency of the engine cycle (η_{cycle}) can be expressed as:

$$\eta_{cycle} = \frac{W_{net}}{Q_u} \tag{1.28}$$

1.6.4 Efficiency of electric generators

An electric generator is a device that converts mechanical energy from a prime mover to electric energy as an output. Prime movers provide the required mechanical energy, and they include heat engines, hydro, and wind turbines. Electromechanical generation of electricity is achieved through the concepts of electrostatic and magnetic induction. Electrets (dielectric materials with embedded charge) are used to generate electricity through electrostatic induction (Wada et al., 2016). These electrets create an electric field which is used by an electret generator to convert mechanical energy into electricity. Nevertheless, electret generators produce power at the microwatt scale, which is suitable for low-power electronic devices such as microphones. In contrast, the CSP technology generates power up to the megawatt scale, and magnetic generators are widely exploited for this purpose. The principle of electromagnetic induction, developed by an English chemist and physicist Michael Farady (1791−867), is applied in magnetic-field generators. In this principle, voltage is induced when: a) a conductor, which intersects lines of a constant magnetic field, moves relative to the field, or b) a magnetic field changes while the conductor intersecting the field is stationary. The induced potential difference increases with the rate of change of the magnetic field.

The net output work from a heat engine (W_{net}) is input to the generator. So, the electric power output (E) from the generator is:

$$E = \eta_{gen} W_{net} \tag{1.29}$$

Where E is the electric output from the generator, and η_{gen} is the efficiency of the generator. Generally, the efficiency of an electric generator is close to $\eta_{gen} = 1.0$.

1.6.5 System efficiency

Ultimately, the overall efficiency of the system is most important. For a given engine cycle, the net output work (W_{net}) is:

$$W_{net} = \eta_{cycle} Q_u \tag{1.30}$$

From Eq. (1.24), $G_{ab} = \eta_{opt} G_{ap}$, and from Eq. (1.27), $Q_u = \eta_{sr} A_{ap} G_{ab}$. Therefore, $E = \eta_{opt} \eta_{rec} \eta_{cycle} \eta_{gen} A_{ap} G_{ap}$. The solar power input to the concentrator is $A_{ap} G_{ap}$. Consequently, the overall ideal efficiency of the system (η_{sys}) can be expressed as a product of efficiencies that correspond to the steps involved in the whole process of converting solar radiation to electricity:

$$\eta_{sys} = \frac{E}{G_{ap}} = \eta_{opt} \eta_{rec} \eta_{cycle} \eta_{gen} = \eta_{opt} \eta_{ecr} \eta_{pb} \tag{1.31}$$

Where $\eta_{pb} = \eta_{cycle} \eta_{gen}$ is the efficiency of the power block.

The optical efficiency of large-scale concentrators often falls in the range of 60%–75%, except for the parabolic dish with $\eta_{opt} > 85\%$ (Kribus, 2017). For receivers, the thermal efficiency decreases with increasing operational temperature. It can decline from 95% at receiver temperature of 62 K to 83% at 773 K. This creates a challenge in receiver design and operation because the efficiency of converting solar radiation to heat decreases with increasing the operating temperature (Blanco & Santigos, 2017). In other words, a higher fraction of the concentrated solar radiation is converted to heat when the receiver is at low temperature. On the other hand, the rate of converting thermal energy to work rises with the operating temperature, but it is constrained by the second law of thermodynamics. It is therefore necessary for the receiver to produce heat of very high grade. However, material properties also restrain the possible achievable temperatures. Most of the CSP plants were based on steam cycles by mid-2020. These plants could achieve thermal efficiencies of up to 42% (Petrakopoulou et al., 2017), with solar-to-electric conversion efficiencies of 8%–35%. This shows that there is need for improving the performance of the CSP technology through research and development.

Example 1.4

A PTC power plant has an optical efficiency of 65%, and it uses Therminol 66 as a heat transfer fluid (with flowrate of 5 kgs^{-1}, inlet temperature of 300 K, and outlet temperature of 653 K). The solar receiver has a thermal efficiency of 80%. If the effective beam radiation on the aperture of the concentrator is 800Wm^{-2}, find the area of the aperture. Assume the specific heat capacity of Therminol 66 is 2207Jkg^{-1}.

Solution

From Eq. (1.27), the area of the aperture can be expressed as:

$$A_{ap} = \frac{Q_u}{\eta_{rec} G_{ab}}$$

But S_{ab} can be computed from Eq. (1.24),

$$G_{ab} = \eta_{opt} G_{ap}$$

From Eq. (1.26), useful heat from the receiver (Q_u) can be given by:

$$Q_u = \dot{m}_{htf} C_{p,htf} \left(T_{htf,out} - T_{htf,in} \right)$$

$$A_{ap} = \frac{\dot{m}_{htf} C_{p,htf} \left(T_{htf,out} - T_{htf,in} \right)}{\eta_{opt} \eta_{sr} G_{ap}}$$

$$= \frac{5 * 2207 * (653 - 300)}{0.65 * 0.8 * 800}$$

$$= 9364 m^2$$

1.7 Common types of concentrating solar power technology

Concentrating solar power technologies are classified into four main types, namely, parabolic trough concentrator (PTC), linear Fresnel reflector (LFR), parabolic dish concentrator (PDC), and solar tower (ST) (Wagner et al., 2018). A summary of selected major characteristics of the different CSP technology types is presented in Table 1.2 (Islam et al., 2018; Zhang et al., 2013). At present, some projects are under construction (projects in which the ground has been broken and construction is still under way), while others are under development (projects that have a signed agreement but actual construction has not yet commenced). Most of the projects under development are ST-based, which shows that the deployment of the ST technology is becoming more attractive (Table 1.3). In contrast, the current status of CSP indicates that the PDC is the most immature CSP technology. The solar-to-electric conversion efficiency of current operational PTC and LFR plants is relatively low (8%−16%), while the ST and PDC exhibit higher efficiencies (20%−35%).

Thermal storage and backup heating are important strategies for elevating the capacity factor of a CSP plant. In this connection, most of the plants equipped with thermal storage, in operation and under construction, are based on the PTC (Table 1.4). This trend may change in future because, in the underdevelopment phase, plants integrated with thermal storage components are

TABLE 1.2 **Selected major characteristics of different concentrating solar power (CSP) technology types. *LFR*, linear Fresnel reflector; *PDC*, parabolic dish concentrator; *PTC*, parabolic trough concentrator, and *ST*, solar tower.**

Parameter	PTC	LFR	ST	PDC
Capacity (MW)	10–200	10–200	10–150	0.01–0.40
Concentration ratio (dimensionless)	25–100	70–80	300–1000	1000–3000
Annual solar-to-electric efficiency (%)	12–16	8–10	20–35	20–35
Operating temperature of solar field (K)	563–823	523–833	523–923	393–1773
Land use (m^2 $MWh^{-1}year^{-1}$)	6–8	4–6	8–12	8–12

All data were compiled from Islam, M. T., Huda, N., Abdullah, A. B., & Saidur, R. (2018). A comprehensive review of state-of-the-art concentrating solar power (CSP) technologies: Current status and research trends. *Renewable and Sustainable Energy Reviews, 91*, 987–1018. https://doi.org/10.1016/j.rser.2018.04.097 except for the annual solar-to-electric efficiency of the PTC (National Renewable Energy laboratory. (2020). *Concentrating solar power projects*. National Renewable Energy laboratory (NREL). https://solarpaces.nrel.gov.), operational temperature of the PDC (Zhang, H. L., Baeyens, J., Degrève, J., & Cacères, G. (2013). Concentrated solar power plants: Review and design methodology. *Renewable and Sustainable Energy Reviews, 22*, 466–481. https://doi.org/10.1016/j.rser.2013.01.032), and ST (SOLGATE. (2005). *Solar hybrid gas turbine electric power system*).

predominantly ST- based. Again, most of the operating power plants with backup heaters are based on the PTC (Table 1.5). However, none of the plants under construction has a backup heater, and only one plant (ST-based) under development is thermally backed up.

TABLE 1.3 **Concentrating solar power (CSP) plants that are operational, under construction and development (National Renewable Energy laboratory, 2020).**

CSP type	Operational	Under construction	Under development
PTC	78	10	4
LFR	6	2	1
ST	20	4	9
PDC	0	0	0

TABLE 1.4 Concentrating solar power plants equipped with thermal storage.

CSP type	Operational	Under construction	Under development	Storage capacity
PTC	36	6	3	0.5–16.0
LFR	4	1	1	0.3–14.0
ST	16	4	9	1.0–17.5

National Renewable Energy laboratory. (2020). *Concentrating solar power projects.* National Renewable Energy laboratory (NREL). https://solarpaces.nrel.gov.

1.7.1 Parabolic trough concentrator

A parabolic trough concentrator utilizes a long-curved mirror (reflector) to focus solar radiation on a tubular receiver located along the focal line of the trough (Fig. 1.19A). This receiver can be made up of two concentric glass tubes with a vacuum between them, and a metallic tube fitted inside the transparent concentric glass tubes. Some tubular receivers are not glazed. The trough and receiver can move together to track the sun during the daytime (García et al., 2011), and multiple reflectors are sometimes linked together to form an array (Fig. 1.19B). Concentrated solar radiation is absorbed by the metallic tube, and a heat transfer fluid, which may be oil, molten salt, or other fluid, flows through the hot metallic tube. The PTC system is cheap and easy to manufacture, but it has a low concentration ratio and can achieve low to medium operating temperatures (Zhang et al., 2013). Nonetheless, this range of temperature is satisfactory for powering the steam turbine.

The first PTC system was built in 1912 in Cairo, Egypt (Ummadisingu & Soni, 2011), and since then, many other PTC power plants have been developed. By mid-2020, most of these plants were in Spain, and the largest PTC

TABLE 1.5 Concentrating solar power plants equipped with backup heaters.

CSP type	Operational	Under construction	Under development
PTC	12	0	0
LFR	0	0	0
ST	1	0	1

National Renewable Energy laboratory. (2020). *Concentrating solar power projects.* National Renewable Energy laboratory (NREL). https://solarpaces.nrel.gov.

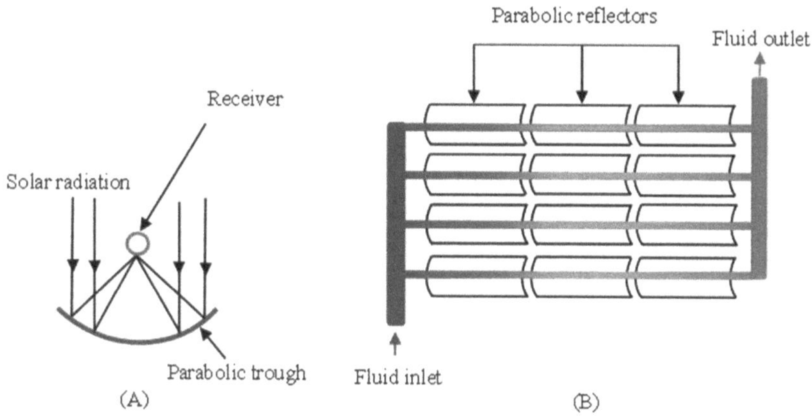

FIGURE 1.19 Schematic representation of parabolic trough concentrator: A) cross-section and B) multiple collectors.

power plants in operation were the Solana Generating Station (start date: 2013), Genesis Solar Energy Project (start date: 2014) and Mojave Solar Project (start date: 2014), with a capacity of 250 MW each, in the United States of America. Both the Genesis Solar Energy and Mojave Solar projects are solar-only without thermal storage, while the Solana Generating Station has storage (capacity of 6 h) and backup-heating systems. At present, the largest CSP power plant (600 MW) in the world is the DEWA CSP Trough Project in the United Arab Emirate (UAE), which is under construction. This power plant is powered by solar radiation only with a thermal storage capacity of 15 h.

1.7.2 Linear Fresnel reflector

An LFR system consists of many curved mirror strips on each side of a tubular receiver (Pelay et al., 2017), as shown in Fig. 1.20, and it has relatively low concentration ratio and operational temperature. One drawback of this system is shading effect between adjacent reflectors, which led to the development of a compact linear Fresnel reflector (CLFR) (Barlev et al., 2011). In the CLFR system, adjacent mirrors are oriented toward two separate absorber tubes located in opposite directions, thereby reducing the size of land required by the solar field.

The first operational LFR plant is the Puerto Errado 1 (1.4 MW) in Germany, which was commissioned in March 2009. Later, the Puerto Errado two power plant (30 MW) was built in Spain and started operating in August 2012.

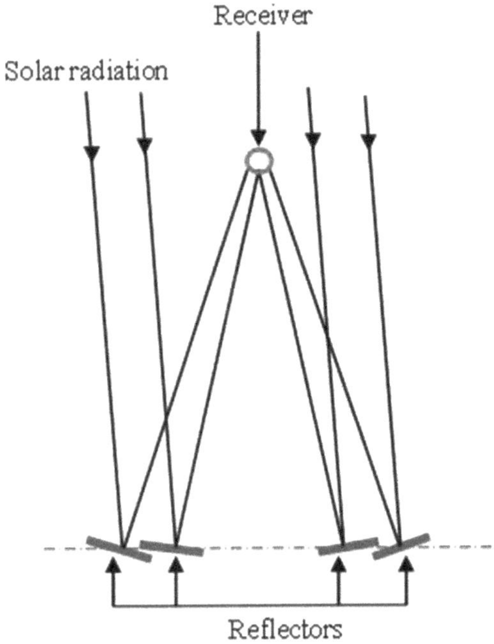

FIGURE 1.20 Schematic representation of a linear Fresnel reflector.

Since then, other LFR plants have been constructed. The largest LRF power plant in operation is the Dhursar power (125 MW), which was commissioned in 2014 in India. This power plant is solar-only without a storage component. By mid-2020, the Dadri ISCC Plant (14 MW) in India and IRESEN 1 MWe plant in Morocco were under construction. Both these plants are solar-only, but the IRESEN plant is equipped with a thermal storage (with a capacity of 20 min). Only the Zhangjiakou 50 MW CSG Fresnel project in China was under development. It is solar-only and equipped with a thermal storage capacity of 14 h.

1.7.3 Solar tower

In an ST, a field of large mirrors is employed to focus direct solar radiation onto a receiver mounted on the tower (Fig. 1.21), where it is converted to heat. These large mirrors, also called heliostats, track the sun individually in two axes. A high amount of solar radiation is reflected by the mirrors onto the receiver. In view of this, the ST technology can achieve higher temperatures than the PTC and LFR, and it can be exploited in the gas turbine technology

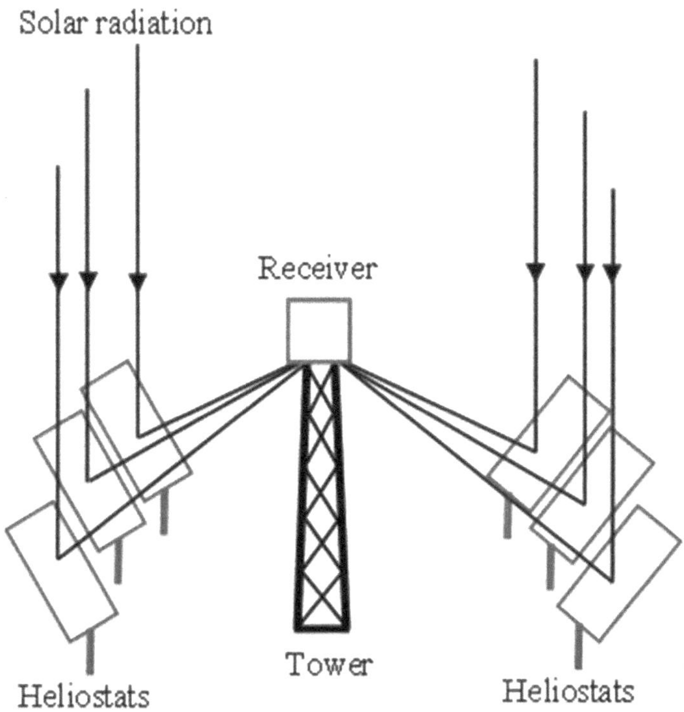

FIGURE 1.21 Schematic representation of a solar tower.

(Madhlopa & Okoroigwe, 2017). Moreover, high temperatures render the use of thermal storage more attractive (Franchini et al., 2013).

The first operating ST power plant is the Planta Solar 10 (11 MW), commissioned in 2007 in Spain. By mid-2020, the largest ST power plant in operation was the Ivanpah Solar Electric Generating System (392 MW) in the United States of America. This power plant started operating in 2014, and it does not have a storage component but a backup heater. All the ST power plants under construction and development are solar-only and equipped with storage components. In terms of capacity, the Tamarugal Solar Energy Project is the largest ST power plant (450 MW), with a storage capacity of 13 h under the development phase in Chile.

1.7.4 Parabolic dish concentrator

A PDC focuses solar radiation on a focal point above a parabolic dish reflector (Fig. 1.22), and the reflector and receiver move together to track the sun

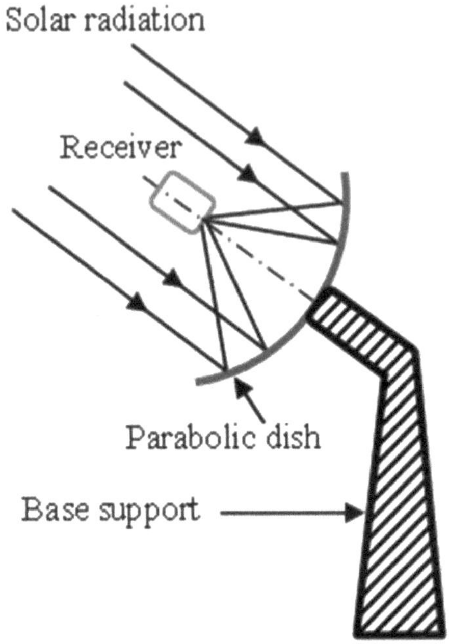

FIGURE 1.22 Schematic representation of a parabolic dish.

(Zhang et al., 2013). This design does not require water for cooling and therefore, it is suitable for exploitation in desert areas. Among the different types of CSP technologies, the PDC exhibits the highest thermodynamic efficiency, but it produces electricity at the highest cost. Another drawback of this type of CSP technology is its low compatibility with thermal storage (Barlev et al., 2011). PDCs have low capacity of power (in the order of kW or less), with each dish generating electricity autonomously. The implication of this is that thousands of dishes are required to build a mega-scale power plant.

There are only two PDC power plants (Maricopa Solar Project and Tooele Army Depot) with a capacity of 1.5 MW each, but none of them is operational (National Renewable Energy laboratory, 2020). The Maricopa Solar and Tooele Army Depot power plants are respectively located in Arizona and Utah, the United States of America. The former plant was commissioned in 2010 as a demonstration system and decommissioned in 2011, while the latter started operating as a commercial system in 2013 but became nonoperational by April 13, 2017. By mid-2020, there were neither plants under construction nor development. This indicates that the level of maturity of the PDC is still very low.

Nomenclature

A area (m^2)
C specific heat capacity ($J\ kg^{-1}\ K^{-1}$)
e internal energy ($J\ kg^{-1}$)
G irradiance ($W\ m^{-2}$)
h coefficient of heat transfer ($Wm^{-2}\ K^{-1}$)
i specific enthalpy ($J\ kg^{-1}$)
K angular modifier (dimensionless)
P pressure (Pa)
Q heat rate (W)
s specific entropy ($J\ kg^{-}\ K^{-1}$)
t time (s)
T temperature (K)
V volume (m^3)
W work (J)
$\dot{m}$ mass flow rate ($kg\ s^{-1}$)

Greek symbols

α absorptance (dimensionless)
γ proportion of reflected solar radiation (dimensionless)
η energy efficiency (dimensionless)
ρ reflectance (dimensionless)
τ transmittance (dimensionless)

References

Aberoumand, S., Ghamari, S., & Shabani, B. (2018). Energy and exergy analysis of a photovoltaic thermal (PV/T) system using nanofluids: An experimental study. *Solar Energy, 165*, 167–177. https://doi.org/10.1016/j.solener.2018.03.028

Ahmad, M., & Karimi, M. N. (2013). Thermodynamic analysis of Kalina cycle. *International Journal of Science and Research, 5*(3), 2244–2249.

Akbari, H., Browne, M. C., Ortega, A., Huang, M. J., Hewitt, N. J., Norton, B., & McCormack, S. J. (2019). Efficient energy storage technologies for photovoltaic systems. *Solar Energy, 192*, 144–168. https://doi.org/10.1016/j.solener.2018.03.052

Ameer, S. A. A., & Shahad, A. K. A. (2017). Characteristics review of optical concentrators. *International Journal of Current Engineering and Technology, 7*(1), 168–182. http://inpressco.com/category/ijcet.

ASHRAE. (2017). *ASHRAE handbook - fundamentals*. American Society of Heating, Refrigerating and Air-Conditioning Engineers (ASHRAE).

Barlev, D., Vidu, R., & Stroeve, P. (2011). Innovation in concentrated solar power. *Solar Energy Materials and Solar Cells, 95*(10), 2703–2725. https://doi.org/10.1016/j.solmat.2011.05.020

Benoit, H., Spreafico, L., Gauthier, D., & Flamant, G. (2016). Review of heat transfer fluids in tube-receivers used in concentrating solar thermal systems: Properties and heat transfer coefficients. *Renewable and Sustainable Energy Reviews, 55*, 298–315. https://doi.org/10.1016/j.rser.2015.10.059

Blanco, M. J., & Santigos, L. R. (2017). In *Advances in concentrating solar thermal research and technology*. Elsevier.

Boyce, M. P. (2002). *Gas turbine engineering handbook* (2nd ed.). Gulf Professional Publishing.

Cameretti, M. C., Langella, G., Sabino, S., & Tuccillo, R. (2015). Modeling of a hybrid solar micro gas-turbine power plant. *Energy Procedia, 82*, 833–840. https://doi.org/10.1016/j.egypro.2015.11.820. Elsevier Ltd.

Cen, Z., Kubiak, P., López, C. M., & Belharouak, I. (2018). Demonstration study of hybrid solar power generation/storage micro-grid system under Qatar climate conditions. *Solar Energy Materials and Solar Cells, 180*, 280–288. https://doi.org/10.1016/j.solmat.2017.06.053

Cohen, H., Rogers, G., & Saravanamuttoo. (1996). *Gas turbine theory*. Longman Group Limited.

De Oliveira-De Jesus, P. M. (2019). Effect of generation capacity factors on carbon emission intensity of electricity of Latin America & the Caribbean, a temporal IDA-LMDI analysis. *Renewable and Sustainable Energy Reviews, 101*, 516–526. https://doi.org/10.1016/j.rser.2018.11.030

Dowling, A. W., Zheng, T., & Zavala, V. M. (2017). Economic assessment of concentrated solar power technologies: A review. *Renewable and Sustainable Energy Reviews, 72*, 1019–1032. https://doi.org/10.1016/j.rser.2017.01.006

Drouot, L. P., & Hillairet, M. J. (1984). The themis program and the 2500-KW themis solar power station at targasonne. *Journal of Solar Energy Engineering*, 83–89. https://doi.org/10.1115/1.3267567

Duffie, J. A., & Beckman, W. A. (2013). Solar engineering of thermal processes: Fourth edition. In *Solar engineering of thermal processes* (4th ed.). John Wiley and Sons. https://doi.org/10.1002/9781118671603

Dunham, M. T., & Iverson, B. D. (2014). High-efficiency thermodynamic power cycles for concentrated solar power systems. *Renewable and Sustainable Energy Reviews, 30*, 758–770. https://doi.org/10.1016/j.rser.2013.11.010

Forman, C., Muritala, I. K., Pardemann, R., & Meyer, B. (2016). Estimating the global waste heat potential. *Renewable and Sustainable Energy Reviews, 57*, 1568–1579. https://doi.org/10.1016/j.rser.2015.12.192

Franchini, G., Perdichizzi, A., Ravelli, S., & Barigozzi, G. (2013). A comparative study between parabolic trough and solar tower technologies in Solar Rankine Cycle and Integrated Solar Combined Cycle plants. *Solar Energy, 98*, 302–314. https://doi.org/10.1016/j.solener.2013.09.033

García, L. I., Álvarez, J. L., & Blanco, D. (2011). Performance model for parabolic trough solar thermal power plants with thermal storage: Comparison to operating plant data. *Solar Energy, 85*(10), 2443–2460. https://doi.org/10.1016/j.solener.2011.07.002

Garg, P., Kumar, P., & Srinivasan, K. (2013). Supercritical carbon dioxide Brayton cycle for concentrated solar power. *Journal of Supercritical Fluids, 76*, 54–60. https://doi.org/10.1016/j.supflu.2013.01.010

Grange, B., Dalet, C., Falcoz, Q., Siros, F., & Ferrière, A. (2014). Simulation of a hybrid solar gas-turbine cycle with storage integration. *Energy Procedia, 49*, 1147–1156. https://doi.org/10.1016/j.egypro.2014.03.124

Green, A., Diep, C., Dunn, R., & Dent, J. (2015). High capacity factor CSP-PV hybrid systems. *Energy Procedia, 69*, 2049–2059. https://doi.org/10.1016/j.egypro.2015.03.218

Güngören Madenoğlu, T. (2017). Effect of operating parameters on synthesis of lithium iron phosphate (LiFePO4) particles in near- and super-critical water. *Journal of Supercritical Fluids, 127*, 103–110. https://doi.org/10.1016/j.supflu.2017.03.031

Ho, C. K. (2017). Advances in central receivers for concentrating solar applications. *Solar Energy, 152*, 38–56. https://doi.org/10.1016/j.solener.2017.03.048

International Energy Agency. (2020). *Data and statistics*. International Energy Agency. https://www.iea.org.

International Renewable Energy Agency. (2013). *Southern African power pool: Planning and prospects for renewable energy.*

International Renewable Energy Agency. (2020a). *Renewable energy statistics 2020.* International Renewable Energy Agency.

International Renewable Energy Agency. (2020b). *Renewable power generation costs in 2019.* International Renewable Energy Agency.

IPCC. (2007). *Climate change 2007: The physical science basis.* Intergovernmental Panel on Climate Change (IPCC).

Islam, M. T., Huda, N., Abdullah, A. B., & Saidur, R. (2018). A comprehensive review of state-of-the-art concentrating solar power (CSP) technologies: Current status and research trends. *Renewable and Sustainable Energy Reviews, 91*, 987–1018. https://doi.org/10.1016/j.rser.2018.04.097

Izquierdo, S., Montañs, C., Dopazo, C., & Fueyo, N. (2010). Analysis of CSP plants for the definition of energy policies: The influence on electricity cost of solar multiples, capacity factors and energy storage. *Energy Policy, 38*(10), 6215–6221. https://doi.org/10.1016/j.enpol.2010.06.009

Kalina, A. I. (2002). *Method of preventing nitridation or carburization of metals. United States Patent 6482272 B2.*

Khan, J., & Arsalan, M. H. (2016). Solar power technologies for sustainable electricity generation - a review. *Renewable and Sustainable Energy Reviews, 55*, 414–425. https://doi.org/10.1016/j.rser.2015.10.135

Kim, E. H., Park, Y. G., & Roh, J. H. (2019). Competitiveness of open-cycle gas turbine and its potential in the future Korean electricity market with high renewable energy mix. *Energy Policy, 129*, 1056–1069. https://doi.org/10.1016/j.enpol.2019.03.014

Korzynietz, R., Brioso, J. A., Del Río, A., Quero, M., Gallas, M., Uhlig, R., Ebert, M., Buck, R., & Teraji, D. (2016). Solugas - comprehensive analysis of the solar hybrid Brayton plant. *Solar Energy, 135*, 578–589. https://doi.org/10.1016/j.solener.2016.06.020

Kribus, A. (2017). Concentrated solar power: Components and materials. In *EPJ Web of Conferences, 00009.* https://doi.org/10.1051/epjconf/201714800009

Labordena, M., Patt, A., Bazilian, M., Howells, M., & Lilliestam, J. (2017). Impact of political and economical barriers for concentrating solar power in Sub-Saharan Africa. *Energy Policy, 102*, 52–72. https://doi.org/10.1016/j.enpol.2016.12.008

Lanchi, M., Montecchi, M., Crescenzi, T., Mele, D., Miliozzi, A., Russo, V., Mazzei, D., Misceo, M., Falchetta, M., & Mancini, R. (2015). Investigation into the coupling of micro gas turbines with CSP technology: OMSoP project. *Energy Procedia, 69*, 1317–1326. https://doi.org/10.1016/j.egypro.2015.03.146

Li, X. (2005). Diversification and localization of energy systems for sustainable development and energy security. *Energy Policy, 33*(17), 2237–2243. https://doi.org/10.1016/j.enpol.2004.05.002

Madhlopa, A. (2018a). Configurations of solar gas turbines. In *Green energy and technology* (pp. 145–162). Springer Verlag. https://doi.org/10.1007/978-3-319-68388-1_6. Issue 9783319683874.

Madhlopa, A. (2018b). Thermodynamic cycles of solar gas turbines. In *Green energy and technology* (pp. 103–143). Springer Verlag. https://doi.org/10.1007/978-3-319-68388-1_5, 9783319683874.

Madhlopa, A., & Okoroigwe, E. (2017). Solar gas turbine systems. In *Encyclopedia of sustainable technologies* (pp. 377–388). Elsevier. https://doi.org/10.1016/B978-0-12-409548-9.10150-2

Mansiri, K., Sukchai, S., & Sirisamphanwong, C. (2014). Investigations to conduct a study about possibilities to use small scale solar dish Stirling engine system in Thailand. *Energy Procedia, 56*(C), 367–377. https://doi.org/10.1016/j.egypro.2014.07.169

Merchant. (2009). *World's first hybrid solar-gas turbine power station is under construction in Israel*. http://www.treehugger.com.

Modi, A., Kærn, M. R., Andreasen, J. G., & Haglind, F. (2016). Thermoeconomic optimization of a Kalina cycle for a central receiver concentrating solar power plant. *Energy Conversion and Management, 115*, 276–287. https://doi.org/10.1016/j.enconman.2016.02.063

National Renewable Energy laboratory. (2020). *Concentrating solar power projects*. National Renewable Energy laboratory (NREL). https://solarpaces.nrel.gov.

Next, -C. S. P. (2016). *High Temperature concentrated solar thermal power plant with particle receiver and direct thermal storage*. Next CSP http://www.next-csp.eu.

Okoroigwe, E. C., & Madhlopa, A. (2019). Advances in biofueling of micro gas turbines for power generation. *Energy Technology, 7*(5). https://doi.org/10.1002/ente.201800689

Olumayegun, O., Wang, M., & Kelsall, G. (2016). Closed-cycle gas turbine for power generation: A state-of-the-art review. *Fuel, 180*, 694–717. https://doi.org/10.1016/j.fuel.2016.04.074

Pelay, U., Luo, L., Fan, Y., Stitou, D., & Rood, M. (2017). Thermal energy storage systems for concentrated solar power plants. *Renewable and Sustainable Energy Reviews, 79*, 82–100. https://doi.org/10.1016/j.rser.2017.03.139

Petrakopoulou, F., Sánchez-Delgado, S., Marugán-Cruz, C., & Santana, D. (2017). Improving the efficiency of gas turbine systems with volumetric solar receivers. *Energy Conversion and Management, 149*, 579–592. https://doi.org/10.1016/j.enconman.2017.07.058

Pillot, B., Muselli, M., Poggi, P., & Dias, J. B. (2019). Historical trends in global energy policy and renewable power system issues in Sub-Saharan Africa: The case of solar PV. *Energy Policy, 127*, 113–124. https://doi.org/10.1016/j.enpol.2018.11.049

Poullikkas, A. (2005). An overview of current and future sustainable gas turbine technologies. *Renewable and Sustainable Energy Reviews, 9*(5), 409–443. https://doi.org/10.1016/j.rser.2004.05.009

Quero, M., Korzynietz, R., Ebert, M., Jiménez, A. A., Del Río, A., & Brioso, J. A. (2014). Solugas - operation experience of the first solar hybrid gas turbine system at MW scale. *Energy Procedia, 49*, 1820–1830. https://doi.org/10.1016/j.egypro.2014.03.193

Rabl, A. (1976). Comparison of solar concentrators. *Solar Energy, 18*(2), 93–111. https://doi.org/10.1016/0038-092X(76)90043-8

Rahmati, A., Varedi-Koulaei, S. M., Ahmadi, M. H., & Ahmadi, H. (2020). Dimensional synthesis of the Stirling engine based on optimizing the output work by evolutionary algorithms. *Energy Reports, 6*, 1468–1486. https://doi.org/10.1016/j.egyr.2020.05.030

Saikku, L., Rautiainen, A., & Kauppi, P. E. (2008). The sustainability challenge of meeting carbon dioxide targets in Europe by 2020. *Energy Policy, 36*(2), 730–742. https://doi.org/10.1016/j.enpol.2007.10.007

Selman, J. R., & Chen, C. C. (2012). Scientific and technical maturity of molten carbonate technology. *International Journal of Hydrogen Energy, 37*(24), 19280–19288. https://doi.org/10.1016/j.ijhydene.2012.06.016

Shan. (2020). A review of Kalina cycle. *International Journal of Smart Energy Technology and Environmental Engineering, 1*(1), 77–107.

SOLGATE. (2005). *Solar hybrid gas turbine electric power system*.

SOLHYCO. (2011). *Solar-hybrid power and cogeneration plants*.

Stein, W. H., & Buck, R. (2017). Advanced power cycles for concentrated solar power. *Solar Energy, 152*, 91–105. https://doi.org/10.1016/j.solener.2017.04.054

Tyagi, V. V., Rahim, N. A. A., Rahim, N. A., & Selvaraj, J. A. L. (2013). Progress in solar PV technology: Research and achievement. *Renewable and Sustainable Energy Reviews, 20*, 443–461. https://doi.org/10.1016/j.rser.2012.09.028

Ummadisingu, A., & Soni, M. S. (2011). Concentrating solar power - technology, potential and policy in India. *Renewable and Sustainable Energy Reviews, 15*(9), 5169–5175. https://doi.org/10.1016/j.rser.2011.07.040

van Vuuren, D. P., Nakicenovic, N., Riahi, K., Brew-Hammond, A., Kammen, D., Modi, V., Nilsson, M., & Smith, K. R. (2012). An energy vision: The transformation towards sustainability-interconnected challenges and solutions. *Current Opinion in Environmental Sustainability, 4*(1), 18–34. https://doi.org/10.1016/j.cosust.2012.01.004

Wada, N., Horiuchi, N., Mukougawa, K., Nozaki, K., Nakamura, M., Nagai, A., Okura, T., & Yamashita, K. (2016). Electrostatic induction power generator using hydroxyapatite ceramic electrets. *Materials Research Bulletin, 74*, 50–56. https://doi.org/10.1016/j.materresbull.2015.10.007

Wagner, M. J., Hamilton, W. T., Newman, A., Dent, J., Diep, C., & Braun, R. (2018). Optimizing dispatch for a concentrated solar power tower. *Solar Energy, 174*, 1198–1211. https://doi.org/10.1016/j.solener.2018.06.093

Xu, J., & Yu, C. (2014). Critical temperature criterion for selection of working fluids for subcritical pressure organic Rankine cycles. *Energy, 74*(C), 719–733. https://doi.org/10.1016/j.energy.2014.07.038

Zhang, H. L., Baeyens, J., Degrève, J., & Cacères, G. (2013). Concentrated solar power plants: Review and design methodology. *Renewable and Sustainable Energy Reviews, 22*, 466–481. https://doi.org/10.1016/j.rser.2013.01.032

Chapter 2

Solar radiation resource

2.1 Introduction

A concentrating solar power (CSP) technology converts solar radiation to heat, which is transferred from the solar receiver to other components of the system. Thus, knowledge about the physics of solar radiation is critical in the development of receivers.

Solar radiation is a free source of energy and the most influential environmental factor that affects solar energy systems (Nafey et al., 2000). Applications of solar technologies fall into photovoltaic (PV) and thermal categories. The former group of solar energy systems directly converts solar radiation into electricity through the photoelectric effect, while the latter generates heat from solar radiation. In the case of a CSP system, this heat is directly (or indirectly) transferred from the receiver to a turbine. In this chapter, fundamentals of solar radiation are discussed and applied to the process of solar conversion in solar receivers. It is shown, in this chapter, that the worldwide technical potential of CSP technology is much higher than the world electricity consumption level. This magnitude of the availability of solar resource is attractive for the advancement of the CSP technology.

2.2 Source of solar radiation

2.2.1 The sun

The sun is the primary source of solar radiation. It emits radiation at an equivalent black body temperature of about 6000 K with a constant intensity outside the earth's atmosphere (Kane, 2005). This radiation travels through space to reach the earth surface.

2.2.2 Solar constant

Irradiance on a surface that is perpendicular to the direction of propagation of solar radiation at an average distance between the earth and the sun outside the atmosphere is known as the solar constant. The solar constant is required in solar radiation analysis as shown in Section 1.5. Consequently, different authors have attempted to establish an accurate value of this constant for

Solar Receivers for Thermal Power Generation. https://doi.org/10.1016/B978-0-323-85271-5.00007-0

application in solar engineering and science. A range of values of 1366–1395 Wm^{-2} have been reported in the literature, but the World Radiation Center adopted a value of 1367 Wm^{-2} (Gueymard, 2004).

2.3 Components of solar radiation

2.3.1 Beam and diffuse radiation

Outside the earth's atmosphere, there is only direct (beam) radiation. However, as solar radiation propagates through the atmospheric matter, it undergoes scattering, absorption, and transmission. The scattering process produces the diffuse component of global solar radiation. So, the total solar radiation that reaches the earth's surface is made up of beam (direct) and diffuse components (Yang et al., 2001). These two components of solar radiation have different optical properties when incident on a surface. Beam radiation travels directly from the sun's disc to a receiver surface, and its rays can be traced from the sun's position and used in the computation of solar angles. The beam component of radiation can also be focused to increase the amount of solar energy intercepted per unit area of an absorber surface. It is, therefore, not surprising that the CSP technology utilizes the direct component of solar radiation.

On the other hand, diffuse radiation comes from the whole sky vault and it cannot be focused. Moreover, its rays are not traceable from the sun's position, and the amount of diffuse solar energy directly received by a given surface depends on the proportion of the sky viewed by the surface (Duffie & Beckman, 2013).

2.3.2 Direct normal irradiance

Direct normal irradiance (G_{dn}) is the amount of solar radiation intercepted per unit area of a surface, which is held at right angles to the rays that propagate in a straight line from the sun at its instantaneous position in the sky. However, the position of the sun varies with time. In view of this, the surface has to be mounted on a solar tracking system. In this case, the surface follows the sun on daily (one-axis) or daily and seasonal (two-axis) basis to maintain normal incidence of the beam radiation.

A sun tracker is an electromechanical device, which ensures that the rays of solar radiation are incident at right angles to the plane of the aperture of a solar collector. Sun trackers can primarily be classified into passive, time-based, and active categories (Sabir & Ali, 2016). In a passive tracker, the sun heats up a low-boiling gas, which creates a pressure gradient causing the gas to flow on one end of the tracker, thereby moving the tracker. The hourly, daily, and yearly positions of the sun are computerized by using an algorithm. At a given time, the tracker compares predetermined and actual values of the tilt and azimuth angles and makes appropriate corrections. Only one axis of the movement of the sun is tracked in single-axis tracking systems. For instance, only the tilt angle may be tracked. In active trackers, sensors are used to establish the position of the sun in the sky, and a reference signal is generated to stimulate DC motor motor-based actuators to properly align the solar collector with the rays of the sun at a given time. Active trackers can be single axis or dual axis. A single-axis solar tracker has a fixed tilt angle, but the azimuth angle changes throughout. This category of active trackers is cheap to apply. In a dual-axis tracker, both the tilt and azimuth angles are varied.

Global (G_{gh}) and diffuse (G_{dh}) components of solar radiation are measured on a horizontal surface at many locations in the world. In this case, direct normal irradiance (G_{dn}) can be calculated from (Kotti et al., 2014):

$$G_{dn} = \frac{G_{gh} - G_{dh}}{\cos\ \theta_z} \tag{2.1}$$

Where G_{gh} is the global irradiance on a horizontal surface, G_{dh} is the diffuse irradiance on a horizontal surface, and θ_z is the solar zenith angle.

The level of direct normal irradiance is a good indicator of the suitability of a site for the deployment of CSP technologies. It varies in space and time, and the mean daily levels of insolation are shown in Fig. 2.1. The recommended minimum annual sum of irradiation for CSP technology to be economically viable is 2000 kWh m^{-2} (Spelling, 2013), and many locations within the sunbelt in the world meet this criterion. So, there is great potential for the exploitation of CSP technology. In 2013, the worldwide technical potential of CSP technology was estimated at 3000000 TWh yr^{-1} (Spelling, 2013), which

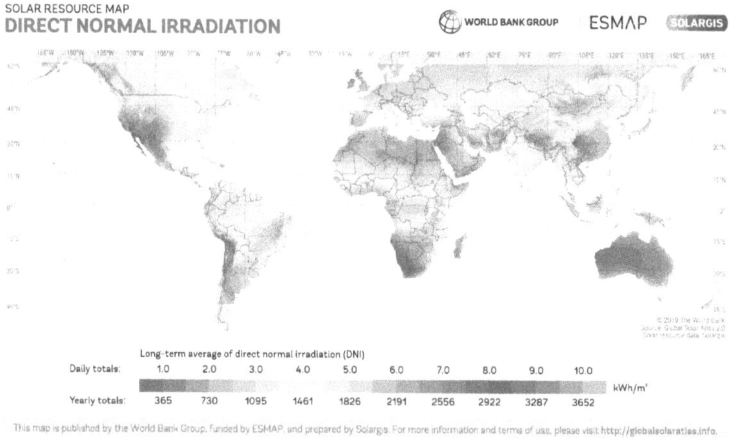

FIGURE 2.1 Distribution of direct normal irradiation worldwide. *From Solargis (2019). Solar resource maps of world: Direct normal irradiation. Solargis. https://solargis.com/maps-and-gis-data/download/world.*

significantly exceeded the world electricity consumption level of 29000 TWh yr^{-1} in 2015 (International Energy Agency, 2017). This solar resource potential indicates the importance of developing receivers that can withstand high levels of solar flux.

2.4 Position of the sun and direction of beam radiation

At any given time and place, knowledge about the position of the sun is required for calculation of the beam component of the solar radiation incident on a surface and for the determination of angular-dependent optical properties of transparent materials. The position of the sun can be specified by the zenith (θ_Z), declination (δ), azimuth (γ_S), and hour (ω) angles (Braun & Mitchell, 1983). Beam radiation makes an angle θ_Z with the normal to a horizontal plane (zenith) and angle θ_I with the normal to an inclined plane. The declination angle (δ) is the angle between the sun's rays and the equatorial plane, and this parameter can be taken to be positive in the northern hemisphere and negative in the southern hemisphere. Further, the solar azimuth angle is measured from south to the horizontal projection of the sun's rays on a horizontal plane, and it is negative in the east of south and positive in the west of south (Duffie & Beckman, 2013). The earth has to turn through an angle ω to bring the meridian of an observer at a specific point directly in line with the rays of the sun, and ω is negative in the morning, zero at solar noon and positive in the afternoon. The various solar and surface angles are shown in Fig. 2.2.

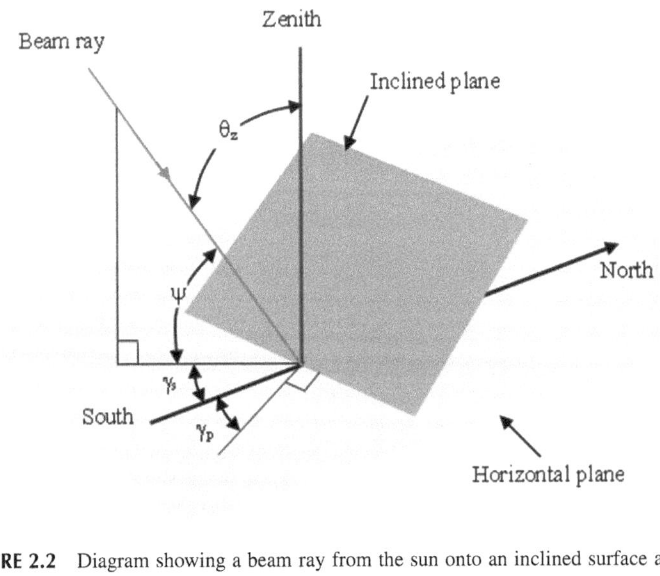

FIGURE 2.2 Diagram showing a beam ray from the sun onto an inclined surface and solar and surface angles (in the northern hemisphere).

The angle of incidence (θ_1) of solar radiation on a surface inclined at angle β to the horizontal plane influences the proportions of solar radiation absorbed and transmitted by the receiving surface. The transmission of solar radiation decreases with increasing the angle of incidence. At solar noon, the angle of incidence is zero or close to zero, while the intensity of solar radiation is optimum. An expression for θ_1 is given by (Howell et al., 1982):

$$\cos\theta_1 = \cos\theta_z \cos\beta + \sin\theta_z \sin\beta \cos\left(\gamma_s - \gamma_p\right) \tag{2.2}$$

Where γ_s is the solar azimuth angle and γ_p is the azimuth angle of the plane.

$$\cos\theta_z = \cos\delta \cos\phi \cos\omega + \sin\delta \sin\phi \tag{2.3}$$

Where ϕ is the latitude of the location.

$$\cos\gamma_s = (\sin\phi \cos\delta \cos\omega - \cos\phi \sin\delta)/\cos\psi \tag{2.4}$$

Where ψ is solar altitude angle ($\psi = 90^{\circ} - \theta_z$)

$$\omega = 15(t_s - 12) \tag{2.5}$$

Where $t_s = t_{ct} + t_{eq} + (\Theta_{sd} - \Theta)/15$, $t_{eq} = 229.2\{0.000075 + 0.001868 \cos \Gamma - 0.032077 \sin \Gamma - 0.014615 \cos (2\Gamma) - 0.04089 \sin (2\Gamma)\}/60$, Θ_{sd} is the standard longitude of the location, and Θ is the longitude of the location.

$$t_s = t_{ct} + t_{eq} + \frac{\Theta_{sd} - \Theta}{15} \tag{2.6}$$

$$\begin{aligned} t_{eq} = 229.2[&0.000075 + 0.001868 \cos\Gamma - 0.032077 \sin\Gamma \\ &- 0.014615 \cos(2\Gamma) - 0.04089 \sin(2\Gamma)]/60 \end{aligned} \tag{2.7}$$

$$\Gamma = 360(N - 1)/365 \tag{2.8}$$

Where N is the day of the year, starting from 1 January (N = 1).

Eqs. (2.3) and (2.4) also involve the declination angle that can be calculated from (Jain, 1988):

$$\begin{aligned} \delta = 0.006918 &- 0.399912 \cos\Gamma + 0.07257 \sin\Gamma - 0.006758 \cos 2\Gamma \\ &+ 0.000907 \sin 2\Gamma - 0.002697 \cos 3\Gamma + 0.00148 \sin 3\Gamma \end{aligned} \tag{2.9}$$

When the sun is directly overhead on a horizontal surface ($\psi = 90^{\circ}$ and $\theta_z = 0$), solar radiation transmittance through the atmosphere is maximum at a given location on the earth's surface. The zenith angle is equal to zero during some time in the year at sites within the tropical zone, and it is always greater than zero at sites outside this region. The magnitude of the zenith angle is smallest, and the potential for stable meteorological conditions is highest around solar noon on a clear day. The values of solar angles and extraterrestrial radiation are required in the computation of solar radiation received by a surface.

Example 2.1

Calculate the angle of incidence on a north-facing surface inclined at 15° from the horizontal at 11:00 on 21st January at Makoka Weather Station (15.533° S, 35.183°E) in Malawi (Madhlopa, 2006).

Solution

The angle of incidence on an inclined surface is computed from Eq. (2.2), which requires equation of time (t_{eq}), solar time (t_s), hour angle (ω), the declination angle (δ), zenith angle (θ_z) and solar (γ_s) and surface (γ_p) azimuth angles, and inclination angle (β). Makoka Weather Station is located in the Southern Hemisphere where latitude is negative. Therefore, ϕ=−15.533°.

Equation of time on 21st January, t_{eq} = 229.2{0.000075 + 0.001868 cos (20*360/365)−0.032077 sin (20*360/365)−0.014615 cos (2*20*360/365)−0.04089 sin (2*20*360/365)}/60 = −0.23282

Solar time

$$\begin{aligned} t_s &= t_{ct} + t_{eq} + (\Phi_{sd} - \Phi)/15 \\ &= 11 - 0.23282 + (30 - 35.183)/15 = 10.4217\ \text{h} \end{aligned}$$

$$\omega = 15(t_s - 12) = 15(11.6995 - 12) = -23.675^\circ$$

$$\cos\theta_1 = \cos\theta_z \cos\beta + \sin\theta_z \sin\beta \cos(\gamma_s - \gamma_p)$$

But $\cos\theta_z = \sin\delta \sin\phi + \cos\delta \cos\phi \cos\omega$

$$\begin{aligned} \delta = {} & 0.006918 - 0.399912 \cos\mathit{\Gamma} + 0.07257 \sin\mathit{\Gamma} - 0.006758 \cos 2\mathit{\Gamma} \\ & + 0.000907 \sin 2\mathit{\Gamma} - 0.002697 \cos 3\mathit{\Gamma} + 0.00148 \sin 3\mathit{\Gamma} \end{aligned}$$

$$\Gamma = 360(21 - 1)/365 = 20 * 360/365 = 19.72603$$

$$\delta = 180[0.006918 - 0.399912 \cos\Gamma + 0.07257 \sin\Gamma - 0.006758 \cos(2\Gamma)$$

$$+ 0.000907 \sin(2\Gamma) - 0.002697 \cos(3\Gamma) + 0.00148 \sin(3\Gamma)]/\pi$$

$$= 180[0.006918 - 0.399912 \cos 19.72603 + 0.07257 \sin 19.72603$$

$$-0.006758 \cos(2 * 19.72603) + 0.000907 \sin(2 * 19.72603)$$

$$-0.002697 \cos(3 * 19.72603) + 0.00148 \sin(319.72603)]/\pi$$

$$= -10.772^\circ$$

Zenith angle, $\theta_z = \cos^{-1}(\sin\delta\sin\phi + \cos\delta\cos\phi\cos\omega) = \cos^{-1}[\sin(-10.772°)\sin(-15.533°) + \cos(-10.772°)\cos(-15.533°)\cos(-23.675°)] = 23.525°$

Solar altitude, $\psi = 90° - \theta z = 90° - 23.525° = 66.475°$.

Solar azimuth angle, $\gamma_s = \cos^{-1}\{(\sin\phi\cos\delta\cos\omega - \cos\phi\sin\delta)/\cos\psi\} = \cos^{-1}\{[\sin(-15°)\cos(-10.772°)\cos(-23.675°) - \cos(-15.533°)\sin(-10.722°)]/\cos(66.475°)\} = 93.552°$

Angle of incidence, $\theta_1 = \cos^{-1}\{\cos\theta_z\cos\beta + \sin\theta_z\sin\beta\cos(\gamma_s - \gamma_p)\} = \cos^{-1}\{\cos(23.525°)\cos(15°) + \sin(23.525°)\sin(15°)\cos(93.552° - 180°)\} = 26.868°$

2.5 Extraterrestrial radiation and solar radiation on inclined surfaces

Extraterrestrial radiation is the theoretical amount of solar energy that would be available on a horizontal plane on the earth's surface if the earth were not surrounded by an atmosphere. In reality, solar radiation undergoes attenuation as it propagates through the atmosphere, resulting in a decrease of its intensity when measured on a plane on the earth's surface. At any specific time within the day, extraterrestrial irradiance (G_o) can be given by (Duffie & Beckman, 2013):

$$G_o = G_{sc}\left[1 + 0.033\cos\left(\frac{360\,N}{365}\right)\right] \tag{2.10}$$

Where G_{sc} is the solar constant.

In modeling and experimental studies of solar energy technologies, hourly (I) and daily (H) solar radiation data are sometimes required. So, integrated values of hourly (I_o) and daily (H_o) extraterrestrial radiation on a horizontal plane are used in such computations:

$$I_o = \frac{43200}{\pi}G_{sc}\left\{1 + 0.033\cos\left(\frac{360\,N}{365}\right)\right\} \left\{\cos\phi\cos\delta\,(\sin\omega_2 - \sin\omega_1)\frac{\pi(\omega_2 - \omega_1)}{180}\sin\phi\sin\delta\right\} \tag{2.11}$$

Where ω_1 is the hour angle at time $t = t_1$ and ω_2 is the hour angle at $t = t_2$.

$$H_o = \frac{86400}{\pi}G_{sc}\left[1 + 0.033\cos\left(\frac{360\,N}{365}\right)\right] \left[\cos\phi\cos\delta\sin\omega_{ss} + \frac{\pi\omega_{ss}}{180}\sin\phi\sin\delta\right] \tag{2.12}$$

Where ω_{ss} is the hour angle at sun set.

The hourly and daily extraterrestrial radiation is influenced by the day of the year and solar angles at a given location. Extraterrestrial radiation can be used to compute the available solar radiation on a horizontal plane on the earth's surface. However, solar radiation on a tilted surface is needed in the design and performance studies of solar collectors. The total irradiance on a tilted plane (G_{gp}) consists of the beam, diffuse, and ground-reflected components, and it can be calculated from:

$$G_{gp} = G_{bp} + G_{dp} + G_{gr} \tag{2.13}$$

Where G_{bp} is the beam irradiance on a tilted plane, G_{dp} is the diffuse irradiance on a tilted plane, and G_{gr} is ground-reflected irradiance.

In practice, global irradiance is commonly measured on a horizontal surface. Models are therefore employed to generate a database of solar radiation received by inclined surfaces. For beam radiation, a geometric factor ($\acute{R}_b$) is used to compute beam irradiance on a tilted surface from the horizontal beam irradiance:

$$\acute{R}_b = \frac{G_{bp}}{G_{bh}} = \frac{\cos \theta_1}{\cos \theta_z} \tag{2.14}$$

For concentrated solar power, diffuse radiation is ignored because it cannot be focused onto a smaller area.

2.6 Available solar radiation on the earth's surface

Solar radiation that reaches the earth's surface is lower than the extraterrestrial radiation. In addition, commonly available solar radiation data are measured on a horizontal surface. Consequently, solar radiation data on an inclined plane are mostly computed from solar radiation on a horizontal plane. For radiation analysis and other applications, it is necessary to measure solar radiation received by a horizontal surface at a given site.

Available global solar radiation on a horizontal surface can be directly measured by using a pyranometer (Thekaekara, 1976), and accurate solar radiation data are needed for simulation, testing, and operation of solar energy technologies (Kudish & Evseev, 2008). In line with this requirement, it is important to calibrate pyranometers by using standard methods in order to capture accurate hemispherical solar radiation data (International Organization of Standardization, 1993). The ratio of the solar radiant flux (including the direct solar radiant flux) received from the sky hemisphere by a given plane to the area of the plane is called hemispherical irradiance (International Organization of Standardization, 1992). According to the ISO 9060 classification standard, there are three classes of pyranometers: secondary standard, first, and second classes, with the secondary standard class being the best (Kipp & Zonen, 2006). First-class pyranometers are also recommended for measuring irradiance for performance evaluation of solar collectors. Nevertheless, it is often desirable to know the individual levels of beam and diffuse radiation for purposes of solar energy system modeling, design, testing, and application. So, these components of global radiation are measured on a given plane.

Beam radiation can be measured by using direct or indirect methods. In the direct approach, a pyrheliometer is used to measure beam radiation at normal incidence, but this instrument is costly (Kudish & Evseev, 2008). Consequently, only a limited number of meteorological sites in the world has pyrheliometers. In the indirect method, beam radiation is determined by subtracting diffuse radiation from the global radiation. In this case, concurrent measurements of global and diffuse radiation are taken in the same plane.

Diffuse radiation can be measured by using a pyranometer fitted with an occulting disk or a shadow ring, with the occulting disk yielding more accurate data. However, the disk requires an expensive sun-tracking system. In view of this, diffuse radiation is commonly measured by using a pyranometer with a stationary ring in the east–west axis. The ring blocks off beam radiation and a small amount of diffuse radiation from reaching the sensor, which results in slightly lower readings of diffuse irradiance. A correction factor is therefore applied to the data to take care of this error. The difference between global and diffuse irradiance yields beam irradiance, which is then used to compute direct normal irradiance required for concentrating solar power.

2.7 Attenuation of solar radiation when incident on opaque and transparent surfaces

For opaque surfaces, part of the incoming solar radiation is reflected, while the remainder is absorbed by the surface. The surface absorbs (α) and reflects (ρ) part of the radiation (Fig. 2.3). The absorbed proportion is converted to heat (unless the surface has photocells that convert part of solar radiation into electricity). For a smooth surface (such as a mirror), the angle of incidence (θ_1) is equal to the angle of reflection (θ_{re}). The absorptance and reflectance of solar radiation are influenced by the wavelength (λ) and direction (θ_i). The implication of this is that the spectral distribution of the solar flux has to be

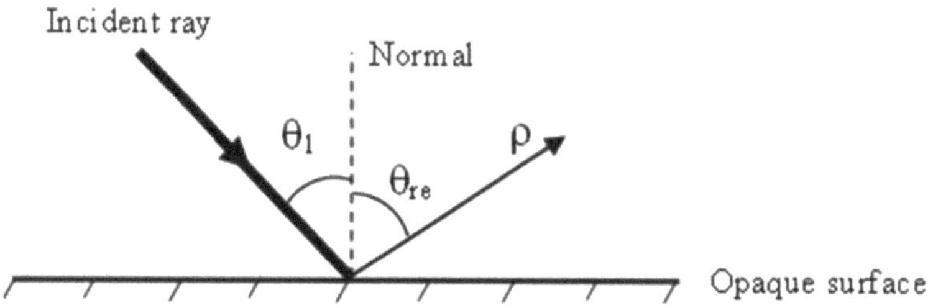

FIGURE 2.3 Attenuation of solar radiation by an opaque surface. θ_1 is the angle of incidence at the boundary between medium 1 and medium 2.

known in order to enable computation of the solar weighted property. To enable comparison of the performance of different solar technologies, it is necessary to use a reference standard solar spectral irradiance distribution (International Organization of Standardization, 1992). Part 1 of the ISO 9845 standard gives a suitable standard spectral irradiance distribution, which can be employed to find the relative performance of solar thermal, photovoltaic, and other systems, and components and materials where the direct and hemispherical irradiance is required. Based on energy balance, the sum of the proportions of reflected and absorbed radiation is equal to one (Eq. 2.15).

$$\alpha(\lambda, \theta_1) + \rho(\lambda, \theta_1) = 1 \tag{2.15}$$

A surface can reflect solar radiation specularly or diffusely (Fig. 2.4). Radiation propagates in one direction from a specular plane reflector, and specular geometric (view) factor and reflectance are required in the computation of reflected beam radiation (Wijeysundera, 1978). This property of radiation is useful in CSP technologies. Only beam radiation can be focused, and therefore, the performance of focusing solar technologies increases with the intensity of beam radiation. For a given set of surface properties, an opaque surface absorbs a maximum level of radiation when the angle of incidence is zero ($\theta_1 = 0$). When $\theta_1 = 0$, the surface is said to receive direct normal irradiance (G_{dn}).

On the other hand, reflected radiation goes in different directions in diffuse reflection, and diffuse view factor and reflectance are required for calculation of the magnitude of reflected diffuse solar radiation. Consequently, a lower proportion of diffuse radiation is desirable for satisfactory performance of CSP technologies. In addition to reflection and absorption, a transparent surface transmits part of the intercepted solar radiation (Fig. 2.5). If the surface is surrounded by air, solar radiation travels through air before it reaches the transparent surface. The surface absorbs (α), reflects (ρ), and transmits (τ) part

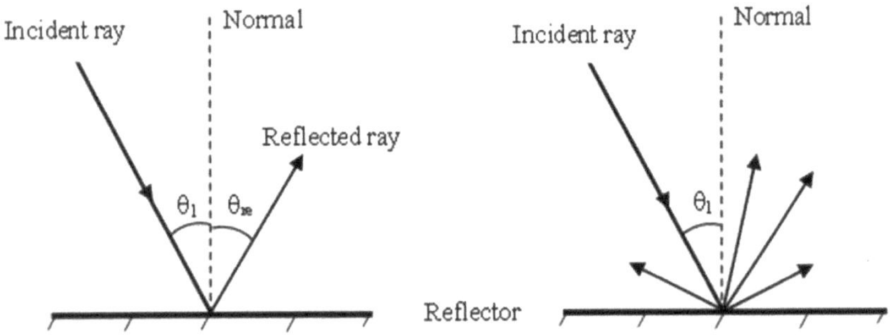

FIGURE 2.4 Attenuation of solar radiation by an opaque surface (A) specular reflection and (B) diffuse reflection.

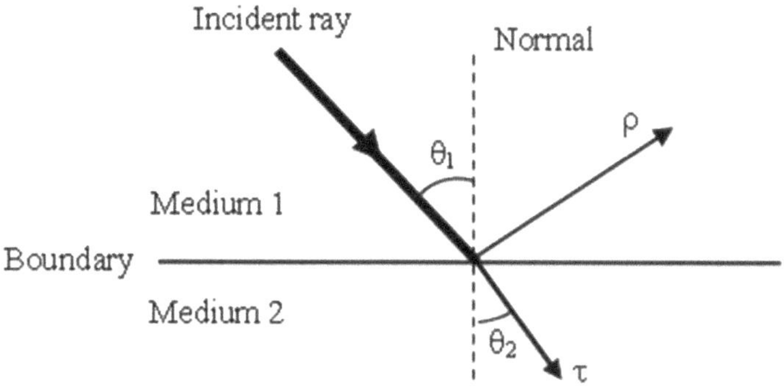

FIGURE 2.5 Attenuation of solar radiation through a transparent surface.

of the radiation. The sum of the proportions of reflected, absorbed, and transmitted radiation is equal to 1, and these fractions are influenced by the optical properties of incoming solar radiation.

$$\alpha(\lambda, \theta_1) + \rho(\lambda, \theta_1) + \tau(\lambda, \theta_1) = 1 \tag{2.16}$$

A ray of light changes direction when it traverses the boundary between two media (medium one and medium 2) that have different values of optical density (n_1 and n_2). This phenomenon is defined as refraction. Snell's law shows that:

$$n_1 \sin\theta_1 = n_2 \sin \theta_2 \tag{2.17}$$

Where n_1 is the refractive index in medium 1, n_2 is the refractive index in medium 2, θ_1 and θ_2 is the angle of refraction.

Example 2.2

A ray of radiation propagates from water (w) to glass (g). If the angle of incidence on the boundary between water and glass is 15°, calculate the angle of refraction. Take $n_w = 4/3$ and $n_g = 3/2$.

Solution

$$n_w \sin\theta_w = n_g \sin\theta_g$$

So, the angle of refraction,

$$\theta_g = \sin^{-1}\left(\frac{n_w \sin\theta_w}{n_g}\right) = \sin^{-1}\left[\frac{(4/3) * \sin(15°)}{(3/2)}\right] = 13.2°$$

Most of the solar radiation received by the earth's surface lies within a wavelength range of $0.29-3.00 \times 10^{-6}$ m (Thekaekara, 1976). Nevertheless, many types of window glazing have a weak spectral selectivity in the solar spectrum (American Society of Heating, Refrigerating and Air-Conditioning Engineers, 2001). In view of this, only their angular dependence can be taken into consideration.

Nomenclature

G irradiance (W m^{-2})
H insolation (J m^{-2})
n refractive index (dimensionless)
$\mathbf{t_{ct}}$ civil time (h)
$\mathbf{t_{eq}}$ equation of time (h)
$\mathbf{t_s}$ solar time (h)

Greek symbols

α absorptance (dimensionless)
β tilt angle (°)
γ azimuthal angle (°)
ϕ latitude of location (°)
λ wavelength (m)
θ angle (°)
Θ longitude (°)
τ transmittance (dimensionless)
ω hour angle (°)
δ declination angle (°)

References

American Society of Heating, Refrigerating and Air-Conditioning Engineers. (2001). *Fundamentals handbook*. American Society of Heating, Refrigerating and Air-Conditioning Engineers (ASHRAE).

Braun, J. E., & Mitchell, J. C. (1983). Solar geometry for fixed and tracking surfaces. *Solar Energy, 31*(5), 439–444. https://doi.org/10.1016/0038-092X(83)90046-4

Duffie, J. A., & Beckman, W. A. (2013). *Solar engineering of thermal processes* (4th ed.). John Wiley & Sons Inc. https://doi.org/10.1002/9781118671603

Gueymard, C. A. (2004). The sun's total and spectral irradiance for solar energy applications and solar radiation models. *Solar Energy, 76*(4), 423–453. https://doi.org/10.1016/j.solener.2003.08.039

Howell, J., Bannerot, R., & Vliet, G. (1982). *Solar-thermal energy systems: Analysis and design*. McGraw_hill Book Company.

International Energy Agency. (2017). *Key World Energy Statistics 2017*. Paris: International Energy Agency. https://doi.org/10.1787/key_energ_stat-2017-en

International Organization of Standardization. (1992). *Standard 9845-1: Solar energy - reference solar spectral irradiance at the ground at different receiving conditions - part 1: Direct normal and hemispherical solar irradiance for air mass 1.5*. International Organization of Standardization (ISO).

International Organization of Standardization. (1993). *Standard 9846: Solar energy - calibration of a pyranometer using a pyrheliometer.* International Organization of Standardization (ISO).

Jain, P. C. (1988). Accurate computations of monthly average daily extraterrestrial irradiation and the maximum possible sunshine duration. *Solar & Wind Technology, 5*, 41–53. https://doi.org/10.1016/0741-983X(88)90088-4

Kane, R. P. (2005). Sun-Earth relation: Historical development and present status - a brief review. *Advances in Space Research, 35*(5), 866–881. https://doi.org/10.1016/j.asr.2005.03.142

Kipp & Zonen. (2006). *Instruction manual: CMP series 11/CMA series pyranometers/albedometer.* Kipp & Zonen.

Kotti, M., Argiriou, A., & Kazantzidis, A. (2014). Estimation of direct normal irradiance from measured global and corrected diffuse horizontal irradiance. *Energy, 70*, 382–392.

Kudish, A. I., & Evseev, E. G. (2008). The assessment of four different correction models applied to the diffuse radiation measured with a shadow ring using global and normal beam radiation measurements for Beer Sheva, Israel. *Solar Energy, 82*(2), 144–156. https://doi.org/10.1016/j.solener.2007.06.006

Madhlopa, Amos (2006). Solar radiation climate in Malawi. *Solar Energy, 80*, 1055–1057. https://doi.org/10.1016/j.solener.2005.08.007

Nafey, A. S., Abdelkader, M., Abdelmotalip, A., & Mabrouk, A. A. (2000). Parameters affecting solar still productivity. *Energy Conversion and Management, 41*(16), 1797–1809. https://doi.org/10.1016/S0196-8904(99)00188-0

Sabir, M. M., & Ali, T. (2016). Optimal PID controller design through swarm intelligence algorithms for sun tracking system. *Applied Mathematics and Computation, 274*, 690–699. https://doi.org/10.1016/j.amc.2015.11.036

Solargis (2019). Solar resource maps of world: Direct normal irradiation. Solargis. https://solargis.com/maps-and-gis-data/download/world.

Spelling, J. (2013). *Hybrid solar gas-turbine power plants: A thermoeconomic analysis.* KTH Royal Institute of Technology.

Thekaekara, M. P. (1976). Solar radiation measurement: Techniques and instrumentation. *Solar Energy, 18*(4), 309–325. https://doi.org/10.1016/0038-092X(76)90058-X

Wijeysundera, N. E. (1978). Geometric factors for plane specular reflectors. *Solar Energy, 20*(1), 81–85. https://doi.org/10.1016/0038-092X(78)90144-5

Yang, K., Huang, G. W., & Tamai, N. (2001). Hybrid model for estimating global solar radiation. *Solar Energy, 70*(1), 13–22. https://doi.org/10.1016/S0038-092X(00)00121-3

Chapter 3

Classification of solar receivers

3.1 Introduction

Generally, classification of things is based on their characteristics, which are features that distinguish one thing from another. Thus, devices with similar characteristics are grouped together, and several levels may be applied for refinement of a classification system. The performance of a solar receiver is affected by design, operational, and environmental parameters. Design and operational parameters are characteristic of a given receiver, while environmental conditions are external factors. Consequently, design and operational parameters constitute a useful basis for classifying receivers. In this vein, two major criteria for classifying receivers are geometric design and adapted heat transfer media.

3.2 Geometric design

Geometry is an important design consideration in the development of receivers. It affects the optical and thermal performance of the receiver. For instance, the amount of reflected solar radiation in a specific direction depends on the shape of the surface. In this regard, it has been found that the distribution of solar radiation and the absorption of heat depend on geometric variables of a cavity receiver (Martínez-Manuel et al., 2021). Heat is also lost from the receiver to the environment through convective and radiative heat transfer, and these two modes of heat transfer are affected by geometric factors (ASHRAE, 2017). Heat convection involves movement of fluid particles from a surface by natural or forced mechanisms, and in each case, the coefficient of heat transfer is affected by the shape of the surface. Similarly, the coefficient of radiative heat transfer is influenced by the optical view factor, which depends on the geometry of surfaces involved in radiation exchange. The amount of exchanged radiative heat increases with the magnitude of the view factor between surfaces. Based on geometry, receivers are classified into: tubular, volumetric, microchannel, linear and point focus, and external and cavity receivers.

Solar Receivers for Thermal Power Generation. https://doi.org/10.1016/B978-0-323-85271-5.00001-X

3.2.1 Tubular receivers

A tubular receiver comprises one or more tubes on which concentrated solar radiation is incident. One major advantage of the circular geometry is its ability to resist pressure. For instance, cylindrical water tanks can withstand the mains pressure. In a tubular receiver, the tubes convert solar radiation to high-temperature heat, which is transferred from the external surfaces to the heat transfer fluid inside the tubes. This variety of receivers utilizes a closed-loop configuration in which the heat transfer fluid is pumped through the irradiated tubes. Some receivers comprise one tube while others have multiple tubes that are assembled into a panel with distribution and collector headers (Fig. 3.1). The cold fluid enters the panel through the distributor header and exits through the collector header, and several panels can be connected together in series or parallel configurations. Higher receiver outlet temperatures are attainable by the series configuration than the parallel one. However, the latter configuration exhibits lower pressure loss.

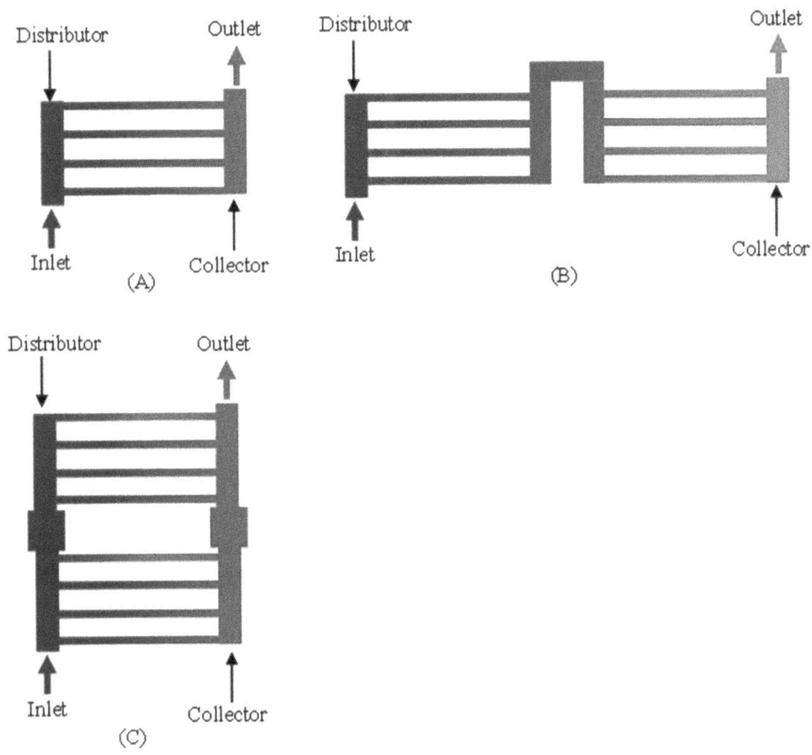

FIGURE 3.1 Schematic illustration of (A) tubular panel, (B) series connection, and (C) parallel connection.

3.2.2 Volumetric receivers

A volumetric solar receiver directly or indirectly absorbs concentrated solar radiation in three dimensions. In direct-absorption volumetric receivers, solar radiation is absorbed by the heat transfer fluid such as a nanofluid. In an indirect absorption volumetric receiver, solar radiation is absorbed by a porous material (honeycomb) or a block of channels. This receiver design often utilizes air as a heat transfer fluid. The fluid is blown through the porous material or block of channels, and heat is transferred from the hot surface of the absorber element to the air. The use of Shell Thermia B oil as a heat transfer fluid has also been explored (Mlatho & McPherson, 2012). Design examples of volumetric receivers are shown in Fig. 3.2.

Ideally, concentrated solar radiation is supposed to propagate deep into the receiver in order to create the hottest zone inside the porous material, farther away from the aperture (Ho, 2017). However, it has been practically observed that the aperture of the volumetric receiver is the hottest region, which results in high radiative heat losses from the receiver to the environment. The air can be heated to temperatures ranging between 1073 and 1773 K, depending on the type of material. The use of microchannels is a recent concept in the development of volumetric receivers, which is aimed at increasing the surface area of contact between the heat transfer fluid and the walls of the channels.

Volumetric receivers can be applied in power plants through two basic principles: (a) Open-loop receiver system with a Rankine cycle and (b) closed-loop receiver system with a Brayton cycle (Ávila-Marín, 2011; Ho, 2017). In the former system, atmospheric air passes through a metal or ceramic volumetric receiver, and then it is utilized to produce steam in a heat recovery steam generator with a separate super-heater. The steam is utilized to drive the Rankine cycle to produce electricity. In the latter system, solar heat is introduced to the gas turbine cycle through pressurized volumetric receivers. In this connection, this variety of receivers has potential for exploitation in combined

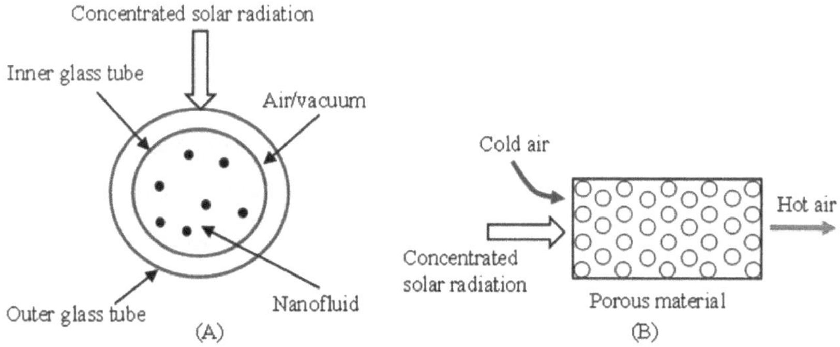

FIGURE 3.2 Volumetric receiver with: (A) nanofluid absorber and (B) porous absorber.

cycle solar gas turbine (CCSGT) systems. Moreover, volumetric receivers are the best alternative to tubular receivers (Ávila-Marín, 2011).

3.2.3 Microchannels

A flow channel is used to distribute a fluid from one point to another, and it may be circular or noncircular. Based on hydraulic diameter (d_h), channels are classified into three categories: (a) conventional channels ($d_h > 3$ mm), (b) minichannels (3 mm $\geq d_h > 200$ μm), and (c) microchannels (200 μm $\geq d_h > 10$ μm) (Kandlikar & Grande, 2003). Microchannels have enormous potential to offer high rates of heat transfer (ASHRAE, 2017). Some of their disadvantages are large pressure drops, high manufacturing cost, dirt clogging, and flow maldistribution, particularly for double-phase flows. Nevertheless, due to the advantage of enhanced rates of heat transfer, microchannels are widely utilized, especially for compact heat exchangers in automotive, aerospace, fuel cell, and high-flux electronic cooling applications. Recently, they have also found application in solar receivers.

A microchannel receiver is made up of enclosed panels or plates with many mirochannels to raise the contact surface area between the heat transfer fluid and the irradiated walls of the receiver. This receiver technology is a recent development. Theoretically, it has been possible to heat supercritical carbon dioxide (with a mass flow of 1 kg s^{-1}) from 803 to 973 K under a solar intensity of 500 kW m^{-2} (Besarati et al., 2015).

3.2.4 Linear and point focus receivers

Receiver pipes (tubes) are located along the focal line of the concentrator, in a linear focus receiver (Fig. 3.3). These tubes absorb heat and conduct it to the heat transfer fluid that is passed through the tubes. Linear receivers find application in the parabolic trough concentrator and linear Fresnel reflector. These receivers can achieve medium to high outlet temperatures (523–833 K).

A point focus solar receiver intercepts concentrated solar radiation on a single point that may be a cylinder, sphere, or other shape. This variety of

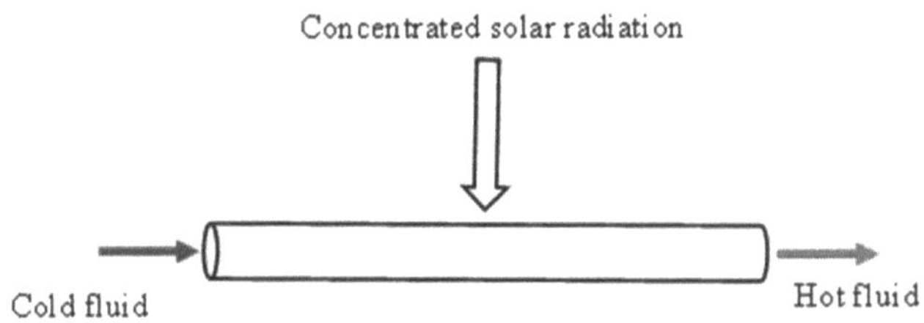

FIGURE 3.3 Distribution of concentrated solar radiation on a line focus receiver.

receivers is capable of attaining very high outlet temperatures (>1000 K), and it finds application in solar tower and parabolic dish power plants.

3.2.5 External and cavity solar receivers

Panels are arranged in cylindrical (or polygonal) shape in external tubular receivers (Fig. 3.4). This type of solar receivers has high operational temperatures and absorber areas, which tend to increase heat losses to the environment through convective and radiative modes of heat transfer.

An external receiver is often mounted at the top of a tower to receive concentrated solar radiation from heliostats. A heat transfer fluid follows a serpentine path, flowing through adjacent up-flow and down-flow panels (Rodriguez-Sanchez et al., 2015). The flow pattern can vary from one receiver to another, depending on the environmental conditions and the operational requirements of the system. One main advantage of this receiver design is its simplicity (Dutta, 2017). It is the most widely applied receiver on solar tower projects across the world because of its simplicity. Nevertheless, it has some inherent limitations: (a) it has a lower limit on the maximum incident flux to avert damage to the receiver material because of the limit on the maximum temperature that it can resist and (b) the heat loss is higher because the receiver surface is exposed to the ambient environment (resulting in higher reflection, convective, and radiative losses).

In a cavity receiver, the absorber tube is fitted along the internal surfaces of the walls of an insulated cavity (Fig. 3.5), and cavity receivers are in a variety of designs such as spherical, cylindrical, conical and cylindrical–conical designs (Bellos et al., 2019). Based on energy flow, cavity receivers are classified into two groups: (a) direct absorption receiver and (b) partially indirect absorption receiver (Wujun et al., 2021). Concentrated solar irradiation is directly absorbed by the absorber element in the direct absorption system,

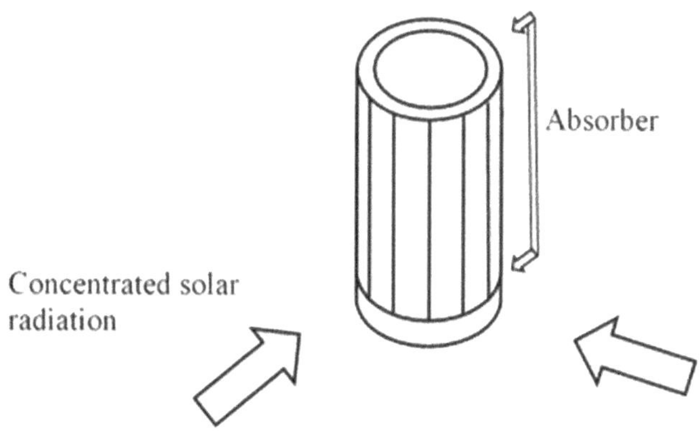

FIGURE 3.4 Cylindrical external solar receiver.

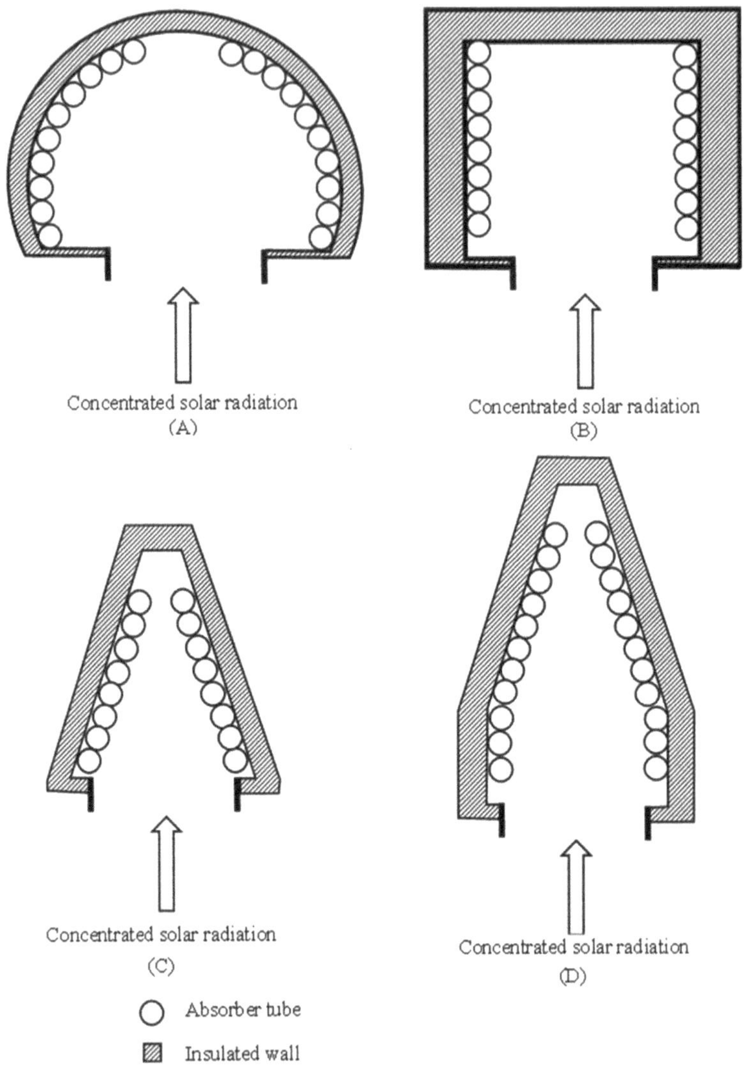

FIGURE 3.5 Cross-section of: (A) spherical, (B) cylindrical, (C) conical, and (D) cylindrical–conical cavity solar receiver.

while part of the concentrated solar radiation is reflected or reradiated by the cavity wall before impinging on the absorber in a partially indirect absorption receiver (Fig. 3.6). Advantages of cavity receivers include (Msaddak et al., 2018): (a) high effective absorption of concentrated solar radiation because of multiple reflection inside the cavity (and higher surface area for heat transfer),

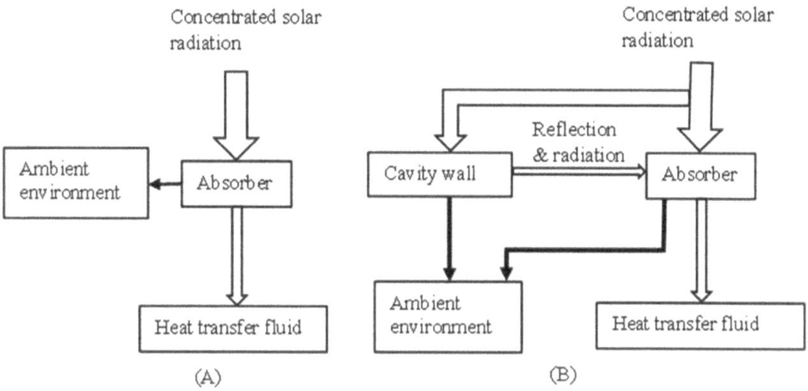

FIGURE 3.6 Two types of cavity receivers: (A) direct absorption receiver and (B) partially indirect absorption receiver.

(b) low requirement in mechanical strength of the wall material because of its limited temperature gradient, and (c) low heat loss from the receiver to the environment (because the cavity is insulated on the exterior part). In addition, this variety of receivers has the ability to handle higher flux at the aperture. Nevertheless, the cavity solar receiver requires a tall tower to enable it "view" the whole field of the heliostats for a given level of power output (Ho, 2017).

3.3 Adaptable heat transfer media

Heat transfer is the process of energy flow from one point to another by virtue of temperature gradient between the points in question, and there are three modes of heat transfer: conduction, convection, and radiation. Heat conduction is the kinetic energy transferred between particles or groups of particles (ASHRAE, 2017), and it can take place in all the three phases of a substance. Heat flows in the direction of declining temperature. The temperatures involved may be transient or steady. Temperature varies with time in a transient process, but it does not change with time in a steady-state process.

Heat convection is the rate of energy transfer between two points in a fluid which involves mixing of the fluid by natural or forced mechanisms. In a natural-convection process, the fluid moves due to the density gradient caused by temperature differences. Forced convection occurs when a moving fluid absorbs heat and transports it away by means of an external pump such as a fan. At the fluid-solid boundary, heat is transferred by means of conduction. Heat may be transferred from a hot solid surface to a cold fluid or from a hot fluid to a cold surface. Within the boundary layer close to the leading edge of the surface, flow is laminar forced convection. The thickness of the boundary layer increases to a critical level as flow continues along the surface. Turbulent forced convection commences after the critical point. Consequently, forced convection can be laminar or turbulent. Flow remains laminar in long tubes or channels with a small hydraulic diameter, and it is said to be fully developed laminar flow at diminished fluid velocity. There is transition from laminar to turbulent flow, and flow is fully developed turbulent at an elevated fluid velocity or in a tube with a large diameter. The movement of fluids in solar receivers is discussed in more details in Chapter 8.

Radiative heat transfer takes place through electromagnetic waves while heat conduction and convection proceeds through media. In contrast, heat radiation does not require a medium for it to take place. In view of this, the use of a vacuum only eliminates convective and conductive heat losses from thermal systems. The nature of the material, microscopic structure, and temperature of the radiator and its surroundings influence the rate of heat radiation. For example, a blackbody absorbs all the radiation incident on it.

Heat is produced by the receiver and then directly or indirectly transferred to the power block by using a heat transfer fluid. The rate of heat transfer from the receiver is highly dependent on the mass flowrate, inlet and outlet temperatures, and nature of the medium (see Section 1.5). Mass flowrate and inlet and outlet temperatures are operational parameters, while thermophysical properties (such as thermal conductivity, specific heat capacity, and viscosity) are natural characteristics of the medium.

In a receiver, heat is transferred by conduction from the outer surface to the inner surface of the wall of the absorber. From the inner surface, heat is transferred to the heat transfer fluid (HTF) via the convection mode. The mean annual efficiency of a concentrating solar power plant is predominantly constrained by the temperature of the working fluid (Benoit et al., 2016). This shows that a heat transfer fluid has a significant impact on the thermal efficiency of the receiver and whole concentrating solar power (CSP) system. The rate of heat convection is influenced by the temperature gradient between the surface and fluid, flow rate, fluid properties, and geometry of the surface in contact with the fluid. Clearly, the characteristics of a given HTF are key in the evacuation of heat from the absorber element. Thus, based on the type of HTF, receivers are classified into gas, liquid, and particle receivers.

3.3.1 Gas solar receivers

Gas receivers use air (or other gases) as a heat transfer fluid. The gas is passed through tubes, honeycombs, channels, or microchannels irradiated with concentrated solar radiation. It may be pressurized or in supercritical form. Thus, gas solar receivers can further be classified into pressurized and supercritical categories. In a pressurized gas receiver, a compressor is employed to compress a gas before it flows into the receiver. Common pressurized gases are air, carbon dioxide (CO_2), helium (He), and hydrogen (H_2). Three types of gas cycles are: (a) conventional, (b) transcritical noncondensing, and (c) supercritical (Garg et al., 2013). Details of these three types of gas cycles are reported in Section 1.5.3. Some supercritical gases used in solar technology are s-CO_2 and s-H_2O. These gases are promising to yield higher efficiencies in advanced thermodynamic cycles.

Direct heating of CO_2 in tubular solar receiver designs is an emerging strategy. This design of receivers can withstand high internal fluid pressures and temperature, and it gives elevated efficiency of thermal-to-electricity conversion close to or higher than 50% (Ho, 2017). A summary of gas receiver designs is shown in Table 3.1.

3.3.2 Liquid solar receivers

Liquid receivers use liquids as heat transfer fluids. These fluids include water, thermal oil (Therminol VP-1, Dowtherm and others), molten salts (such as solar salt, Hitec, Hitec XL, etc), and liquid metals (liquid sodium and lead bismuth eutectic). The liquids are passed through panels of tubes that are irradiated by concentrated solar radiation. Water is commonly used for direct generation of steam to drive the Rankine cycle, and most of the operating CSP plants in the world are steam based. For indirect heating, these CSP plants utilize molten salts as a heat transfer fluid. Liquid receivers can be used to indirectly preheat the working fluid in solar gas turbine systems.

3.3.3 Particle solar receivers

In a particle receiver, small solid particles are suspended in a fluid (usually gas). The suspension has higher density ($>1000\ \mathrm{kg\ m^{-3}}$) and capable of achieving a wider range of temperatures than the present molten salt or direct-steam generation. These particles are irradiated with concentrated solar radiation, and it is possible to raise their temperature above 1273 K (Ho, 2017; Ho et al., 2017). One main advantage of particle receivers is the option of direct irradiation of the particles, which overcomes the challenge of heat-transfer resistance and flux limitations associated with indirect heating through tube walls. Nevertheless, these receivers can only heat up the working fluid of an engine cycle through a heat exchanger, which would reduce the efficiency of solar conversion to useful heat. Particle receiver designs can further be grouped into free-falling particle receiver, obstructed particle receiver, rotary receiver, fluidized bed receiver, and gravity driven (Ho, 2017).

3.3.3.1 Free-falling particle receivers

Basically, free-falling receivers comprise a top hopper, a cavity, and a bottom hopper. The solid particles are released through an opening at the base of the top hopper above the cavity (Fig. 3.7). The free-falling particles form a curtain and are directly irradiated by concentrated solar radiation in the cavity. The flow of particles can be regulated by using a sliding gate or some other mechanism to match the levels of irradiance and still attain the desired outlet

TABLE 3.1 Summary of gas receiver designs (Ho, 2017).

Receiver design	Outlet temperature/ thermal efficiency	Advantages	Challenges/ research needs
Tubular gas receivers	973–1073 K/~90% (modeling), up to 85% (testing)	Ability to use features and geometry (such as cavity) to increase trapping of radiation.	Resistance to heat transfer through tube walls, fatigue, and high-cost materials to contain high-pressure, high-temperature flow, no direct storage.
Volumetric air receivers	>1073 K/50% –90%	Can use ambient air which is cheap, potential for volumetric (interior) heating to minimize heat losses near the aperture.	Flow instabilities, high temperatures, and radiative heat losses near the aperture due to limited light penetration, no direct storage.
Volumetric air receivers	>1073 K/50% –90%	Can use ambient air which is cheap, potential for volumetric (interior) heating to minimize heat losses near the aperture.	Flow instabilities, high temperatures, and radiative heat losses near the aperture due to limited light penetration, no direct storage.
Microchannel gas	973 K/up to 90% (scaled up from bench-scale tests)	Augmented surface area and heat transfer to gas, ability to add up microchannel modules for larger capacities.	Potential for increased pressure drop, potential manufacturing complexity and costs, no direct storage.

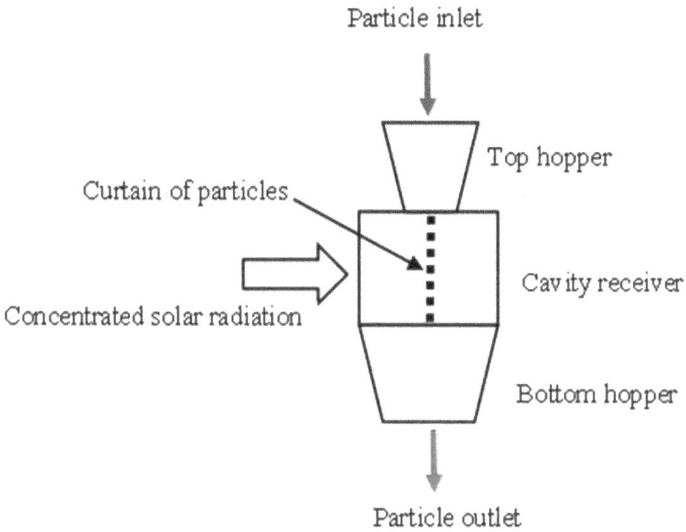

FIGURE 3.7 Schematic representation of a free-falling particle solar receiver.

temperature of the particles. The hot particles are directly used to heat the working fluid of the power cycle or they can be stored for later use.

The concept of free-falling particle curtain has evolved to pilot scale because of its great potential for application in the new generation of CSP plants that are expected to operate at high temperature. In this context, there has been some improvements to models for this concept and validation by experiments at pilot scale (Siegel et al., 2010). On-sun testing of a prototype of the receiver was performed with an input power of 1.58−2.5 MW_{th} at the National Solar Thermal Test Facility in Albuquerque, United States of America. Aluminosilicate particles containing 7% Fe_2O_3 with 697 mm mean diameter were utilized in this study. Another option is the concept of face-down solid particle receiver in which the particle curtain lines the inner wall of a cylinder closed on its upper part, with the bottom part facing the incoming concentrated solar radiation. A comparison of the solid particle solar power plant with standard options of generation technologies shows that a particle-receiver solar tower with a combined cycle yields the lowest levelized cost of electricity (about 0.10 EUR kWh^{-1}) (Giuliano et al., 2011).

3.3.3.2 Obstructed particle receivers

In the free-falling receivers, residence time can be achieved by using a sliding gate. Another technique of augmenting the residence time of particles within the zone of concentrated solar radiation is to block the flow with porous structures or an array of obstacles that mechanically obstruct their descent and reduce velocity of their descent while still permitting direct absorption of the radiation.

Sandia introduced the early concepts of obstructed flow designs in the 1980s by employing ceramic structures suspended from the back wall to slow down the particles (Hruby, 1986). Advances in the development of this category of receivers have included use of: (a) a spiral ramp along which particles flow under the influence of gravity and mechanically induced vibration (Xiao et al., 2014), (b) a staggered array of porous mesh structures (Ho et al., 2014), and (c) interconnected porous structures to decelerate the flow of particles (Lee et al., 2015).

3.3.3.3 Rotary/centrifugal receivers

A rotary kiln is a long tube that is tilted (to the horizontal plane) and rotated, and the diameter of the kiln may be fixed or variable. Low tilt angles (3–4°) are utilized in this variety of kilns (Esence et al., 2020). The tilt angle and the rotation allow solid reactants to move down the tube. These two factors also affect the time of retaining solid particles in the kiln. Based on flow pattern, rotary kilns are classified into co-current and counter-current kilns. In a co-current rotary kiln, flue gas flows in the same direction of the solid reactants. On the other hand, flue gas flows in the opposite direction of the solid reactants in a counter-current rotary kiln. Rotary kilns are applied in several industrial processes (such as cement and steel industries) due to their ability to supply well-mixed particles, thereby promoting the uniform distribution of temperature, and they can achieve very high temperatures of up to 2273 K (Gallo et al., 2016). It is therefore not surprising that rotary kilns have been adopted in the development of solar receivers.

In a rotary receiver, particles are introduced into a rotating cylinder, which is irradiated with concentrated solar radiation (Fig. 3.8). A transparent cover may befitted on the front part of the cylinder to allow transmission of solar

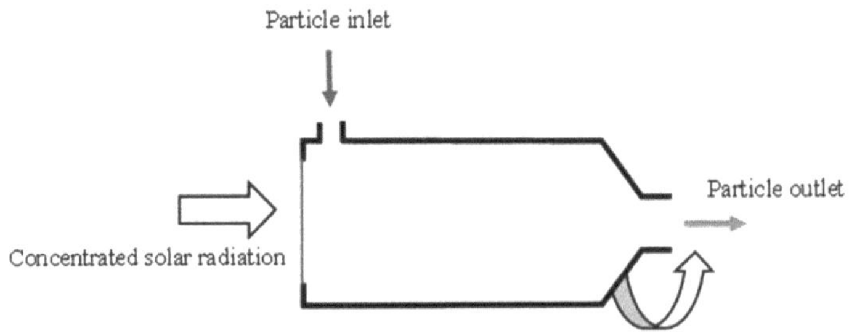

FIGURE 3.8 Cross-section of a rotary particle solar receiver.

radiation into the cavity. A gas is passed through the cylinder in a co-current or counter current flow pattern.

3.3.3.4 Confined fluidized bed receivers

Fluidization is a process in which particles of a solid material are caused to flow like a fluid by blowing a gas (or liquid) upward through a reactor that is filled with solid particles (Fig. 3.9). This process creates a fluidized bed that provides the best zone for contact between the fluid and solid particles (well-mixed blend), thereby offering excellent heat transfer between the fluid and solid particles (Cocco et al., 2014).

The behavior of solid particles in a fluid is influenced by their density, size, shape, and velocity of the fluid. In this vein, there are eight regimes of fluidization (Khawaja, 2015): (a) fixed bed, (b) minimum fluidization, (c) smooth fluidization, (d) bubbling bed, (e) axial slugging bed, (f) flat slugging bed, (g) turbulent fluidization, and (h) pneumatic transport. In a fixed bed, a few particles vibrate within the same height as the bed at rest when the flow of a fluid through a bed of solid particles is elevated continually. As the velocity of the fluid increases, a point is attained where the drag force exerted by the upward moving fluid equals the weight of the particles, and the void space of the bed slightly rises. Fluidization commences at this point, and it is called minimum fluidization. In certain cases, the fluidized bed expands uniformly once the fluid is passed through the granular particles. This state is called uniform fluidization, and it often occurs when very small particles are fluidized by gas at relatively low superficial velocities, or when granular materials are fluidized by a liquid. A further increase in the velocity of the fluid starts creating bubbles, resulting in a bubbling fluidized bed. A granular material, made of larger particles such as sand with a particle diameter $> 250\ \mu m$, will start bubbling without initially passing through the uniform fluidization regime, when the gas velocity exceeds the minimum fluidizing velocity. When the velocity of the fluid is increased further, the bubbles in a bubbling fluidized bed merge and expand as they rise. Slugging sets in when the size of the bubbles becomes comparable to the diameter of the bed. Slugs may have round or flat edges. The axial and flat slugging fluidization beds may generally be referred to as slugging fluidization bed, thereby yielding seven main regimes of fluidization beds. As fluid velocity increases, the upper boundary of the bed becomes less distinct, and bubbles begin to break up. At this point, the upper surface of the bed disappears and a turbulent motion of solid clusters and voids of gas of different sizes and shapes starts. Finally, the fluidized bed becomes an entrained bed in which there is dispersed lean phase fluidized bed as the fluid velocity increases further.

In the context of fluidization, powders are classified into four groups, A, B, C, and D (Geldart, 1973). These powder categories are referred to as Geldart

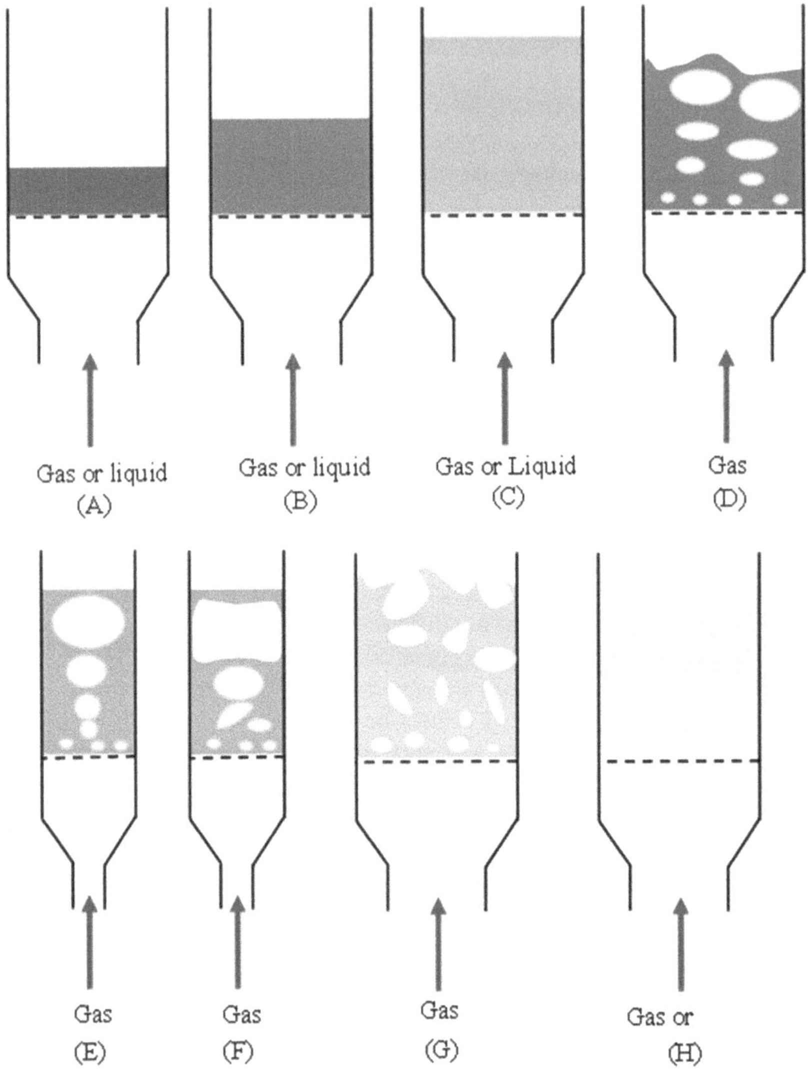

FIGURE 3.9 Regimes of fluidization: (A) Fixed bed, (B) minimum fluidization, (C) bubbling bed, (D) slugging bed, (E) turbulent, and (F) pneumatic transport bed. *Adapted from Khawaja (2015).*

Groups, and this classification is based on their location on the diagram of solid-fluid density difference and particle size.

Group A: This group has a small mean particle size with a density of less than 1400 kg m^{-3}. The particles can easily be aerated and fluidized. Before a bubbling bed phase sets in, the bed of this group of particles expands considerably at minimum fluidization because of diminished bulk density. Group A powders are exploited in most of powder-catalyzed beds.

Group B: In this group, the range of particle size is 40–500 μm, and the particle density varies between 1400 and 4000 kg m^{-3}. Bubbles start forming at the onset of minimum fluidization. So, the minimum fluidization velocity and the minimum bubbling velocity are equal. Group B powders fluidize easily and find application in a wide range of fluidized systems. In this context, most of the fluidized-bed combustors and pyrolysis units utilize coal powder with properties of this group (Cocco et al., 2014).

Group C: This group has a particle size of less than 40 μm. It contains very fine and, therefore, the most cohesive particles. In view of this, these particles fluidize under conditions that are extremely difficult to attain, and they may require the use of an external force, such as mechanical shakeup.

Group D: Powders in this group have the largest particle sizes (above 500 μm) and densities. So, fluidization of this group requires a large amount of gas and is typically associated with high levels of abrasion. Enormous bubbles are formed during the fluidization of Group D powders. These powders are therefore sometimes fluidized in spouting beds which have lower requirements of gas. A spouting bed is a kind of fluidized bed in which the gas flows principally in the center of the bed.

In a basic fluidized bed solar receiver, solid particles are fluidized by using a gas, and the bed is irradiated with concentrated solar radiation (Fig. 3.10). A gas is blown through the bed of solid particles, and heat is evacuated from the receiver by the gas or the solid–gas mixture, depending on the fluidization regime.

3.3.3.5 Gravity-driven particle flow through enclosures

In a gravity-driven particle receiver, particles are indirectly heated as they flow downward under the force of gravity around exterior surfaces of tubes fitted within an enclosure. Concentrated solar radiation is incident on the interior surfaces of the tubes, and heat is transferred to the exterior surfaces of the tubes by conduction and then to the particles (Fig. 3.11).

Modeling and experimental studies conducted on a small-scale have shown that heat transfer to the particles is limited in locations around the tubular structures where the particles lost contact with the heated wall surfaces (Ho, 2017). Other possible drawbacks include difficulties in maintaining a sufficient mass flow and achieving a significant penetration and uniform flux of concentrated solar radiation within the tubular cavities. Advantages of this design include absence of particle loss through an open aperture and lower heat losses than those exhibited by an open cavity receiver.

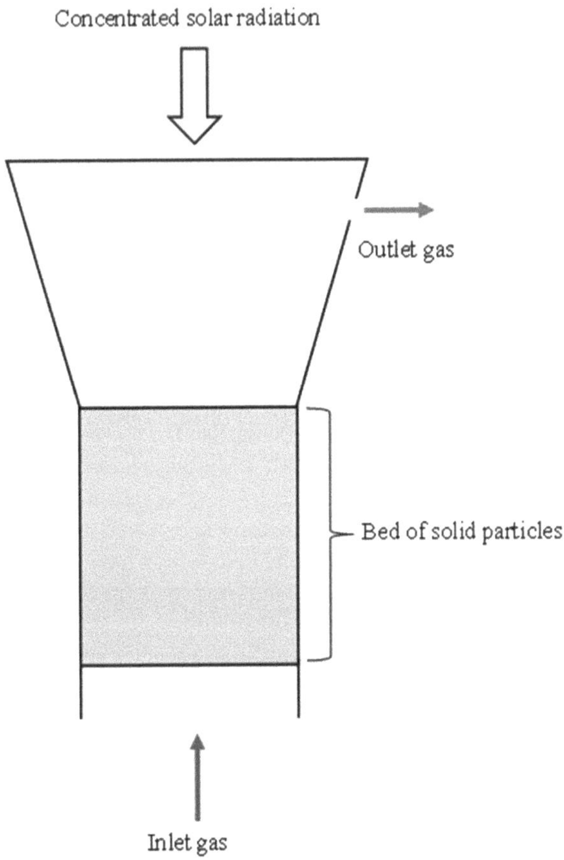

FIGURE 3.10 A fluidized bed solar receiver.

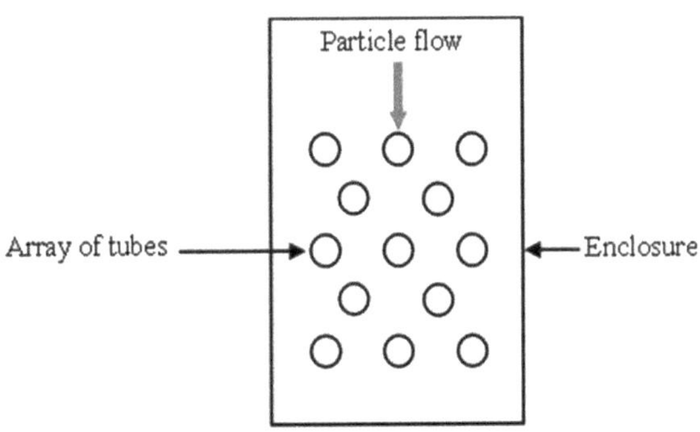

FIGURE 3.11 Indirect particle receiver with particles flowing inside an enclosure.

References

ASHRAE. (2017). *ASHRAE handbook - fundamentals*. Atlanta: American Society of Heating, Refrigerating and Air-Conditioning Engineers (ASHRAE).

Ávila-Marín, A. L. (2011). Volumetric receivers in solar thermal power plants with central receiver system technology: A review. *Solar Energy, 85*(5), 891–910. https://doi.org/10.1016/j.solener.2011.02.002

Bellos, E., Bousi, E., Tzivanidis, C., & Pavlovic, S. (2019). Optical and thermal analysis of different cavity receiver designs for solar dish concentrators. *Energy Conversion and Management, X*, 2. https://doi.org/10.1016/j.ecmx.2019.100013

Benoit, H., Spreafico, L., Gauthier, D., & Flamant, G. (2016). Review of heat transfer fluids in tube-receivers used in concentrating solar thermal systems: Properties and heat transfer coefficients. *Renewable and Sustainable Energy Reviews, 55*, 298–315.

Besarati, S. M., Goswami, D. Y., & Stefanakos, E. K. (2015). Development of a solar receiver based on compact heat exchanger technology for supercritical carbon dioxide power cycles. *Journal of Solar Energy Engineering, Transactions of the ASME, 137*(3). https://doi.org/10.1115/1.4029861

Cocco, R., Karri, S. B. R., & Knowlton, T. (2014). Introduction to fluidization. *Chemical Engineering Progress*, 21–29. https://www.aiche.org/publications/cep.

Dutta, P. (2017). High temperature solar receiver and thermal storage systems. *Applied Thermal Engineering, 124*, 624–632. https://doi.org/10.1016/j.applthermaleng.2017.06.028

Esence, T., Guillot, E., Tessonneaud, M., Sans, J. L., & Flamant, G. (2020). Solar calcination at pilot scale in a continuous flow multistage horizontal fluidized bed. *Solar Energy, 207*, 367–378. https://doi.org/10.1016/j.solener.2020.06.098

Gallo, A., Roldán, M. I., Alonso, E., & Fuentealba, E. (2016). Considerations for using a rotary kiln for high temperature industrial processes with and without thermal storage. In *EuroSun 2016/ISES conference proceedings*. Palma de Mallorca: International Solar Energy Society. https://doi.org/10.18086/eurosun.2016.02.04

Garg, P., Kumar, P., & Srinivasan, K. (2013). Supercritical carbon dioxide Brayton cycle for concentrated solar power. *Journal of Supercritical Fluids, 76*, 54–60. https://doi.org/10.1016/j.supflu.2013.01.010

Geldart, D. (1973). Types of gas fluidisation. *Powder Technology, 7*(5), 285–292. https://doi.org/10.1016/0032-5

Giuliano, S., Buck, R., & Eguigen, S. (2011). Analysis of solar-thermal power plants with thermal storage hybrid operation strategy. *ASME Journal of Solar Energy Engineering, 133*(3), 031007. https://doi.org/10.1115/1.4004246

Ho, C. K. (2017). Advances in central receivers for concentrating solar applications. *Solar Energy, 152*, 38–56. https://doi.org/10.1016/j.solener.2017.03.048

Ho, C. K., Christian, J. M., Ortega, J. D., Yellowhair, J., Mosquera, M. J., & Andraka, C. E. (2014). Reduction of radiative heat losses for solar thermal receivers. In *Proceedings of the SPIE optics + photonics solar energy + technology high and low concentrator systems for solar energy applications IX, San Diego, August-17–21, 2014*.

Ho, C. K., Christian, J. M., Romano, D., Yellowhair, J., Siegel, N., Savoldi, L., & Zanino, R. (2017). Characterization of particle flow in a free-falling solar particle receiver. *Journal of Solar Energy Engineering, Transactions of the ASME, 139*(2). https://doi.org/10.1115/1.4035258

Hruby, J. M. (1986). *A technical feasibility study of a solid particle solar central receiver for high temperature applications, SAND86-8211*. Livermore, CA: Sandia National Laboratories.

Kandlikar, S. G., & Grande, W. J. (2003). Evolution of microchannel flow passages-thermohydraulic performance and fabrication technology. *Heat Transfer Engineering, 24*(1), 3–17. https://doi.org/10.1080/01457630304040

Khawaja, H. A. (2015). Review of the phenomenon of fluidization and its numerical modelling techniques. *International Journal of Multiphysics*, 397–407. https://doi.org/10.1260/1750-9548.9.4.397

Lee, T., Lim, S., Shin, S., Sadowski, D. L., Abdel-Khalik, S. I., Jeter, S. M., & Al-Ansary, H. (2015). Numerical simulation of particulate flow in interconnected porous media for central particle-heating receiver applications. *Solar Energy, 113*, 14–24. https://doi.org/10.1016/j.solener.2014.12.024

Martínez-Manuel, L., Wang, W., Laumert, B., & Peña-Cruz, M. I. (2021). Numerical analysis on the optical geometrical optimization for an axial type impinging solar receiver. *Energy, 216*. https://doi.org/10.1016/j.energy.2020.119293

Mlatho, J. S. P., & McPherson, M. (2012). Performance optimization of solar receivers that use oil as a heat transfer fluid. *International Journal of Sustainable Energy*. https://doi.org/10.1080/14786451.2012.716060

Msaddak, A., Sediki, E., & Ben Salah, M. (2018). Assessment of thermal heat loss from solar cavity receiver with Lattice Boltzmann method. *Solar Energy, 173*, 1115–1125. https://doi.org/10.1016/j.solener.2018.08.059

Rodríguez-Sánchez, M. R., Sánchez-González, A., Marugan-Cruz, C., & Santana, D. (2015). Flow patterns of external solar receivers. *Solar Energy, 122*, 940–953. https://doi.org/10.1016/j.solener.2015.10.025

Siegel, N., Ho, C. K., Khalsaz, S. S., & Kolb, G. (2010). Development and evaluation of a prototype solid particle receiver: On-sun testing and model validation. *ASME Journal of Solar Energy Engineering, 132*, 021008. https://doi.org/10.1115/1.4001146

Wujun, W., Wangzhong, M., Fei, Y., Silvia, T., Joydeep, D., & Björn, L. (2021). Solar selective reflector materials: Another option for enhancing the efficiency of the high-temperature solar receivers/reactors. *Solar Energy Materials and Solar Cells, 224*, 110995. https://doi.org/10.1016/j.solmat.2021.110995

Xiao, G., Guo, K. K., Ni, M. J., Luo, Z. Y., & Cen, K. F. (2014). Optical and thermal performance of a high-temperature spiral solar particle receiver. *Solar Energy, 109*, 200–213.

Chapter 4

Optical properties of materials for solar receivers

4.1 Introduction

The performance of solar receivers is affected by optical factors. In this connection, some receivers are covered with transparent surfaces, while others are unglazed. For glazed receivers, concentrated solar radiation is transmitted through the cover before imping on the absorber element. The absorber is opaque to solar radiation, and it emits thermal radiation. Consequently, knowledge about the optical properties of both transparent and opaque materials is necessary for the advancement concentrating solar power.

An optical property of a material is the interaction of the material with electromagnetic radiation (especially visible light). An electromagnetic wave is made up of electric and magnetic fields that oscillate at right angles to each other, and amplitude and wavelength are the most fundamental parameters for describing waves. Solar radiation is electromagnetic energy from the sun. When radiation is incident on a material, it can undergo reflection, absorption, and transmission such that the intensity of the incident radiation (I) is equal to the sum of the reflected, absorbed, and transmitted intensities:

$$I = I_r + I_a + I_t \tag{4.1}$$

Where I_r is the reflected intensity, I_a is the absorbed intensity, and I_t is the transmitted intensity.

For transparent covers of receivers, it is desirable that the transmittance should be high, but low levels of reflectance and absorptance are required. Some materials are transparent to visible radiation but opaque to other types of electromagnetic waves. Thus, selective transmission is useful in strategies that are aimed at increasing the optical and thermal efficiencies of receivers. Opaque materials do not allow solar radiation to propagate through them ($I_t = 0$). Some opaque materials find application in reflectors and absorbers. Opaque materials used in reflector components (such as solar concentrators) need to have elevated levels of reflectance (low absorptance), while opaque materials with high absorptance are required in absorber elements. Another important optical property of materials is polarization (the filtration of

Solar Receivers for Thermal Power Generation. https://doi.org/10.1016/B978-0-323-85271-5.00002-1

electromagnetic waves so that only one direction of the oscillations propagates through a polarizing material). This property affects the reflectance, absorptance, and transmittance of some materials (see Section 4.2.1).

The spectrum of electromagnetic radiation is made up of different wavelength bands (including ultraviolet, visible, infrared, and others, in order of increasing wavelength λ). The visible region (λ = 0.38−0.78 μm) lies between the ultraviolet and infrared bands. Important wavelengths in solar energy and its applications fall in the ultraviolet to near-infrared (from 0.29 to ~ 25 μm). Outside the atmosphere, the wavelength range of solar radiation is 0.25−3.00 μm which changes to 0.29−2.50 μm when the radiation reaches the ground (Duffie & Beckman, 2013). So, the solar radiation wavelength band stretches from ultraviolet through visible to infrared, but most of the energy is concentrated in the visible band of the solar spectrum. All electromagnetic waves travel at the speed of light (v) and therefore have different wavelengths.

$$V = \lambda f \tag{4.2}$$

Where f is the frequency the radiation.

4.2 Transmission of radiation through transparent materials

Knowledge about transmission, reflection, and absorption of solar radiation by different parts of a solar receiver is critical in its development. Some receivers have transparent covers to reduce heat loss to the environment, thereby augmenting the thermal efficiency of the system. The transmittance, reflectance, and absorptance of a cover material depend on the incident radiation, and thickness, refractive index (n), and extinction coefficient (K) of the material. The refractive index and the extinction coefficient are influenced by the wavelength of the radiation, generally referred to as the spectral dependence. In some applications, the spectral influence may be ignored depending on the nature of the material. For example, optical properties of glass do not significantly vary with the wavelength of light. Nevertheless, optical properties of other cover materials are significantly affected by the wavelength of the radiation.

4.2.1 Reflection and absorption of beam radiation

It has been reported in Section 2.3.1 that rays of beam radiation can be traced from the sun's position and utilized in the computation of solar angles of beam radiation. When a ray of light traverses the boundary between two media (medium 1 and medium 2) that have different values of refractive index, it changes direction (Fig. 4.1). This phenomenon is known as refraction and it is introduced in Section 2.7. The angle of refraction depends on the angle of incident and the refractive index of the involved media. In this regard, the refractive index (n) of a medium is the ratio of the speed of light in a vacuum to the speed of light in the medium. It is an optical property of the medium. The speed of light in a vacuum is constant (about 3 x 10^8 m s^{-1}). Thus, from Eq. (4.3), the speed of a ray of light decreases as the refractive index increases. The refracted ray bends towards the normal when $n_1<n_2$, but away from the normal when $n_1>n_2$. Refractive indices of selected materials are presented in Table 4.1.

$$n = c/v \quad (4.3)$$

Where c is the speed of light in a vacuum and v is the speed of light in a specific medium.

Solar radiation undergoes slight polarization as it propagates through a transparent material. Unpolarized radiation has perpendicular and parallel components that are polarized differently by the material, which results in differences in the reflectance, absorptance, and transmittance of the material for the two components. Fresnel formulated some useful expressions for the reflection of radiation on a smooth surface. Values of perpendicular (r_n)

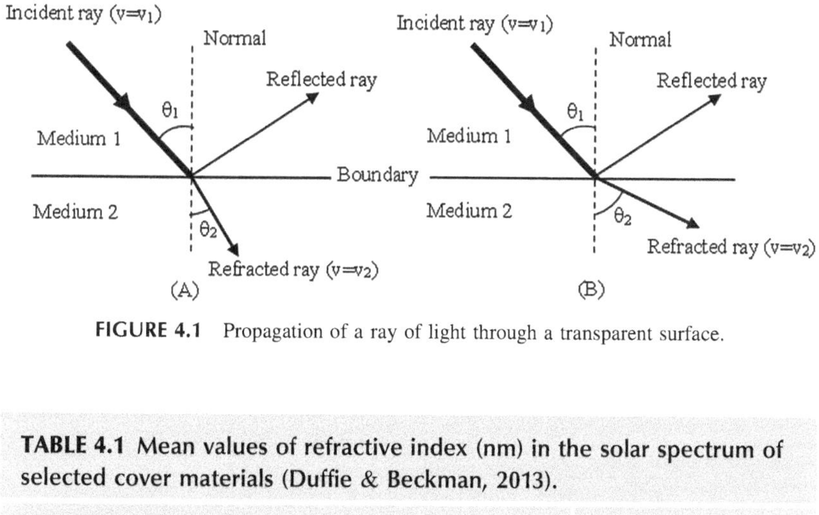

FIGURE 4.1 Propagation of a ray of light through a transparent surface.

TABLE 4.1 Mean values of refractive index (nm) in the solar spectrum of selected cover materials (Duffie & Beckman, 2013).

Cover material	n_m
Glass	1.526
Polycarbonate	1.6
Polyfluorinated ethylene propylene	1.34
Polymethyl methacrylate	1.49
Polytetrafluoroethylene	1.37
Polyvinyl fluoride	1.45

and parallel (r_{pr}) reflectance are calculated as follows (Duffie & Beckman, 2013):

$$r_n = \frac{\sin^2(\theta_2 - \theta_1)}{\sin^2(\theta_2 + \theta_1)} \tag{4.4}$$

$$r_{pr} = \frac{\tan^2(\theta_2 - \theta_1)}{\tan^2(\theta_2 + \theta_1)} \tag{4.5}$$

$$r = \frac{I_r}{I_i} = \frac{r_n + r_{pr}}{2} \tag{4.6}$$

Where I_r is the intensity of reflected radiation and I_i is the intensity of incident radiation.

Eq. (4.6) gives the average reflectance of the material. At normal incidence, $\theta_1 = \theta_2 = 0$, and combining Eqs. (4.3) and (4.6) yields the reflectance at normal incidence (r_o):

$$r_o = \left(\frac{n_1 - n_2}{n_1 + n_2}\right)^2 \tag{4.7}$$

Example 4.1

A ray of light from air is incident at an angle of 10° on a transparent cover of a particle solar receiver. If the angle of refraction is 6.5^o, calculate the reflectance of the cover.

Solution

The reflectance (r) of the cover can be calculated from Eq. (4.6),

$$r = \frac{r_n + r_{pr}}{2}$$

$$\text{But } r_n = \frac{\sin^2(\theta_2 - \theta_1)}{\sin^2(\theta_2 + \theta_1)} = \frac{\sin^2(6.5^o - 10^o)}{\sin^2(6.5^o + 10^o)} = 0.0462$$

$$r_{pr} = \frac{\tan^2(\theta_2 - \theta_1)}{\tan^2(\theta_2 + \theta_1)} = \frac{\tan^2(6.5^o - 10^o)}{\tan^2(6.5^o + 10^o)} = 0.0426$$

Therefore,

$$r = \frac{0.0462 + 0.0426}{2} = 0.04$$

In solar applications, the transparent cover system is a slab or film of material with two interfaces per cover that create optical losses through reflection. For each component of polarization, the radiation reflected at an interface is different when the angle of incidence is not normal, and the transmitted and reflected radiation becomes partially polarized. In this case, each component of polarization needs to be considered independently. Within a transparent layer, a ray of light undergoes multiple reflections as shown in Fig. 4.2.

Assuming that the absorptance is negligible, the reflection of one component of the unpolarized light can be considered at a time. For the perpendicular component, $(1-r_n)$ of the incident beam is transmitted to the interface 2. Of this amount, $(1-r_n)^2$ is transmitted through the second interface and $(1-r_n)r_n$ is reflected back to the first interface. This process of internal reflection is repeated until the energy in the reflected ray is negligible. Transmitted terms (through the second interface) can be summed up to obtain the transmittance of the perpendicular component of polarization (Duffie & Beckman, 2013):

$$\tau_n = (1 - r_n)^2 \sum_{i=0}^{i=\infty} r_n^{2i} = \frac{(1 - r_n)^2}{1 - r_n^2} = \frac{1 - r_n}{1 + r_n} \tag{4.8}$$

A similar expression (τ_{pr}) is obtained for the parallel component of polarization. It should be noted that r_n and r_{pr} are not equal except at normal incidence. The transmittance of unpolarized light (τ_r) is equal to the mean value of τ_n and τ_{pr}.

$$\tau_r = \frac{1}{2}\left(\frac{1 - r_n}{1 + r_n} + \frac{1 - r_{pr}}{1 + r_{pr}}\right) \tag{4.9}$$

In Eq. (4.9), the subscript r is used as a reminder that only reflection losses were taken into account in the derivation of the expression. In reality, the

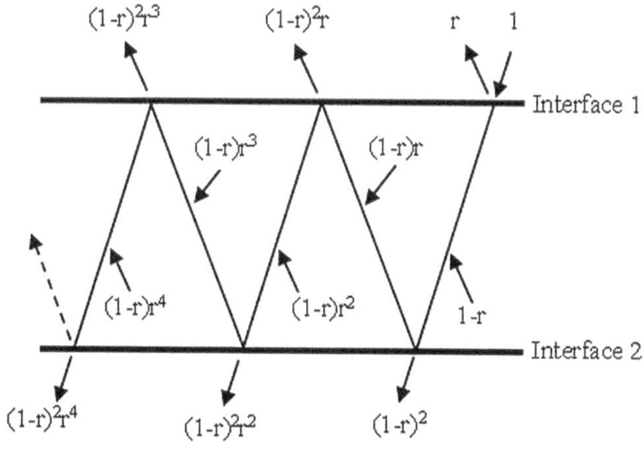

FIGURE 4.2 Transmission of radiation through a nonabsorbing transparent layer. *Adapted from Duffie and Beckman (2013).*

transparent layer absorbs part of the incident radiation. Some solar collectors comprise multiple (N) transparent covers. In such systems, a similar analysis gives the following expression of the transmittance of unpolarized light (τ_{rN}):

$$\tau_r = \frac{1}{2}\left[\frac{1 - r_n}{1 + (2N - 1)r_n} + \frac{1 - r_{pr}}{1 + (2N - 1)r_{pr}}\right] \tag{4.10}$$

Example 4.2

Calculate the transmittance of unpolarized light through two transparent covers with a refractive index of 1.526 at normal incidence and at 65° (Assuming the average refractive index of air is 1.000).

Solution

$$\tau_r = \frac{1}{2}\left(\frac{1 - r_n}{1 + (2N - 1)r_n} + \frac{1 - r_{pr}}{1 + (2N - 1)r_{pr}}\right)$$

At normal incidence, the reflectance at one interface is as follows:

$$r_o = \left(\frac{n_1 - n_2}{n_1 + n_2}\right)^2 = \left(\frac{1.526 - 1.000}{1.526 + 1.000}\right)^2 = 0.0434$$

In addition, $r_n = r_{pr}$ at normal incidence.
Therefore,

$$\tau_r(0) = \frac{2}{2}\left[\frac{1 - r_o}{1 + (2N - 1)r_o}\right] = \frac{1 - 0.0434}{1 + 3 * 0.0434} = 0.85$$

The angle of refraction is computed by using Snell's law (Eq. 2.17). So, the angle of refraction on the first cover interface is:

$$\theta_{co} = \sin^{-1}\left(\frac{n_a \sin \theta_a}{n_{co}}\right) = \sin^{-1}\left(\frac{\sin 65^o}{1.526}\right) = 36.44^o$$

As radiation passes through a partially transparent medium, part of it is absorbed by the medium, and this phenomenon can be described by the Bouguer's law (Siegel & Howell, 1971; Zaversky et al., 2018). This principle is premised on the assumption that the amount of absorbed radiation is proportional to the local intensity (I) and the distance (x) covered in the medium.

$$dI = -KI\,dx \tag{4.11}$$

Where K is a constant of proportionality known as the extinction coefficient.

K is assumed to be constant in the solar spectrum, and so, Eq. (4.10) can be integrated along the actual path length (L) in the medium (from $x = 0$ to $x = L/\cos\theta_2$) to obtain the transmittance (τ_a):

$$\tau_a = \frac{I_{transmitted}}{I_{incident}} = \exp\left(-\frac{KL}{\cos\theta_2}\right) \tag{4.12}$$

Where the subscript a indicates that only absorption losses have been considered in the derivation of the mathematical expression for transmittance.

Values of τ_a for different materials of specific path length are normally reported in literature at normal incidence. For glass, the coefficient of extinction falls approximately within the range 4–32 m^{-1}. Glass with low values of K is preferable for use as a transparent cover over the window of a receiver.

4.2.2 Optical properties of transparent covers

In Section 4.2.1, the transmittance of a transparent cover was analyzed without considering both reflection and absorption losses. However, these losses occur in a reality, and so, it is necessary to include them in the expressions for calculating the transmittance of a material. These losses can be determined by ray-tracing or net radiation techniques (Siegel & Howell, 2002). As shown in the preceding section, there are perpendicular and parallel components of polarization. For the perpendicular component, transmittance (τ_n), reflectance (ρ_n), and absorptance (α_n) of the cover can be expressed as follows (Duffie & Beckman, 2013):

$$\tau_n = \frac{\tau_a(1-r_n)^2}{1-(r_n\tau_a)^2} = \tau_a\frac{1-r_n}{1+r_n}\left[\frac{1-r_n^2}{1-(r_n\tau_a)^2}\right] \tag{4.13}$$

$$\rho_n = r_n + r_n\frac{\tau_a^2(1-r_n)^2}{1-(r_n\tau_a)^2} = r_n(1+\tau_a\tau_n) \tag{4.14}$$

$$\alpha_n = (1-\tau_a)\left(\frac{1-r_n}{1-r_n\tau_a}\right) \tag{4.15}$$

Similar equations of transmittance, reflection, and absorptance are found for the parallel component. In view of this, the optical properties of the cover are found by taking the mean of the two components for the incident unpolarized radiation. For practical covers, the parameter τ_a is rarely less than 0.90 and r is of the order of 0.1. Consequently, the last term in Eq. (4.13) is approximately equal to 1.00, and this equation can be simplified to:

$$\tau_n \cong \tau_a\frac{1-r_n}{1+r_n} \tag{4.16}$$

For the parallel component, the transmittance is as follows:

$$\tau_{pr} \cong \tau_a \frac{1 - r_{pr}}{1 + r_{pr}} \tag{4.17}$$

Using Eq. (4.9), this implies that the transmittance of the cover is approximately:

$$\tau = \frac{1}{2}\left(\tau_n + \tau_{pr}\right) \cong \frac{1}{2}\tau_a \left(\frac{1 - r_n}{1 + r_n} + \frac{1 - r_{pr}}{1 + r_{pr}}\right) = \tau_a \tau_r \tag{4.18}$$

Similarly, an approximate value of the absorptance of a solar collector cover can be found by letting the last term in Eq. (4.15) be equal to 1.00, which yields the following:

$$\alpha \cong (1 - \tau_a) \tag{4.19}$$

It should be noted that the ignored term in Eq. (4.13) is smaller than the ignored term in Eq. (4.15). Nevertheless, the transmittance is much larger than absorptance so that the whole accuracy of the two approximations is basically the same. For a single cover, the reflectance is:

$$\rho = 1 - \alpha - \tau \tag{4.20}$$

which is approximately the following:

$$\rho \cong \tau_a (1 - \tau_r) = \tau_a - \tau \tag{4.21}$$

Where the single term τ_r accounts for the process of polarization in the approximate equations rather than the use of the more complex expressions for each optical property.

To reduce heat loss, multicovers are fitted on some solar collectors. However, the effective transmittance of a system declines with increasing the number of covers (except at high levels of the angle of incidence where $\tau \cong 0$ for any number of covers). Consequently, an optimal number of covers is desirable. The ray-tracing method can be applied to derive appropriate equations for computing the transmittance, reflectance, and absorptance of multicover systems. Advantages of the ray-tracing technique include the following: (a) ability to capture optical phenomena such as multiple scattering at low computational cost and (b) flexibility in use of arbitrary geometries (Lambrick et al., 2018). The method can be applied to systems with any number of covers (Whillier, 1953). These covers may be identical or different. For N identical covers, the transmittance without considering absorptance can be expressed as follows:

$$\tau_a = \exp\left(-\frac{NKL}{\cos\theta_2}\right) \tag{4.22}$$

The following equations can be used to calculate the transmittance and reflectance of a system with two covers that are not identical:

$$\tau = \frac{1}{2}(\tau_n + \tau_{pr}) = \frac{1}{2}\left[\left(\frac{\tau_1\tau_2}{1-\rho_1\rho_2}\right)_n + \left(\frac{\tau_1\tau_2}{1-\rho_1\rho_2}\right)_{pr}\right] \quad (4.23)$$

$$\rho = \frac{1}{2}(\rho_n + \rho_{pr}) = \frac{1}{2}\left[\left(\rho_1 + \frac{\tau\tau_1\rho_2}{\tau_2}\right)_n + \left(\rho_1 + \frac{\tau\tau_1\rho_2}{\tau_2}\right)_{pr}\right] \quad (4.24)$$

Where the subscripts 1 and 2 refer to the outer and inner covers, respectively.

Example 4.3

A ray of light is incident at incidence angles of 65° for two glass covers. Calculate the solar transmittance if the thickness of each cover is 2.5 mm with an extinction coefficient of 15 m^{-1}, and a refractive index of 1.526.

Solution

$$\tau = \tau_a \tau_r$$

For two glass layers,

$$\tau_a = \exp\left(-\frac{2KL}{\cos\theta_2}\right)$$

From Example 4.2, $\theta_2 = 36.44°$ and $\tau_r = 0.70$.
Therefore,

$$\tau_a = \exp\left(-\frac{2 * 15 * 0.0025}{\cos(36.44^o)}\right) = 0.91$$

$$\tau = 0.91 * 0.70 = 0.64$$

4.2.3 Transmission of diffuse radiation

The optical analysis presented in Sections 4.2.1 and 4.2.2 is based on the characteristics of beam radiation. However, solar radiation comprises beam and diffuse components that have different optical characteristics (see Chapter 2). Moreover, beam radiation can be reflected specularly or diffusely. Thus, it is necessary to treat each component separately.

Solar radiation incident on a collector includes diffuse radiation from the sky and, sometimes, radiation reflected from the ground. Generally, the solar radiation reflected by the ground is diffuse in nature. Consequently, the amount of this radiation component is also diffuse. The radiation that propagates through a cover system can be computed by integrating the transmitted radiation over all the angles but, unlike beam radiation, diffuse radiation comes from different angles. Thus, in general, the angular distribution of diffuse radiation is not known. Anisotropic sky models are commonly used to analyze the angular distribution of radiation. These models include the Hay and Davies model, the Klucher and Reindl model (Reindl et al., 1990), and the Perez model (Perez et al., 1987). They break down solar radiation into components that correspond to different geometric regions of the virtual surface of the sky dome, namely: (a) beam (parallel rays directly from the sun's disc), (b) sky isotropic (as a uniform dome), (c) circumsolar (a point source or cone following the sun), (d) horizon brightening (thin band or a ring at the horizon), and ground-reflected albedo (extension of the dome) (Arias-Rosales & LeDuc, 2020). The angular incidence distribution is different for each anisotropic component distribution. Consequently, a given component is expected to exhibit different characteristics of transmittance.

For isotropic diffuse radiation, the integration is simple. It can be performed by identifying an equivalent angle for beam radiation that yields the same transmittance as for diffuse radiation. An equivalent angle of 60° has been established as suitable for a wide spectrum of conditions that prevail in the application of solar collectors (Duffie & Beckman, 2013). This means that the transmittance for isotropic diffuse radiation is the same as that for beam radiation incident at an angle of 60°. The angle of incidence of circumsolar diffuse radiation can be assumed to be the same as the angle of incidence of the beam radiation. Often, diffuse radiation from the horizon contributes a small proportion to the total radiation, and it can be considered as having approximately the same angle of incidence as the isotropic component. The equivalent angle of incidence ($\theta_{e,d}$) for all the diffuse radiation can be assumed to be the same, and the ground reflected radiation can be taken to have an equivalent angle of incidence ($\theta_{e,g}$) given by the following:

$$\theta_{e,d} = 59.7 - 0.1388\beta + 0.001497\,\beta^2 \tag{4.25}$$

$$\theta_{e,g} = 90 - 0.5788\beta + 0.002693\,\beta^2 \tag{4.26}$$

4.2.4 Transmittance–absorptance product

Part of transmitted fraction (τ) of solar radiation that reaches the absorber surface (with an absorptance of α) is absorbed ($\alpha\tau$) and converted to heat, while the remainder $(1-\alpha)\tau$ is reflected to the cover surface. Usually, the absorber plate is not smooth to produce specular reflection. Consequently, the reflected radiation from the absorber plate is assumed to be diffuse (and not polarized). The cover reflects a fraction $(1-\alpha)\tau_{\rho d}$ of this radiation back to

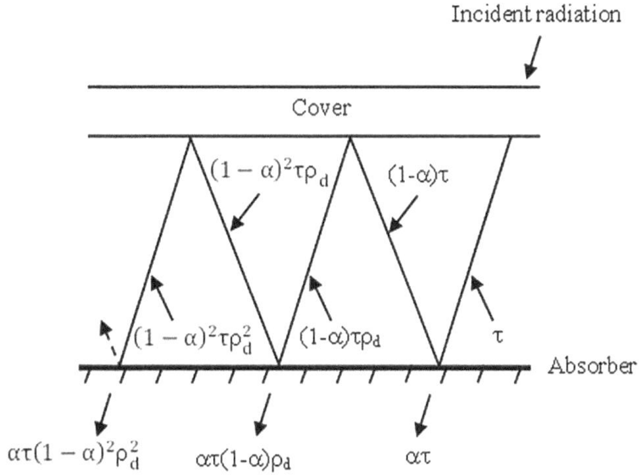

FIGURE 4.3 Transmission and absorption of solar radiation in a two-cover system. *Adapted from Duffie and Beckman (2013).*

absorber plate. These attenuation processes continue until the energy of the reflected ray is negligible (Fig. 4.3).

The reflectance ρ_d is a little bit different from the reflectance of the incident diffuse solar radiation. For a two-cover system presented in Fig. 4.3, multiple reflection of radiation continues to such an extent that the fraction of absorbed solar radiation can be given by the following expression:

$$(\alpha\tau) = \alpha\tau \sum_{i=0}^{i=\infty} [(1-\alpha)\rho_d]^i = \frac{\alpha\tau}{1-(1-\alpha)\rho_d} \tag{4.27}$$

Solar receivers intercept concentrated beam radiation. Nevertheless, in general, the absorber surface of the receiver reflects diffusely. Consequently, knowledge about the reflection of diffuse radiation is necessary for the development of receiver systems.

4.2.5 Spectral dependence of transmittance

Transmittance is influenced by the wavelength (λ) and direction (θ_i) of radiation. Therefore, knowledge about the spectral distribution of the solar flux is important to enable computation of the solar weighted property. In this vein, a reference standard solar spectral irradiance distribution is required to compare the performance of different solar technologies (see Section 1.7).

For most of the glazing materials, transmittance is influenced by the wavelength of the incident radiation. Glass, the most widely used cover material, absorbs a small fraction of the solar energy spectrum when its concentration of iron (III) oxide (Fe_2O_3) is diminished. Infrared radiation is absorbed at elevated levels of Fe_2O_3 concentration. The maximum intensity of the solar spectrum lies in the range 0.3–2.0 μm (D'Alessandro et al., 2021). Moreover, glass is significantly opaque to radiation of wavelengths above approximately 3 μm and it is taken as opaque to long-wave radiation. Thus, glass is transparent to the solar spectrum except for the type that absorbs heat (long wavelength). Low-iron (clear) glass has transmittance values of 88% −92% at normal incidence (Greenwald & McHugh, 1985). Other advantages of glass are: (a) long life (if it is properly fitted) and (b) transmission of a small proportion of long wave radiation (Brow & Schmitt, 2009).

Transmittances of some transparent materials are more wavelength dependent than low-iron glass. In such cases, it is necessary to compute the individual transmittances of monochromatic radiation and then sum over the whole solar spectrum. If the transmittance of monochromatic radiation is independent of the angle of incidence, the transmittance for incident radiation of a specific spectral distribution is calculated by using the following equation:

$$\tau = \sum_{j=1}^{j=n} \tau_{\lambda,j} \Delta f_j \tag{4.28}$$

Where is $\tau_{\lambda,j}$ the transmittance of monochromatic radiation and Δf_j is the change of energy in the blackbody spectrum (details for computation of this parameter are reported in Section 4.3.4).

Otherwise, if the transmittance of the incident radiation of a given spectral distribution is angular dependent, then:

$$\tau(\theta) = \frac{\int_0^\infty \tau_\lambda(\theta) I_{\lambda i}(\theta) d\lambda}{\int_0^\infty I_{\lambda i}(\theta) d\lambda} \tag{4.29}$$

Where $I_{\lambda,i}$ (θ) is the monochromatic radiation incident at angle θ and $\tau_\lambda(\theta)$ is the transmittance of monochromatic radiation incident at angle (θ).

In some cases, both the solar transmittance of the cover system and the solar absorptance of the absorber plate depend on the wavelength and angle of incidence of the radiation. In such cases, the fraction of solar radiation absorbed by an absorber plate is computed from:

$$\alpha\tau(\theta) = \frac{\int_0^\infty \alpha_\lambda(\theta)\ \tau_\lambda(\theta) I_{\lambda i}(\theta) d\lambda}{\int_0^\infty I_{\lambda i}(\theta) d\lambda} \tag{4.30}$$

Where $\alpha_\lambda(\theta)$ is the absorptance of monochromatic radiation incident at angle (θ).

4.2.6 Transparent selective surface

For solar thermal collectors, a transparent cover should ideally have high transmittance over the incident solar spectral region while also preventing

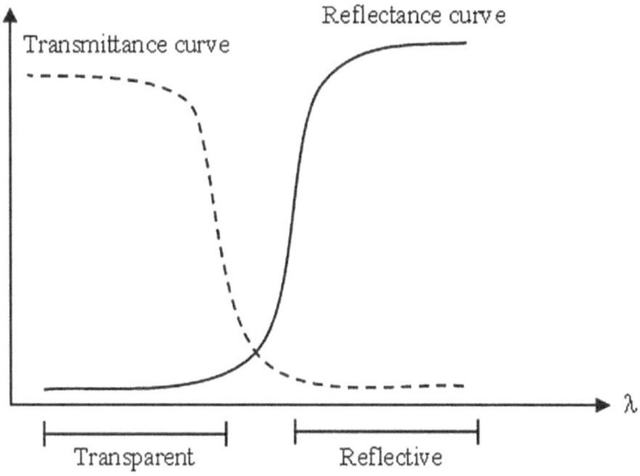

FIGURE 4.4 Ideal optical properties of a transparent heat mirror. *Adapted from Duffie and Beckman (2013).*

infrared radiation (~2.5 μm < λ < ~25 μm) from escaping. However, transparent materials that naturally possess a combination of these properties are scarce. In general, there are naturally no materials with such a combination of ideal optical properties. Conventional transparent materials, such as glass and organic thin films, have good transmittance in the visible range, but the infrared-red emissivity of these materials is very high, which leads to the dissipation of a large amount of energy through thermal radiation (Li et al., 2020). Consequently, research attention has been given to the development of materials that can strongly transmit solar radiation and be very reflective to infrared radiation. Surfaces with such optical properties are called transparent heat mirrors (D'Alessandro et al., 2021). Fig. 4.4 depicts the transmittance and reflectance of a typical transparent heat mirror as functions of the wavelength of the incident radiation.

Heat mirror coatings are applied on a plastic or glass substrate. Plastics are used as cover materials in applications of low temperature and therefore they are not suitable for receivers. In view of this limitation, glass substrates are commonly exploited in heat mirrors for high temperature applications. Two types of transparent heat mirrors have practically been developed (Rostami et al., 2020): (a) ultrathin metal films that are sandwiched between thin-film dielectric antireflective coatings and (b) doped transparent conducting oxides. Some of the conventional transparent coating materials are fluorine-doped tin dioxide (SnO_2:F), antimony-doped tin dioxide (SnO_2:Sb), tin-doped indium sulfur trioxide (In_2SO_3:Sn), and aluminum-doped zinc oxide (ZnO:Al) (Kennedy, 2002). These thin films allow transmission of solar radiation and supress the propagation of infrared radiation through them. Nevertheless, the transmittance of heat mirrors made of thin metal films is insufficient for solar thermal applications.

Another technique for achieving transparent heat mirrors is emerging to manufacture them in the form of photonic structures. These structures can be one-, two-, or three dimensional. Research findings show that at 1000 K, the transparent photonic crystal heat mirror reflects more radiation back to the absorber surface of a receiver than a conventional transparent mirror based on tin-doped indium trioxide film (Rostami et al., 2020). The performance of parabolic dish receivers can be improved by 100% when transparent photonic crystal heat mirrors are used. Consequently, this emerging technology of transparent heat mirrors is promising to significantly improve the performance of concentrating solar power plants.

4.3 Opaque materials

4.3.1 Absorptance and emittance

When solar radiation is incident on an opaque surface, part of the radiation is absorbed, while the remainder is reflected by the surface. The absorptance of monochromatic radiation in a given direction is a property of the surface, and it is defined as the proportion of the incident radiation of wavelength (λ) from the direction (μ, γ) that is absorbed by the surface. It can be expressed as follows:

$$\alpha_\lambda(\mu,\gamma) = \frac{I_{\lambda,a}(\mu,\gamma)}{I_{\lambda,i}(\mu,\gamma)} \tag{4.31}$$

Where $I_{\lambda,a}(\mu,\gamma)$ is the absorbed radiation, $I_{\lambda,i}(\mu,\gamma)$ is the incident radiation, μ is the cosine of the polar angle, and γ is the azimuthal angle.

The fraction of all the radiation over the entire spectrum of wavelengths from the direction (μ,γ) that is absorbed by a surface is known as the directional absorptance and it is described by the following equation:

$$\alpha(\mu,\gamma) = \frac{\int_0^\infty \alpha_\lambda(\mu,\gamma)\, I_{\lambda,i}(\mu,\gamma)d\lambda}{\int_0^\infty I_{\lambda,i}(\mu,\gamma)d\lambda} = \frac{1}{I_i(\mu,\gamma)}\int_0^\infty \alpha_\lambda(\mu,\gamma)\, I_{\lambda,i}(\mu,\gamma)d\lambda \tag{4.32}$$

The parameter $\alpha(\mu,\gamma)$ is not a characteristic of the surface because it depends on the wavelength distribution of the incident radiation and contains an unspecified function $I_{\lambda,i}(\mu,\gamma)$. In addition, surfaces emit thermal radiation. In this regard, knowledge about the emittance of a material is essential for the computation of radiative heat transfer (Granados et al., 2020). To analyze the emissive energy, it is necessary to understand the concept of a blackbody. A blackbody is an ideal body that absorbs all incident radiation, irrespective of wavelength and direction. It emits diffusely and is a perfect emitter. A real

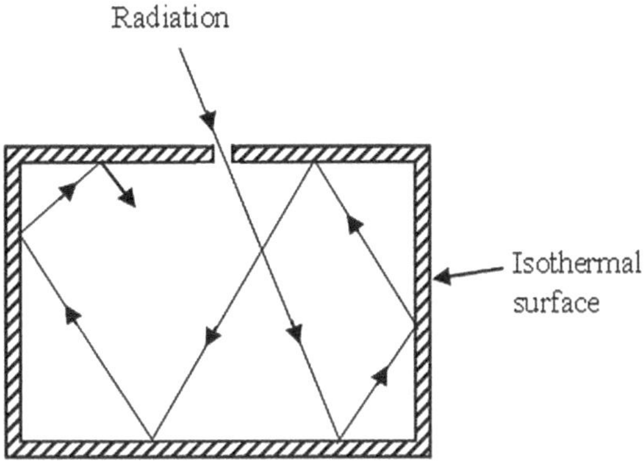

FIGURE 4.5 Characteristics of a blackbody cavity.

surface cannot emit more energy than a blackbody at a given temperature and wavelength. However, the nearest approximation to a blackbody is obtained by a cavity that has its inner surface at a uniform temperature. Radiation is expected to undergo multiple reflections before reemergence if it enters the cavity via a tiny aperture (Fig. 4.5). So, a blackbody is a standard against which radiative properties of real surfaces may be compared.

The monochromatic directional emittance of a surface (ε_λ) is defined as the ratio of the intensity of monochromatic radiation that is given off by a surface in a specific direction to the intensity of monochromatic radiation that would be radiated by a blackbody at the same temperature. The parameter ε_λ is a property of the surface and it is given by the following:

$$\varepsilon_\lambda(\mu,\gamma)=\frac{I_\lambda(\mu,\gamma)}{I_{\lambda,b}(\mu,\gamma)} \tag{4.33}$$

Where $I_\lambda(\mu,\gamma)$ is radiation emitted by a real surface and $I_{\lambda b}(\mu,\gamma)$ is radiation emitted by a blackbody.

The directional emittance is defined as the ratio of the emitted total intensity in the direction (μ,γ) to the blackbody intensity. It is also a property of the surface expressed as follows:

$$\varepsilon(\mu,\gamma)=\frac{\int_0^\infty \varepsilon_\lambda(\mu,\gamma)\, I_{\lambda,b}(\mu,\gamma)d\lambda}{\int_0^\infty I_{\lambda,b}(\mu,\gamma)d\lambda}=\frac{1}{I_b}\int_0^\infty \varepsilon_\lambda(\mu,\gamma)\, I_{\lambda,b}(\mu,\gamma)d\lambda \tag{4.34}$$

Hemispherical properties can be defined based on the definitions of the directional absorptance and emittance of a surface. In this context, the monochromatic hemispherical absorptance and emittance are found through integration over the enclosing hemisphere:

$$\alpha_\lambda = \frac{\int_0^{2\pi}\int_0^1 \alpha_\lambda\,(\mu,\gamma)\; I_{\lambda,i}(\mu,\gamma)\mu d\mu\; d\gamma}{\int_0^{2\pi}\int_0^1 I_{\lambda,i}(\mu,\gamma)\mu d\mu d\gamma} \tag{4.35}$$

$$\varepsilon_\lambda = \frac{\int_0^{2\pi}\int_0^1 \varepsilon_\lambda\,(\mu,\gamma)\; I_{\lambda,b}(\mu,\gamma)\mu\; d\mu\; d\gamma}{\int_0^{2\pi}\int_0^1 I_{\lambda,b}(\mu,\gamma)\mu d\mu d\gamma} = \frac{\int_0^{2\pi}\int_0^1 \varepsilon_\lambda\,(\mu,\gamma)\; I_{\lambda,b}(\mu,\gamma)\mu d\mu d\gamma}{E_{\lambda,b}} \tag{4.36}$$

For a blackbody radiator, the monochromatic energy ($E_{\lambda,b}$) can be computed by using Planck's law:

$$E_{\lambda,b} = \frac{C_1}{\lambda^5[\exp(C_2/\lambda T) - 1]} \tag{4.37}$$

Where $C_1 = 2\pi h \quad C_o^2 = 3.7405 \times 10^8\ W\ \mu m^4\ m^{-2}$ and $C_2 = h\quad C_o\ k^{-1} = 14{,}387.8\ \mu m\ K$ are respectively Planck's first and second radiation constants, h is Planck's constant, and k is Boltzmann's constant.

For a given temperature, there is a wavelength (λ_{max}) that corresponds to the maximum intensity of blackbody radiation. This is defined by Wien's displacement law:

$$\lambda_{max} T = C_3 \tag{4.38}$$

Where C_3 is a constant = 2897.8 μm K.

It should be noted that the parameter α_λ is not a property, while ε_λ is a property. Expressions for hemispherical absorptance (α) and emittance (ε) are derived through integration:

$$\alpha = \frac{\int_0^\infty\int_0^{2\pi}\int_0^1 \alpha_\lambda\,(\mu,\gamma)\; I_{\lambda,i}(\mu,\gamma)\mu d\mu d\gamma d\lambda}{\int_0^\infty\int_0^{2\pi}\int_0^1 I_{\lambda,i}(\mu,\gamma)\mu d\mu d\gamma d\lambda} \tag{4.39}$$

$$\varepsilon = \frac{\int_0^\infty\int_0^{2\pi}\int_0^1 \varepsilon_\lambda\,(\mu,\gamma)\; I_{\lambda,b}(\mu,\gamma)\mu d\mu d\gamma d\lambda}{\int_0^\infty\int_0^{2\pi}\int_0^1 I_{\lambda,b}(\mu,\gamma)\mu d\mu d\gamma d\lambda} = \frac{\int_0^{2\pi} \varepsilon_\lambda E_{\lambda,b} d\lambda}{E_b} \tag{4.40}$$

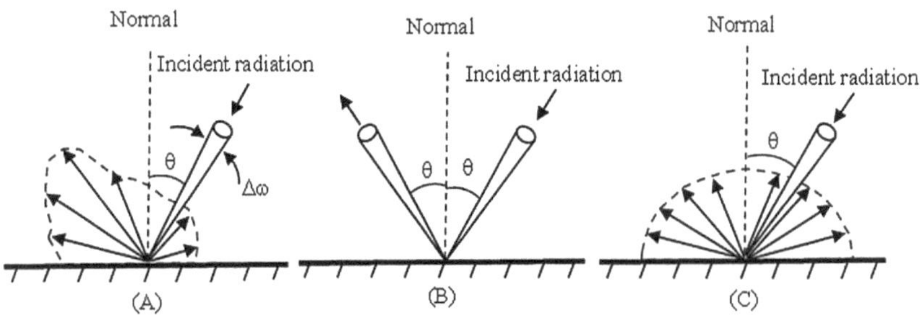

FIGURE 4.6 Reflection of radiation by an opaque surface: (A) general reflection, (B) specular reflection, and (C) diffuse reflection. θ is the angle of incidence and Δω is a small solid angle of the radiation.

Eq. (4.39) can be simplified when the monochromatic radiation is independent of the direction (when $\alpha_\lambda(\mu, \gamma) = \alpha_\lambda$):

$$\alpha = = \frac{\int_0^\infty \alpha_\lambda \, q_{\lambda,i} d\lambda}{\int_0^\infty q_{\lambda,i} d\lambda} \tag{4.41}$$

Where $q_{\lambda,i}$ is the incident monochromatic radiant energy.

4.3.2 Reflectance

When radiation is incident on an opaque surface, a fraction of it is reflected specularly or diffusely (Fig. 4.6). Radiation propagates in one direction from a specular plane reflector ($\Delta\omega$ is a small solid angle of radiation). On the other hand, reflected radiation is scattered in different directions in diffuse reflection. In a practical system, a surface is neither all specular nor all diffuse. For a given surface, the intensity of radiation in a specific direction depends on the wavelength and the spatial distribution of the incident radiation.

The monochromatic reflectance can be expressed as a function of direction:

$$\rho_\lambda(\mu_r, \gamma_r, \mu_i, \gamma_i) = \lim_{\Delta\omega_i \to 0} \frac{\pi I_{\lambda,r}(\mu_r, \gamma_r)}{I_{\lambda,i} \; \mu_i \Delta\omega_i} \tag{4.42}$$

Where $I_{\lambda,r}$ (μ_r, γ_r) is the reflected intensity in the direction (μ_r, γ_r), $I_{\lambda,i}$ $\mu_i \Delta\omega_i$ is the incident intensity from the direction (μ_i, γ_i), and $\Delta\omega_i$ is a small solid angle of the radiation.

There are two types of hemispherical reflectances, namely: (a) angular hemispherical reflectance and (b) monochromatic hemispherical reflectance. In an angular hemispherical reflectance, radiation with a thin solid angle is incident on a surface and all the reflected radiation is collected in a specified direction when the surface is irradiated from different directions. A monochromatic hemispherical reflectance $\rho_\lambda(\mu_i, \gamma_i)$ is the ratio of the monochromatic radiant energy reflected in all directions to the radiant flux incident on the surface within a small solid angle $\Delta\omega$. The monochromatic radiant energy reflected in all directions ($q_{\lambda,r}$) can be computed from:

$$q_{\lambda,r} = \frac{1}{\pi} \int_0^{2\pi} \int_0^1 \rho_\lambda \, (\mu_r, \gamma_r, \mu_i, \gamma_i) I_{\lambda,i} \mu_i \Delta\omega_i \mu_r d\mu_r d\gamma_r \tag{4.43}$$

So, the monochromatic angular hemispherical reflectance is:

$$\rho_\lambda(\mu_i, \gamma_i) = \frac{q_{\lambda,r}}{I_{\lambda,i} \mu_i \Delta\omega_i} = \frac{1}{\pi} \int_0^{2\pi} \int_0^1 \rho_\lambda \, (\mu_r, \gamma_r, \mu_i, \gamma_i) \mu_r d\mu_r d\gamma_r \tag{4.44}$$

From Eq. (4.44), it is noted that $\rho_\lambda(\mu_i, \gamma_i)$ is a property of the surface. Integration over all the wavelengths yields the angular hemispherical reflectance $\rho(\mu_i, \gamma_i)$ which is not a property because it is a function of the wavelength distribution of the incident radiation.

Another parameter is the monochromatic hemispherical-angular reflectance. This parameter is the ratio of the reflected monochromatic energy in the direction (μ_r, γ_r) to the monochromatic energy from all directions divided by π. Integration of the incident energy intensity over the hemisphere gives the monochromatic incident energy:

$$q_{\lambda,i} = \int_0^{2\pi}\int_0^1 I_{\lambda,i}\mu_i \, d\mu_i d\gamma_i \tag{4.45}$$

and hemispherical-angular reflectance is:

$$\rho_\lambda(\mu_r,\gamma_r) = \frac{I_{\lambda,r}(\mu_r,\gamma_r)}{q_{\lambda,i}/\pi} \tag{4.46}$$

The hemispherical-angular reflectance, presented in Eq. (4.46), can also be expressed as a reflectance function:

$$\rho_\lambda(\mu_r,\gamma_r) = \frac{\int_0^{2\pi}\int_0^1 \rho_\lambda\,(\mu_r,\gamma_r,\mu_i,\gamma_i)\; I_{\lambda,i}\mu_i d\mu_i d\gamma_i}{\int_0^{2\pi}\int_0^1 I_{\lambda,i}\mu_i d\mu_i d\gamma_i} \tag{4.47}$$

The monochromatic hemispherical-angular reflectance $\rho_\lambda(\mu_r, \gamma_r)$ is not a property of the surface because it depends on the angular distribution of the incident intensity. There is a special case when the incident radiation is diffuse. In this case, the monochromatic hemispherical-angular reflectance is equal to the monochromatic angular hemispherical reflectance of a surface. It is important for $\rho_\lambda(\mu_i, \gamma_i)$ and $\rho_\lambda(\mu_r, \gamma_r)$ to be equal when $I_{\lambda,i}$ is uniform because the measurement of $\rho_\lambda(\mu_r, \gamma_r)$ is simpler than that of $\rho_\lambda(\mu_i, \gamma_i)$. Integration over all wavelengths of both $\rho_\lambda(\mu_i, \gamma_i)$ and $\rho_\lambda(\mu_r, \gamma_r)$ gives the total for each reflectance. For angular hemispherical reflectance, the total is:

$$\rho(\mu_i,\gamma_i) = \frac{\int_0^\infty q_{\lambda,r}d\lambda}{\int_0^\infty I_{\lambda,r}\mu_i\Delta\omega_i d\lambda} = \frac{1}{\pi I_i}\int_0^\infty\int_0^{2\pi}\int_0^1 \rho_\lambda\,(\mu_i,\gamma_i,\mu_r,\gamma_r)\; I_{\lambda,i}\;\mu_r d\mu_r d\gamma_r d\lambda \tag{4.48}$$

It is possible to express the reflected monochromatic energy in terms of the reflection function:

$$q_{\lambda,r} = \int_0^{2\pi}\int_0^1\left[\int_0^{2\pi}\int_0^1 \frac{\rho_\lambda\,(\mu_i,\gamma_i,\mu_r,\gamma_r)}{\pi} I_{\lambda,i}\mu_i d\mu_i d\gamma_i\right]\mu_r d\mu_r d\gamma_r \tag{4.49}$$

Dividing Eq. (4.49) by Eq. (4.45) gives the monochromatic hemispherical reflectance. For a diffuse surface, the magnitude of the monochromatic hemispherical reflectance is equal to that of the reflectance function and it is not affected by the spatial distribution of the intensity of the incident radiation. Integration of Eqs. (4.45) and (4.49) over all the wavelengths yields the total incident energy and total reflected energy, respectively. The hemispherical reflectance (ρ) is equal to the ratio of the total reflected energy (q_r) to the total incident energy (q_i). The parameter ρ depends on the angular and wavelength distributions of the incident radiation, and it is given by the following:

$$\rho = \frac{q_r}{q_i} = \frac{\int_0^\infty q_{\lambda,r} d\lambda}{\int_0^\infty q_{\lambda,i} d\lambda} \tag{4.50}$$

A distinct form of the hemispherical reflectance is useful for low-temperature applications that exclude solar radiation. Eq. (4.50) is the distinct form that is founded on the assumption that the reflection function does not depend on direction (diffuse approximation) nor wavelength (gray approximation). In a gray surface, the spectral absorptivity and emissivity are independent of wavelength within the spectral regions of irradiation and emission of the surface. Combining these two approximations makes the surface reflectance to become independent of all variables except maybe the temperature of the surface. Nonetheless, even the surface temperature is often ignored.

4.3.3 Functional relationships among absorptance, emittance, and reflectance

The intensity of monochromatic radiation in a direction (μ, γ) from a tiny area of the surface is made up of emitted radiation ($I_{\lambda,e}$) and reflected radiation ($I_{\lambda,r}$) that should be equal to the intensity ($I_{\lambda b}$) of monochromatic radiation from a blackbody. Thus,

$$I_{\lambda b} = I_{\lambda,e}(\mu,\lambda) + I_{\lambda,r}(\mu,\lambda) \tag{4.51}$$

But:

$$I_{\lambda,e}(\mu,\ \lambda) = \varepsilon_\lambda(\mu,\lambda) I_{\lambda b} \tag{4.52}$$

$$I_{\lambda,r}(\mu,\ \lambda) = \rho_\lambda(\mu_r,\lambda_r) I_{\lambda b} \tag{4.53}$$

For diffuse radiation, $\rho_\lambda(\mu_r,\ \gamma_r) = \rho_\lambda(\mu_i,\ \gamma_i)$. Consequently, Eq. (4.51) can be expressed as follows:

$$I_{\lambda b} = \varepsilon_\lambda(\mu,\lambda) I_{\lambda b} + \rho_\lambda(\mu_i,\lambda_i) I_{\lambda b} \tag{4.54}$$

This implies that:

$$\varepsilon_\lambda(\mu,\lambda) = 1 - \rho_\lambda(\mu_i,\lambda_i) \tag{4.55}$$

Kirchhoff's law shows that:

$$\varepsilon_\lambda(\mu,\lambda) = \alpha_\lambda(\mu,\lambda) = 1 - \rho_\lambda(\mu_i,\lambda_i) \tag{4.56}$$

Eq. (4.56) indicates that knowledge about the monochromatic angular hemispherical reflectance can be used to calculate both the monochromatic emittance and absorptance of a surface. Monochromatic or total energy from all directions is either absorbed or reflected from an opaque surface, such that:

$$\alpha_\lambda + \rho_\lambda = \varepsilon_\lambda + \rho_\lambda = 1 \tag{4.57}$$

$$\alpha + \rho = 1 \tag{4.58}$$

4.3.4 Selective absorber surfaces

It is desirable that the absorber element of a solar receiver should have high absorptance for radiation in the solar energy spectrum. Unfortunately, this component also loses energy through different mechanisms that include thermal radiation from the absorbing surface. So, it important to have low long-wave emittance of the surface in order to curtail heat losses. These absorber elements operate at relatively lower temperature compared to the effective surface temperature of the sun (approximately 6000 K). For example, most concentrating solar power plants operate at temperatures below (1000 K) as of 2021 (see Section 1.7). In view of this, the wavelength range of the radiation emitted by practical absorber surfaces overlaps by a small margin with the solar spectrum. Under these circumstances, it is possible to develop surfaces that have high solar absorptance but low long-wave emittance. A material that has high absorptance in the solar spectrum and low emittance in the infrared band is known as a selective solar absorber.

A selective absorber is an idealized surface with high spectral absorptance in the short-wavelength band and high spectral reflectance in the long-wavelength band. In general, there are no materials with such a combination of ideal optical properties. Consequently, selective absorbers can practically be produced by combining layers of different materials. A common selective absorber type is the absorber–reflector tandem, which is produced by combining two surfaces (Tesfamichael, 2000). One of the surfaces has high absorptance in the solar spectral band and the other one with high reflectance in the infrared band. Surface coating is a common way of producing selective absorbers, whereby the coating layer absorbs the solar spectrum and the substrate has low emittance in the long-wavelength band. Most of the coating materials in use are metal oxides and substrates are metals, such as copper oxide coated on copper. Black chromium (Cr/Cr_2O_3) is one of the most widely used selective coatings in solar collectors for the efficient conversion of solar radiation to heat (Nunes et al., 2018). The coatings are produced by electroplating, and laboratory specimens of this variety of coatings have exhibited high absorptances of up to 0.96 and low emittances down to 0.08.

Based on principles of design and absorption, selective solar absorbing coatings can be categorized into five different groups (Xu et al., 2020): (a)

Intrinsic absorber, (b) Semiconductor absorber, (c) Multilayer interference stacks, (d) Cermet absorber, and (e) Textured surface. An intrinsic selective absorber comprises one material with intrinsic selective properties. There is no naturally occurring material that has intrinsically ideal solar-selective properties, only some estimates. Transition metals (such as tungsten metal) and semiconductors (such as molybdenum trioxide doped with molybdenum, silicon doped with boron, hafnium carbide, and tin (IV) oxide) possess intrinsic properties. However, these materials require significant modification for use as intrinsic absorbers. A semiconductor layer is sandwiched between a substrate and antireflecting layers in a semiconductor absorber. The semiconductor material absorbs solar radiation and suppresses emission of infrared radiation due to the band gap (Kennedy, 2002; Rostami, 2019). In a multilayer configuration, dielectric and metallic layers are stacked up on a substrate. This category uses the interference property of light to absorb close to the peak of the solar radiation instead of within the infrared band. The selective effect rises from the multiple reflections that pass through the bottom dielectric layer and it does not depend on the selectivity of the dielectric. The cermet absorber is a composite of ceramic and metal materials. In this selective surface, metal nanoparticles are dispersed within the ceramic matrix. A textured surface has modified surface topographical features at micro- and nanoscales. It is a common method of producing spectral selectivity by optically capturing solar energy. The five classes of coatings are schematically presented in Fig. 4.7.

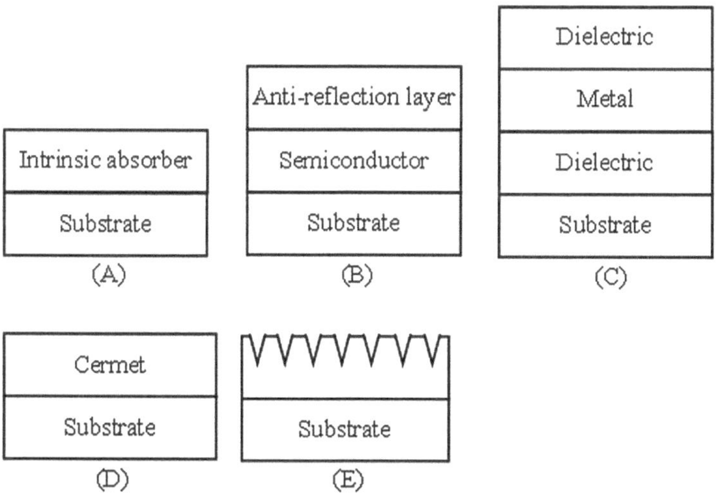

FIGURE 4.7 Schematic representation of five solar absorber coatings: (A) Intrinsic absorber, (B) Semiconductor absorber, (C) Multilayer interference stacks, (D) Cermet absorber, and (E) Textured surface. *Adapted from Xu et al. (2020).*

The multilayer interference stack and cermet coatings are exploited most widely. Based on the absorbing dielectrics, cermet coatings are divided into three major groups: double cermet solar selective coatings, transition metal nitride multilayer coatings, and transition metal oxide multilayer coatings. From substrate to surface, a typical double cermet coating comprises a metal infrared reflective layer, high and low metal volume fraction cermet solar absorption layers (HMVF and LMVF), and a ceramic antireflection layer (AR). The structure of the transition metal nitride/oxide multilayer coatings is similar to that of the cermet. It is made up of ceramic-top and metallic-bottom layers with interspaced gradient layers. A coating that has absorptance higher than 0.90 in the solar spectrum region and thermal emittance ε lower than 0.10 the infrared wavelength range is deemed to perform well (Prado & Sowmy, 2016). Most of the selective absorbers, developed by different researchers, meet this required (Tables 4.2 and 4.3).

4.3.5 Computation of absorptance and emittance

Measurements of monochromatic reflectance (ρ_λ) are commonly taken in practice. So, data on ρ_λ is often available. From Eq. (4.56), this parameter is related to absorptance (α_λ) and emittance (ε_λ). Consequently, this data can be utilized to determine the absorptance (α) and emittance (ε) of the surface. This computation can be done by dividing the spectrum (the incident energy spectrum for absorptance or blackbody spectrum for emittance) into intervals (Fig. 4.8) and numerically integrating to find α or ε.

Integration of Eq. (4.37) from 0 to some wavelength λ yields the total energy emitted per unit area by a blackbody:

$$E_{0-\lambda b} = \int_0^\lambda E_{\lambda,b} d\lambda \tag{4.59}$$

Substituting Eq. (4.37) into Eq. (4.59) and dividing by σT^4 gives a function of λT:

$$f_{0-\lambda T} = \frac{E_{0-\lambda T}}{\sigma T^4} = \int_0^{\lambda T} \frac{C_1 d(\lambda T)}{\sigma(\lambda T)^5 [\exp(C_2/\lambda T) - 1]} \tag{4.60}$$

$f_{0\text{-}\lambda T}$ is the fraction of the blackbody energy from zero to λT. Values of the integral are available in tabular form (Table 4.4). The integration can also be done numerically on a computer by using Eq. (4.61):

$$f_{0-\lambda T} = \frac{15}{\pi^4} \sum_{n=1}^{n=\infty} \left[\frac{e^{-na}}{n} \left(a^3 + \frac{3a^2}{n} + \frac{6a}{n^2} + \frac{6}{n^3} \right) \right] \tag{4.61}$$

Where $a = \frac{C_2}{\lambda T}$. Summation of 10 terms is practically accurate enough (Table 4.4).

TABLE 4.2 High-temperature double cermet solar selective coatings (Xu et al., 2020).

Substrate	Coating	Annealing (°C/h/ atmosphere)	α	ε
Si	W/W−Al2O3(H)/W −Al2O3(L)/Al2O3	580/720/V^{-1}	0.937	0.093 (673 K)
SS	W/W−Al2O3(H)/W −Al2O3(L)/SiO2	400/2496/A^{-1}	0.950	0.106 (673 K)
SS	W/W−Al2O3(H)/W −Al2O3(L)/Al2O3	400/2496/A^{-1}	0.922	0.073 (673 K)
SS	Mo/Pt−Al2O3(H)/Pt −Al2O3(L)/Al2O3	800/24/V^{-1}	0.938	0.10 (355 K)
Si	Ag/Al2O3/Ag−Al2O3(H)/Ag −Al2O3(L)/Al2O3	500/12/N^{-1}	0.729	–
Si	AgAl/Al2O3/AgAl −Al2O3(H)/AgAl−Al2O3(L)/ Al2O3	500/1002/N^{-1}	0.950	0.11 (673 K)
SS	W/Al2O3/Wti-Al2O3/W −Al2O3/Al2O3	600/840/V^{-1}	0.928	0.103 (773 K)
SS	Mo/AlNi−Al2O3(H)/AlNi −Al2O3(L)/Al2O3	500/138/V^{-1}	0.950	0.078 (353 K)
SS	W/Al2O3/Wni-Al2O3(H)/ Wni-Al2O3(L)/Al2O3/SiO2	600/168/V^{-1}	0.908	0.057 (355 K)
SS	Mo/MoSi2−Al2O3(H)/MoSi2 −Al2O3(L)/Al2O3	400/1140/V^{-1}	0.950	0.07 (353 K)
SS	WC-Al2O3(H)/WC- Al2O3(L)/Al2O3	600/5/V^{-1}	0.940	0.10 (355 K)
SS	Co-WC-Al2O3/Al2O3	600/168/A^{-1}	0.898	0.172 (355 K)
SS	W/WNi-YSZ(H)/WNi-YSZ(L)/ YSZ/SiO2	600/168/V^{-1}	0.930	0.085 (773 K)
SS	W/AlSiOx:W(H)/ AlSiOx:W(L)/AlSiOx	580/1050/V^{-1}	0.951	0.113 (673 K)
SS	Ag/MoeSi3N4(H)/ MoeSi3N4(L)/Si3N4	–	0.926	0.055 (673 K)
SS	Ag/MoSi2eSi3N4 (H)/ MoSi2eSi3N4(L)/Si3N4	600/200/V^{-1}	0.880	0.020 (298 K)
SS	W/TiNeSiO2(H)/ TiNeSiO2(L)/SiO2	600/168/V^{-1}	0.946	0.036 (355 K)

Continued

TABLE 4.2 High-temperature double cermet solar selective coatings (Xu et al., 2020).—cont'd

Substrate	Coating	Annealing (°C/h/ atmosphere)	α	ε
Ann-SS	W/TiNeSiO2(H)/ TiNeSiO2(L)/SiO2	$600/168/V^{-1}$	0.949	0.098 (355 K)
$LaAlO_3$	W/TiNeSiO2(H)/ TiNeSiO2(L)/SiO2	$700/168/V^{-1}$	0.942	0.080 (355 K)
SS(573 K)	ZrC-ZrOx/Al2O3	$500/100/V^{-1}$	0.880	0.13 (355 K)

Ann-SS, annealed stainless steel, *SS*, stainless steeel, *α*, absorptance, *ε*, emittance.

To compute the emittance ε of a surface, the jth segment of the blackbody spectrum at the surface temperature T is examined (Fig. 4.8). Corresponding to this segment is a monochromatic emittance $\varepsilon_{\lambda,j}$ which is the ratio of the shaded area to the total area of the segment under the curve of the blackbody spectrum. The ratio of monochromatic energy (E_λ) emitted by a surface to the monochromatic energy ($E_{\lambda,b}$) emitted by a blackbody is considered as characteristic of the segment at a suitable wavelength (normally midpoint of the interval). The change of energy (Δf_j) in the blackbody spectrum can be computed by using tabulated data of $f_{0-\lambda T}$. Δf_j is the difference in $f_{0-\lambda T}$ at the lower and upper wavelength limits of the interval. The jth increment in $f_{0-\lambda T}$ has a corresponding increment in ε given by Eq. (4.62):

$$\Delta\varepsilon_j = \varepsilon_j \Delta f_j \tag{4.62}$$

Therefore,

$$\varepsilon = \sum_{j=1}^{j=n} \varepsilon_j \Delta f_j \tag{4.63}$$

Eq. (4.63) can be expressed in terms of reflectance:

$$\varepsilon = \sum_{j=1}^{j=n} (1 - \rho_j) \Delta f_j = 1 - \sum_{j=1}^{j=n} \rho_j \Delta f_j \tag{4.64}$$

In cases where the energy increments Δf_j are equal, Eq. (4.64) can be simplified to:

TABLE 4.3 High-temperature transition metal nitride multilayer coatings (Xu et al., 2020).

Substrate	Coating	Annealing (°C/h/ atmosphere)	α	ε
SS	W/WAlN/WAlON/Al2O3	500/2/A^{-1}	0.92	0.11 (355 K)
SS	W/Ag/WN-AlN/AlN/SiO2	400/696/V^{-1}	0.951	0.06 (673 K)
Cu	NbAlN/NbAlON/Si3N4	500/2/A^{-1}	0.943	0.10 (355 K)
Cu	CrAlN/CrAlON/Al2O3	550/100/A^{-1}	0.922	0.08 (355 K)
Cu	CrAlN/CrAlON/Si3N4	450/100/A^{-1}	0.918	0.12 (355 K)
SS	Cr/AlCrN/AlCrON/AlCrO	500/2/A^{-1}	0.940	0.14 (373 K)
SS	TiAlN/TiAlSiN/Si3N4	500/2/A^{-1}	0.913	0.094 (348 K)
Cu	TiAlN/TiAlON/Si3N4	600/2/A^{-1}	0.920	0.08 (355 K)
SS	W/AlTiN/AlTiON/AlTiO	600/100/V^{-1}	0.950	0.08 (355 K)
SS	Ti/AlTiN/AlTiON/AlTiO	450/1004/V^{-1}	0.926	0.16 (673 K)
SS	Al/Ti0.5Al0.5N/ Ti0.25Al0.75N/AlN	400/192/A^{-1}	0.911	0.06 (355 K)
SS	Pt/AlN/TiAlN/AlN/SiO2	900/3/V^{-1}	0.920	0.15 (355 K)
Si	Cu/Zr0.3Al0.7N/ Zr0.2Al0.8N/Al34O60N6	400/192/A^{-1}	0.952	0.074 (673 K)
SS	AlMoN(H)/AlMoN(L)	600/435/V^{-1}	0.921	0.13 (355 K)
SS	W/AlHfN(H)/AlHfN(L)	580/260/V^{-1}	0.933	0.06 (355 K)
SS	W/AlSixN/AlSiOyNx/ AlSiOx	400/1200/A^{-1}	0.940	0.089 (673 K)
SS	Al/NbTiSiN/NbTiSiON/ SiO2	550/100/V^{-1}	0.910	0.13 (673 K)

Continued

TABLE 4.3 High-temperature transition metal nitride multilayer coatings (Xu et al., 2020).—cont'd

Substrate	Coating	Annealing (°C/h/ atmosphere)	α	ε
SS	Cu/TiN/TiSiN/SiN	700/2/V^{-1}	0.922	0.05
SS	Mo/ZrSiN/ZrSiON/SiO2	500/500/V^{-1}	0.930	0.13 (773 K)
SS	Cr/AlCrSiN/AlCrSiON/ AlCrO	600/600/A^{-1}	0.924	0.16
Cu	TiAlSiN/TiAlSiON/SiO2	–	0.96	0.05 (373 K)
SS	W/CrAlSiNx/CrAlSiNxOy/ SiAlOx	400/650/A^{-1}	0.951	0.097 (673 K)
SS	W/CrAlSiNx/CrAlSiNxOy/ SiAlOx	600/650/V^{-1}	0.95	0.094 (673 K)
SS	W/WSiAlNx/WSiAlOyNx/ SiAlOx	450/400/A^{-1}	0.96	0.096 (673 K)
SS	W/WSiAlNx/WSiAlOyNx/ SiAlOx	600/300/V^{-1}	0.958	0.102 (673 K)
SS	W/TiAlC/TiAlCN/TiAlSiCN/ TiAlSiCO/TiAlSiO	650/100/V^{-1}	0.955	0.07 (355 K)
SS	Al/NbTiN/NbTiON/SiO2	550/100/V^{-1}	0.866	0.14 (673 K)
SS(Fe_3O4)	Mo/TiZrN/TiZrON/SiON	500/300/V^{-1}	0.92	0.10 (353 K)
Cu	NbTiON(H)/NbTiON(L)/ SiON	500/40/V^{-1}	0.94	0.08 (353 K)
SS	NbTiON(H)/NbTiON(L)/ SiON	600/40/V^{-1}	0.9	0.13 (353 K)
SS	CrMoN(H)/CrMoN(L)/ CrON	400/71/A^{-1}	0.913	0.12 (355 K)
Al	CrMoN(H)/CrMoN(L)/ CrON	265/600/A^{-1}	0.896	0.08 (355 K)
SS	Al/NbMoN/NbMoON/ SiO2	500/240/V^{-1}	0.937	0.13 (673 K)
SS	HfMoN(H)/HfMoN(L)/ HfON/Al2O3	650/100/V^{-1}	0.941	0.14 (355 K)

SS, stainless steeel, α, absorptance, ε, emittance.

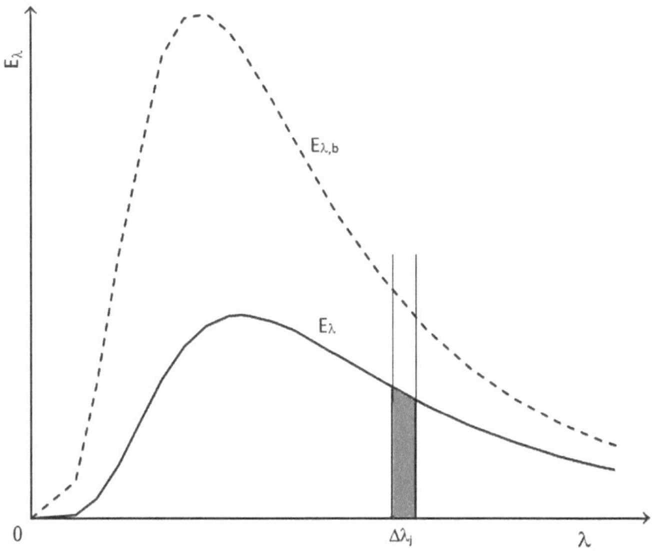

FIGURE 4.8 Segmentation of an emission spectrum.

$$\varepsilon = \frac{1}{n}\sum_{j=1}^{j=n} \varepsilon_j = 1 - \frac{1}{n}\sum_{j=1}^{j=n} \rho_j \tag{4.65}$$

The absorptance of an opaque surface can be calculated using a similar procedure, except for the need to specify the incident radiation. The incident radiation will be generally not blackbody radiation. Consequently, it is necessary to use other available information to facilitate the computational process.

Example 4.4

Calculate the fraction of blackbody radiation at $\lambda T = 2000$ μm K.

Solution

$$f_{0-\lambda T} = \frac{15}{\pi^4}\left[\frac{e^{-a}}{1}\left(a^3 + \frac{3a^2}{1} + \frac{6a}{1^2} + \frac{6}{1^3}\right)\right] + \ldots + \frac{15}{\pi^4}\left[\frac{e^{-a}}{10}\left(a^3 + \frac{3a^2}{10} + \frac{6a}{10^2} + \frac{6}{10^3}\right)\right]$$

$a = C_2/\lambda T = 14387.8/2000 = 7.1939$
$f_{0-2000} = 0.06673$.
From tables, $f_{0-2000} = 0.0667$.

TABLE 4.4 Fraction ($f_{0-\lambda T}$) of blackbody radiant energy between zero and λT (Duffie & Beckman, 2013).

λT (μm K)	$f_{0-\lambda T}$	λT (μm K)	$f_{0-\lambda T}$	λT (μm K)	$f_{0-\lambda T}$	λT (μm K)	$f_{0-\lambda T}$	λT (μm K)	$f_{0-\lambda T}$
1,000	0.0003	3,100	0.2958	5,200	0.6579	7,300	0.8244	9,400	0.9005
1,100	0.0009	3,200	0.3181	5,300	0.6693	7,400	0.8295	9,500	0.903
1,200	0.0021	3,300	0.3401	5,400	0.6803	7,500	0.8343	9,600	0.9054
1,300	0.0043	3,400	0.3617	5,500	0.6909	7,600	0.839	9,700	0.9076
1,400	0.0077	3,500	0.3829	5,600	0.701	7,700	0.8436	9,800	0.9099
1,500	0.0128	3,600	0.4036	5,700	0.7107	7,800	0.8479	9,900	0.912
1,600	0.0197	3,700	0.4238	5,800	0.7201	7,900	0.8521	10,000	0.9141
1,700	0.0285	3,800	0.4434	5,900	0.7291	8,000	0.8562	11,000	0.9318
1,800	0.0393	3,900	0.4624	6,000	0.7378	8,100	0.8601	12,000	0.945
1,900	0.0521	4,000	0.4809	6,100	0.7461	8,200	0.8639	13,000	0.955
2,000	0.0667	4,100	0.4987	6,200	0.7541	8,300	0.8676	14,000	0.9628
2,100	0.083	4,200	0.516	6,300	0.7618	8,400	0.8711	15,000	0.9689
2,200	0.1009	4,300	0.5327	6,400	0.7692	8,500	0.8745	16,000	0.9737
2,300	0.12	4,400	0.5488	6,500	0.7763	8,600	0.8778	17,000	0.9776

TABLE 4.4 Fraction ($f_{0-\lambda T}$) of blackbody radiant energy between zero and λT (Duffie & Beckman, 2013).—cont'd

λT (μm K)	$f_{0-\lambda T}$	λT (μm K)	$f_{0-\lambda T}$	λT (μm K)	$f_{0-\lambda T}$	λT (μm K)	$f_{0-\lambda T}$	λT (μm K)	$f_{0-\lambda T}$
2,400	0.1402	4,500	0.5643	6,600	0.7831	8,700	0.881	18,000	0.9807
2,500	0.1613	4,600	0.5793	6,700	0.7897	8,800	0.8841	19,000	0.9833
2,600	0.1831	4,700	0.5937	6,800	0.7961	8,900	0.8871	20,000	0.9855
2,700	0.2053	4,800	0.6075	6,900	0.8022	9,000	0.8899	30,000	0.9952
2,800	0.2279	4,900	0.6209	7,000	0.808	9,100	0.8927	40,000	0.9978
2,900	0.2506	5,000	0.6337	7,100	0.8137	9,200	0.8954	50,000	0.9988
3,000	0.2732	5,100	0.6461	7,200	0.8191	9,300	0.898	∞	1

4.3.6 Measurement of surface radiation

Most of the reflectance data available in the literature have been acquired by using a technique that was devised by Gier et al. (1954). In this technique, a cool sample is exposed to blackbody radiation from a cavity source at a temperature of about 1273 K, and the monochromatic radiation reflected from the surface of the sample is matched with the monochromatic blackbody radiation from the cavity (Fig. 4.9). Radiation A is the blackbody radiation reflected from the sample, while radiation B is the blackbody radiation emitted by the cavity inner surface. The fraction $A\lambda/B\lambda$ is equal to $\rho\lambda(\mu,\gamma)$. This instrument is known as a hohlraum. Different hohlraum designs have been developed, such as the cylindrical, rugby, elliptical, and spherical hohlraums (Lan et al., 2016). Cylindrical hohlraums are most widely used in inertial fusion studies. The data captured by using a hohlraum is hemispherical-angular monochromatic reflectance.

In most hohlraum systems, the angle of the measured radiation is usually fixed at such a low value that measurements can only be made at one angle. It is possible to rotate the sample in order to measure all angles in some designs of hohlraums.

4.3.7 Angular dependence of absorptance of solar radiation

For a given surface finish, the absorption of solar radiation is affected by the angle of incidence on the surface. Nevertheless, data on the angular dependence of solar absorptance are scarce for most surfaces exploited in solar collectors. For common blackened surfaces, the directional absorptance for solar radiation depends on the incidence angle of the radiation. Based on limited available data, it has been suggested that selective surfaces may show similar optical properties. The ratio of solar absorptance (α) at a given incident angle (θ) to solar absorptance (α_n) at normal incidence for a flat black surface can be given by Eq. (4.66) (Duffie and Beckman, 2013):

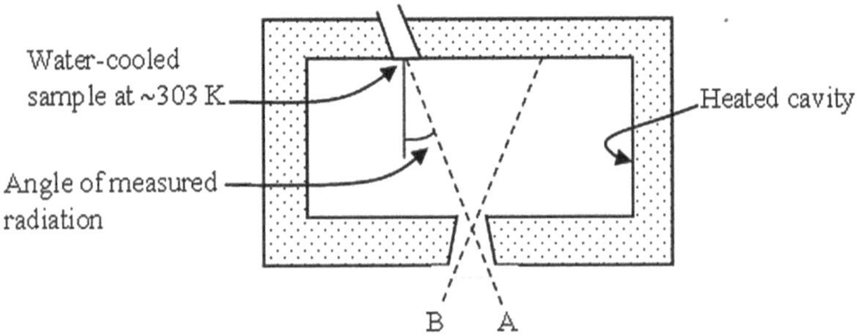

FIGURE 4.9 Schematic representation of a hohlraum for measuring monochromatic hemispherical-angular reflectance.

$$\alpha/\alpha_n = 1 - 1.5879 \times 10^{-3}\theta + 2.7314 \times 10^{-4}\theta^2 - 2.3026 \times 10^{-5}\theta^3 + 9.0244 \times 10^{-7}\theta^4 - 1.8000 \times 10^{-8}\theta^5 + 1.7734 \times 10^{-10}\theta^6 - 6.9937 \times 10^{-13}\theta^7 \quad (4.66)$$

4.3.8 Absorptance of external and cavity solar receivers

Surfaces can exchange thermal radiation in a wide spectrum of wavelengths. In this vein, the magnitude of radiation exchange is a strong function of the geometries and orientations of surfaces, and their radiative properties and temperatures (Bergman et al., 2011). One of the important geometric parameters in radiation exchange is the view factor. It is the ratio of radiation leaving one surface that is intercepted by another surface. A flat (or convex) surface can only exchange radiation with other surfaces (Fig. 4.10), while a concave surface can exchange radiation with itself and other surfaces within its vicinity. Several flat surfaces may be combined to form an enclosure that exchanges radiation with itself only.

External solar receivers are made up of panels that are arranged in cylindrical (or polygonal) shape in such a way that the panels do not exchange radiation. Thus, the absorptance of each panel can be treated as a flat absorber. On the other hand, the geometry of a cavity receiver is close to a blackbody cavity. So, cavity receivers can absorb concentrated solar irradiation with a very high optical efficiency. They can operate at a high temperature (1000−1600 K) and pressure (0.3−3.0 MPa) required in gas turbines for long periods (Wang & Laumert, 2017). Some cavity receivers are designed without transparent covers. For this category of receivers, the proportion of incident radiation that is absorbed by the cavity depends on the absorptance of the inner surfaces of the cavity and the ratio of the areas of the cavity aperture and inner surfaces. Approximately, the effective absorptance of a cavity receiver (α_{eff}) is described by Duffie and Beckman (2013):

$$\alpha_{eff} = \frac{\alpha_i}{\alpha_i + \left(A_{ap}/A_i\right)(1 - \alpha_i)} \quad (4.67)$$

Where A_{ap} is the area of the aperture of the cavity, A_i is the area of the inner surface of the cavity, and α_i is the absorptance of the inner surface of the cavity. The parameter α_i can be estimated at the effective angle of incidence of diffuse radiation ($\theta \sim 60^o$). From Eq. (4.67), it is observed that, for a given level of surface absorptance (α_i), the effective absorptance of the cavity receiver increases with decreasing the ratio A_{ap}/A_i. Usually, the surface absorptance is enhanced by using a selective coating, and properties of promising substrate and coating materials for application at high temperature are presented in Table 4.5.

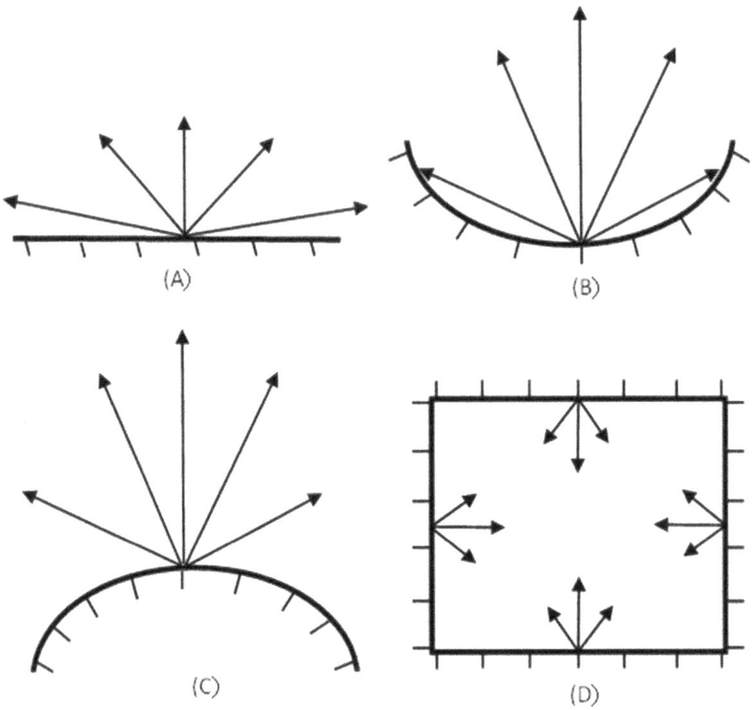

FIGURE 4.10 Schematic representation of radiation leaving: (A) flat surface, (B) concave surface, (C) convex, and (D) enclosure.

TABLE 4.5 Properties of selected promising substrate and coating materials for application at high temperature (Wang & Laumert, 2017).

Material	Maximum operating temperature (K)	Solar absorptance
Inconel 600	1368	0.9
253 MA	1423	0.8
Kanthal APM	1523	0.7
YSZ TBC coating	1373	0.4
Pyromark 2500	1366	0.95

Nomenclature

A area (m^2)
c speed of light ($m\ s^{-1}$)
E flux ($W\ m^{-2}$)
f frequency (Hz)
I radiation intensity ($W\ m^{-2}$)
K coefficient of extinction (m^{-1})
L length (m)
n refractive index (dimensionless)
T temperature (K)
v speed ($m\ s^{-1}$)

Greek symbols

α absorptance (dimensionless)
β tilt angle (^{o})
γ azimuthal angle (^{o})
ε emittance (dimensionless)
θ angle (^{o})
λ wavelength (m)
ρ reflectance (dimensionless)
τ transmittance (dimensionless)

References

Arias-Rosales, A., & LeDuc, P. R. (2020). Modeling the transmittance of anisotropic diffuse radiation towards estimating energy losses in solar panel coverings. *Applied Energy, 268*, 114872. https://doi.org/10.1016/j.apenergy.2020.114872

Bergman, T. L., Lavine, A. S., Incropera, F. P., & Dewitt, D. P. (2011). *Fundamentals of heat and mass transfer.*

Brow, R. K., & Schmitt, M. L. (2009). A survey of energy and environmental applications of glass. *Journal of the European Ceramic Society, 29*(7), 1193–1201. https://doi.org/10.1016/j.jeurceramsoc.2008.08.011

D'Alessandro, C., De Maio, D., Musto, M., De Luca, D., Di Gennaro, E., Bermel, P., & Russo, R. (2021). Performance analysis of evacuated solar thermal panels with an infrared mirror. *Applied Energy, 288*, 116603. https://doi.org/10.1016/j.apenergy.2021.116603

Duffie, J., & Beckman. (2013). *Solar engineering of thermal processes.* John Wiley & Sons Inc.

Gier, J. T., Dunkle, R. V., & Bevans, J. T. (1954). Measurement of absolute spectral reflectivity from 10 to 15 microns. *Journal of the Optical Society of America, 558.* https://doi.org/10.1364/JOSA.44.000558

Granados, L., Takamure, N., Bing, J., Huang, S., Merhvarz, H., McKenzie, D. R., & Ho-Baillie, A. (2020). Direct determination of total hemispherical emittance of perovskite and silicon solar cells. *Cell Reports Physical Science, 1*(1), 100008. https://doi.org/10.1016/j.xcrp.2019.100008

Greenwald, M. L., & McHugh, T. K. (1985). *Practical solar energy technology.* Prentice-Hall, Inc.

Kennedy, C. E. (2002). *Review of mid-to high-temperature solar selective absorber materials.* National Renewable Energy Laboratory.

Lambrick, S. M., Bergin, M., Jardine, A. P., & Ward, D. J. (2018). A ray tracing method for predicting contrast in neutral atom beam imaging. *Micron, 113*, 61–68. https://doi.org/10.1016/j.micron.2018.06.014

Lan, K., Liu, J., Li, Z., Xie, X., Huo, W., Chen, Y., Ren, G., Zheng, C., Yang, D., Li, S., Yang, Z., Guo, L., Li, S., Zhang, M., Han, X., Zhai, C., Hou, L., Li, Y., Deng, K., ... He, X. (2016). Progress in octahedral spherical hohlraum study. *Matter and Radiation at Extremes, 1*(1), 8–27. https://doi.org/10.1016/j.mre.2016.01.003

Li, S., Xu, Q., Zhou, Z., Zhao, W., Li, X., Wen, Z., Yao, Y., Xu, H., Deng, H., Dai, N., & Hao, J. (2020). Flexible transparent heat mirror for thermal applications. *Nanomaterials, 10*, 2479. https://doi.org/10.3390/nano10122479

Nunes, R. A. X., Costa, V. C., Sade, W., Araújo, F. R., & Silvaa, G. M. (2018). Selective surfaces of black chromium for use in solar absorbers. *Materials Research, 21*(1). https://doi.org/10.1590/1980-5373-mr-2017-0556

Perez, R., Seals, R., Ineichen, P., Stewart, R., & Menicucci, D. (1987). A new simplified version of the perez diffuse irradiance model for tilted surfaces. *Solar Energy, 39*(3), 221–231. https://doi.org/10.1016/S0038-092X(87)80031-2

Prado, R. T. A., & Sowmy, D. S. (2016). Innovations in passive solar water heating systems. In *Advances in solar heating and cooling* (pp. 117–150). Elsevier Inc. https://doi.org/10.1016/B978-0-08-100301-5.00007-2

Reindl, D. T., Beckman, W. A., & Duffie, J. A. (1990). Evaluation of hourly tilted surface radiation models. *Solar Energy, 45*(1), 9–17. https://doi.org/10.1016/0038-092X(90)90061-G

Rostami, M. (2019). *The design of novel transparent heat mirrors for enhanced solar thermal applications*. MSc Thesis. York University.

Rostami, M., Pirvaram, A., Talebzadeh, N., & O'Brien, P. G. (2021). Numerical evaluation of one-dimensional transparent photonic crystal heat mirror coatings for parabolic dish concentrator receivers. *Renewable Energy, 171*, 1202–1212. https://doi.org/10.1016/j.renene.2021.03.007

Rostami, M., Talebzadeh, N., & O'Brien, P. G. (2020). Transparent photonic crystal heat mirrors for solar thermal applications. *Energies, 13*(6), 1464. https://doi.org/10.3390/en13061464

Siegel, R., & Howell, J. R. (1971). *Thermal radiation heat transfer – radiation transfer with absorbing, emitting, and scattering Media: Vol. III, (NASA SP-164)*. National Aeronautics and Space Administration.

Siegel, R., & Howell, J. R. (2002). *Thermal Radiation Heat Transfer* (4th). New York: Taylor & Francis.

Tesfamichael, T. (2000). *Characterization of Selective Solar Absorbers: Experimental and Theoretical Modeling*. Uppsala: Acta Universitatis Upsaliensis. ISBN 9155447724.

Wang, W., & Laumert, B. (2017). Effect of cavity surface material on the concentrated solar flux distribution for an impinging receiver. *Solar Energy Materials and Solar Cells, 161*, 177–182. https://doi.org/10.1016/j.solmat.2016.12.008

Whillier, A. (1953). *Solar energy collection and its utilization for house heating*. ScD. Thesis. Massachusetts Institute of Technology (MIT).

Xu, K., Du, M., Hao, L., Mi, J., Yu, Q., & Li, S. (2020). A review of high-temperature selective absorbing coatings for solar thermal applications. *Journal of Materiomics, 6*(1), 167–182. https://doi.org/10.1016/j.jmat.2019.12.012

Zaversky, F., Aldaz, L., Sánchez, M., Ávila-Marín, A. L., Roldán, M. I., Fernández-Reche, J., Füssel, A., Beckert, W., & Adler, J. (2018). Numerical and experimental evaluation and optimization of ceramic foam as solar absorber – single-layer vs multi-layer configurations. *Applied Energy, 210*, 351–375. https://doi.org/10.1016/j.apenergy.2017.11.003

Chapter 5

Characteristics of heat transfer media

5.1 Introduction

A solar receiver is a photothermal converter that absorbs concentrated solar radiation and converts it into heat. The solar radiation is directly or indirectly absorbed by a heat transfer fluid (HTF). In indirect absorption systems, heat is transferred from the walls of the absorber surface of the receiver to the fluid. One or more flow loops of fluids may be used in a system. In case of one loop, the same fluid is used as a heat transfer fluid and a working fluid. Some systems have separate flow loops for heat transfer and working fluids. In this case, heat is transmitted from the heat transfer fluid to the working fluid of the engine cycle via a heat exchanger. Part of the heat collected by the working fluid is converted to mechanical energy through a thermodynamic cycle (see Section 1.5). Thus, the heat transfer fluid is a key element of concentrating solar power (CSP) systems.

There are different thermodynamic cycles, with variations in the range of their working temperatures. In view of this, properties of heat transfer fluids should: (a) match the containment materials and storage media, (b) be capable of operating in the desired temperature range, (c) collect and transmit heat with little effort, and (d) flow well in confined spaces such as tubes (Benoit et al., 2016).

5.1.1 Required characteristics of heat transfer fluids

To achieve a high performance of CSP systems, it is necessary for fluids that are exploited as heat transfer media in solar thermal devises to possess most of the characteristics of a good heat transfer fluid. In brief, the major required characteristics of heat transfer media are as follows (Benoit et al., 2016; Bignon, 1980; Ho & Iverson, 2014):

5.1.1.1 Working temperature range and thermal stability

The efficiency of CSP power plants is predominantly restricted by the working fluid temperature. Consequently, a heat transfer fluid should have a wide range

Solar Receivers for Thermal Power Generation. https://doi.org/10.1016/B978-0-323-85271-5.00005-7

of the working temperature. This implies that the candidate fluid should have a low melting point in order to avert its solidification in pipes or other flow channels, and a high upper limit of the working temperature that permits the exploitation of efficient thermodynamic cycles.

Application of heat to a substance can cause physical and chemical changes to the properties of the material. A physical modification does not produce a fundamentally different substance, while a chemical change transforms a given substance into a chemically new substance. Both types of changes can alter the desirable properties of a fluid. For example, phase change is a physical transformation that affects the efficiency of thermodynamic cycles. Some important parameters of phase change are the critical temperature and pressure of a material (see Section 1.5). Therefore, the thermal stability of materials is a critical factor in the selection of heat transfer fluids. Thermal stability is defined as the ability of a material to resist the action of heat and to maintain its properties at a given temperature. Solar receivers operate at elevated temperatures, and therefore, heat transfer fluids with high thermal stability are required.

5.1.1.2 Heat transfer properties

Properties of a heat transfer fluid play a vital role in receiving, transferring, and storing heat. In this context, good properties promote moderate temperature difference between the receiving surface and the fluid, thereby permitting it to attain elevated temperatures without creating excessive thermal stresses on the surface.

The rate of heat evacuation by a fluid increases with the mass flow rate (see Section 1.5). Consequently, density is an import property of the heat transfer fluid. Fluids with high density and thermal conductivity are preferred. A high thermal conductivity of the fluid is critical for the augmentation of the heat transfer coefficient. Even though indirect methods of thermal storage with a secondary medium are available, a high level of the specific heat capacity is desirable to enable direct storage of energy. Pumping power increases with the viscosity of a fluid. Consequently, a low viscosity is favorable to decrease pressure drop and pumping power.

5.1.1.3 Working pressure

When a fluid stream passes through a conduit (such as pipe, duct, etc), it exerts forces on the walls of the conduit. The magnitudes of these forces increase with the pressure of the fluid. In a receiver, heat transfer fluids are commonly pumped through tubes and other components of a CSP system. Consequently, it is necessary for the walls of the tubes to have a certain level of mechanical strength to withstand the pressure.

Tubes with thin walls can be utilized when the working pressure is low, which decreases the wall temperature gradient and consequently the thermal stress. Moreover, the cost of tubes increases with the thickness of the tube walls. The critical pressure of the fluid has great influence on the efficiency of a thermodynamic cycle. Fluids with low levels of critical pressure can easily be exploited in the supercritical state, thereby achieving high thermodynamic efficiency. Therefore, fluids that can be used at a low working pressure are preferred.

5.1.1.4 Operational aspects

Operational aspects, such as safety and corrosion behavior of materials, play a key role. Diminished hazard properties and sound compatibility of materials are highly advocated. According to ASHRAE Standard 34, the toxicity and flammability classification of refrigerants gives six safety groups, A1, A2, A3, B1, B2, and B3, in order of increasing hazard (ASHRAE, 2017).

Another important operational aspect is the impact of a working fluid on the environment. In this connection, the main concern is about the global warming potential (GWP) and the ozone depletion potential (ODP) of a fluid. Working fluids with low magnitudes of GWP and ODP are preferred.

5.1.1.5 Affordability of materials

A working fluid should be affordable to assist in reducing the cost of electricity generation. Generally, technologies with lower cost are assigned higher shares in the generation mix, and the levelized cost of electricity (LCOE) is one of the most widely used metrics of the economic performance of generation technologies. The LCOE is a unit cost of energy generated by a power plant over its lifetime (Larsson et al., 2014). Advantages of the LCOE method include: (i) absence of restrictions on project scale, (ii) its independence of power generation technologies, and (iii) its application even when energy technologies are different. This metric is a good indicator of the economic competitiveness of emerging energy technologies (Clauser & Ewert, 2018). So, it has been extensively applied in the development of renewable energy technologies, including CSP.

Carbon-neutral electricity can be competitive on the market if its costs are low. In this vein, IRENA (2013) argues that carbon-neutral electricity can compete against the largest carbon emitter (coal power) if its generation costs are lower than approximately 0.08 USD/kWh. In 2019, the global weighted-average levelized cost of CSP-based electricity was 0.182 USD/kWh, but for CSP projects that were due to be commissioned in 2020 and 2021, the weighted-average price of electricity was between 0.075 and 0.094 USD kWh^{-1} (IRENA, 2020). So, there is a decreasing trend in the generation cost of the CSP technology.

From the perspective of heat transfer fluids, some important strategies that can contribute to cost reduction are: (a) exploitation of high HTF temperatures and (b) switching synthetic oil with molten salt. The utilization of high HTF temperature can contribute to the reduction of storage costs (IRENA, 2020) and increase the thermodynamic efficiency of the system, thereby decreasing the overall generation costs. Switching of synthetic oil with molten salt allows achievement of higher temperatures, which also helps in raising the efficiency of the system and in reducing the levelized cost of electricity.

5.1.2 Wall-to-fluid coefficient of convective heat transfer

In conventional receiver designs, concentrated solar radiation is absorbed and converted to heat by opaque absorber walls. Heat is transferred from the hot walls to the fluid through convection, and the coefficient of convective heat transfer (h) from the walls to the fluid is influenced by fluid properties. High coefficients of heat transfer are desirable to achieve a good performance of solar receivers. For natural convection, the coefficient of heat transfer (h_c) can be calculated from:

$$h_c = \mathrm{Nu}\ k/S \tag{5.1}$$

Where k is the thermal conductivity of the fluid, N_u is the Nusselt number (the ratio of convective to conductive heat transfer coefficients within a fluid), and S is the characteristic length or equivalent spacing. The characteristic length is the height, length, diameter, and 0.5 * diameter for vertical plates or pipes, horizontal plates, horizontal pipes, and spheres, respectively.

The Nu can be expressed as a function of the Grashof (Gr) and Prandtl (Pr) numbers or as a function of the Rayleigh number:

$$\mathrm{Nu} = f(\mathrm{Gr},\ \mathrm{Pr}) = f(\mathrm{Ra}) \tag{5.2}$$

Where Gr is the Grashof number (the ratio of buoyancy to viscous force in a fluid), Pr is the Prandtl number (the ratio of kinematic viscosity to thermal diffusivity), and Ra is the Rayleigh number.

The Gr and Pr numbers are also dimensionless parameters given by:

$$\mathrm{Pr} = \frac{\nu}{\xi} = \frac{\mu C_p}{k} \tag{5.3}$$

$$\xi = \frac{k}{\rho C_p} \tag{5.4}$$

$$\mathrm{Gr} = \frac{g\ S^3\ \beta\ \rho^2(\Delta T)}{\mu^2} \tag{5.5}$$

Where g is the acceleration due to gravity, ΔT is the difference between surface and the fluid temperatures, C_p is the specific heat capacity at constant pressure, ξ is the thermal diffusivity, β is the coefficient of thermal expansivity, ρ is density, μ is dynamic viscosity, and ν is the kinematic viscosity. Materials with high values of diffusivity respond quickly to variations in their thermal environment, while materials with low values of diffusivity respond more slowly, and consequently, they take longer to attain a new equilibrium state.

Generally, fluid properties depend on temperature and are computed at the mean of the fluid and wall temperatures. For gases, these properties are significantly affected by the pressure of the gas. Tables are available for thermophysical properties of different fluids at specified temperature and pressure. In addition, some correlations have been established for estimating thermophysical properties of different substances (Benoit et al., 2016; Sohal et al., 2013). ASHRAE (2017) and many other authors provide a summary of models used for calculating the Nu for different geometries. For instance, for a horizontal cylinder, the Nu for natural convection can be calculated from:

$$\mathrm{Nu} = \left\{ 0.6 + \frac{0.387\mathrm{Ra}^{1/4}}{\left[1 + (0.559/\mathrm{Pr})^{9/16}\right]^{4/9}} \right\}^2, \text{ for } 10^9 < \mathrm{Ra} < 10^{13} \tag{5.6}$$

Heat transfer and working fluids are often circulated through the system components by forced convection. When a fluid flows under forced convention, a boundary layer is created between the fluid and the surface. Flow within the boundary layer close to the leading edge of the surface is laminar forced convection. As the flow continues along the surface, there is a rise in the thickness of the boundary layer to a critical level. Thereafter, turbulent forced convection commences. Therefore, forced convection can be laminar or turbulent. Flow conditions can be described by using a dimensionless parameter known as the Reynolds number (Re). At a very low fluid velocity, flow remains laminar in long tubes or channels with a small hydraulic diameter, and it is said to be fully developed laminar flow ($\mathrm{Re} < 2300$). At a high fluid velocity or in a tube with a large diameter, transition to turbulent flow takes place, and flow is fully developed turbulent ($\mathrm{Re} > 2300$). The characteristic length for internal flow in a circular tube is the inside diameter of the tube. The hydraulic diameter (D_h) is taken as the characteristic length of the conduit for noncircular tubes:

$$D_h = 3 \times \frac{\text{Cross sectional area for flow}}{\text{Total wetted perimeter}} \tag{5.7}$$

The Nu of forced convection is a function of Re and Pr:

$$\mathrm{Nu} = f(\mathrm{Re}, \mathrm{Pr}) \tag{5.8}$$

For circular tubes, flow is described as laminar when Re is less than 2300 (Bellos et al., 2017), and, in such cases, the Re can be calculated from:

$$\mathrm{Re} = \frac{64}{\vartheta} \tag{5.9}$$

Where ϑ = friction factor.

When the Re is greater than 2300, the flow regime becomes turbulent, and Re can be calculated from (ASHRAE, 2017):

$$Re = \frac{4\ \dot{m}}{\pi\ \mu\ D} \tag{5.10}$$

Where D is the internal diameter of the conduit and $\dot{m}$ is the mass flow rate of the fluid. When the velocity of the working fluid inside the flow channel is high, the fluid flow is characterized as turbulent. The value of ϑ is almost constant for turbulent flow, varying only from about 0.01 to 0.05.

In case of internal flow in a tube where the working fluid exchanges heat with walls of the tube through convection, the Nu can be given by:

$$Nu = \frac{h_c D}{k} \tag{5.11}$$

For fully developed laminar internal flow (Re < 2300), the Nu can be computed from the following:

$$Nu = \begin{cases} 3.66, \text{ for uniform surface temperature} \\ 4.36, \text{ for unform heat flux} \end{cases} \tag{5.12}$$

For fully developed turbulent internal flow, the Nu can be found from Eq. (5.13).

$$Nu = 0.023\ \left(Re^2\ Pr\right)^{0.4}, \text{ for cooling fluid with } Re \geq 10,000 \tag{5.13}$$

Example 5.1

The mean temperature of a surface-based tubular air receiver (radius = 0.025 m) is 700 K, and the flow rate of the air through the receiver is 0.2 kgs^{-1}. Calculate the average wall-to-fluid coefficient of heat transfer. Assume pressure is atmospheric.

Solution

$$h_c = k\ Nu/D$$

From tables of the thermal properties of air at T = 700 K in Appendix C, $k = 0.0524\ W\ m^{-1}K^{-1}$, $\mu = 3.388 \times 10^{-5}$ Pa s, and Pr = 0.695

$$\text{But } Re = \frac{4\dot{m}}{\pi\mu D} = \frac{4 * 0.2}{\pi * 3.388 * 10^{-5} * 2 * 0.025} = 150323$$

Since Re > 10,000, flow is fully developed turbulent.

$$Nu = 0.023 * \left(150323^2 * 0.695\right)^{0.4} = 269.50$$

Therefore, $h_c = k\ Nu/D = 0.0524*269.50/(2*0.025) = 282\ Wm^{-2}K^{-1}$

Conventionally, thermal oil, molten salts, air at atmospheric pressure, and water are exploited as heat transfer fluids in CSP plants, with thermal oil being the commonest heat transfer fluid (Fig. 5.1). Thermal oil is used as a primary heat transfer fluid to receive heat from the absorber element and transfer it to a secondary fluid or storage medium. The majority of CSP plants are based on the Rankine cycle (see Section 1.5). Consequently, water is commonly used as a secondary fluid, whereby air, thermal oil, molten salt, or liquid sodium are used as primary heat transfer fluids that indirectly heat the water to produce steam. There is no single fluid that meets all the important requirements of a heat transfer fluid. Nevertheless, research and development activities are ongoing to discover fluids with properties that yield higher coefficients of heat transfer. So far, there has been some advancement in the development of heat transfer fluids for application in the CSP and other sectors. In this regard, some of the conventional and advanced fluids are presented and discussed in Sections 5.2 and 5.3.

5.2 Conventional heat transfer media

5.2.1 Water/steam

Water (H_2O) is an inorganic dipolar substance with negative and positive poles on opposite sides of the molecule. Consequently, water molecules form hydrogen bonds with one another. The dipolar structure of the molecule and the hydrogen type of bonding greatly affect the density, vapor pressure, freezing point, boiling point, surface tension, phase changes, specific heat, dielectric constant, viscosity, cohesion, adhesion, and capillarity of water

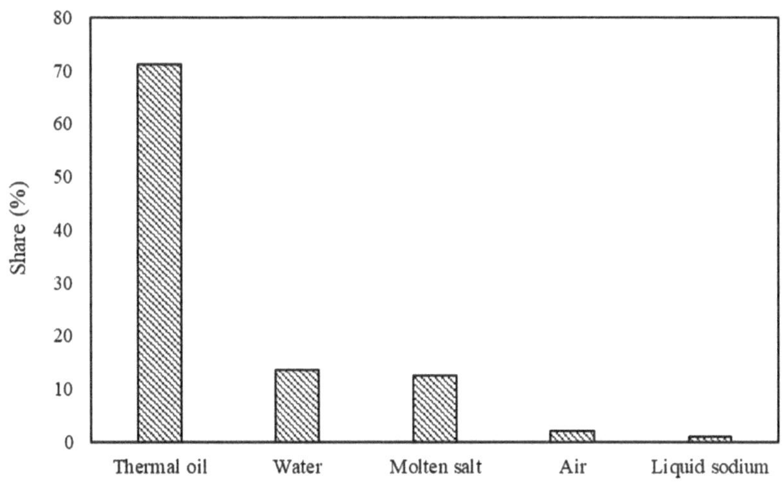

FIGURE 5.1 Share of operating concentrating solar power plants based on primary heat transfer fluid around the world in 2021. *Drawn by author using data from NREL (2021).*

(Boyd, 2015). Some of these properties are critical for selecting water as a heat transfer fluid.

Water is safe, noncorrosive, inexpensive, and readily available in many parts of the world. It also has some good properties that are affected by temperature. However, there are some drawbacks of using it as a working fluid. Water has possible instability and a challenge of managing it at high temperature/high pressure. It undergoes phase change from liquid to gas although it is chemically stable up to very high temperatures. The saturation temperature increases with the pressure. Nevertheless, there is a limit to the operational pressure. For instance, a high pressure requires thick tube walls, which in turn augments the resistance to heat transfer and the cost of the system.

In the majority of water/steam receivers, the liquid water heating and steam generation component is separated from the steam superheating component due to differences in their heat transfer characteristics. Another challenge in the steam generation component of the process is control of the boiling process to prevent energy losses that arise from the water recirculation. It is imperative to achieve this control without reaching the complete dryness point where an abrupt fall in the heat transfer coefficient would trigger a great tube temperature increase that can threaten the piping system. Water can be exploited as a heat transfer fluid in the saturated state for thermodynamic cycles up to its critical point ($T = 647.04$ K and $P = 22.1$ MPa), which is the upper limit before the water attains the supercritical phase. At $P < 22.1$ MPa, steam superheating permits raising the working temperature. For instance, molten salt-based receivers of solar tower plants typically utilize superheated steam at $T = 823$ K and $P = 12.6$ MPa. Steam can be directly or indirectly produced by a receiver.

Direct steam generation (DSG) is a process in which steam is directly generated in a receiver and delivered to a power block for production of electricity. The power block is based on the Rankine cycle Fig. 5.2. Initially, the DSG system was developed for a parabolic trough concentrator (PTC) power plant, but recent studies indicate consideration of hybridization of the PTC plant with other types of CSP plants (such as a solar tower, linear Fresnel reflector, and parabolic dish concentrator) in order to reduce the cost of steam and electricity generation (Hachicha, 2017; Islam et al., 2018). In the DSG system, water is used as a heat transfer fluid, and it is possible to utilize the water as both working fluid and storage media.

At present, most the commercial power plants are based on the PTC system, which predominantly uses synthetic oil as a heat transfer fluid. One of the drawbacks of synthetic oil is its low range of operational temperature. Advantages of the DSG system over a conventional synthetic-oil-based PTC system are (Fernández-García et al., 2010): (a) absence of environmental risk of fire and leakage, (b) achievement of maximum temperature for the thermodynamic cycle of more than 673 K, that is hard to attain by using synthetic oil, (c) higher overall plant efficiency because of the absence of an oil/steam

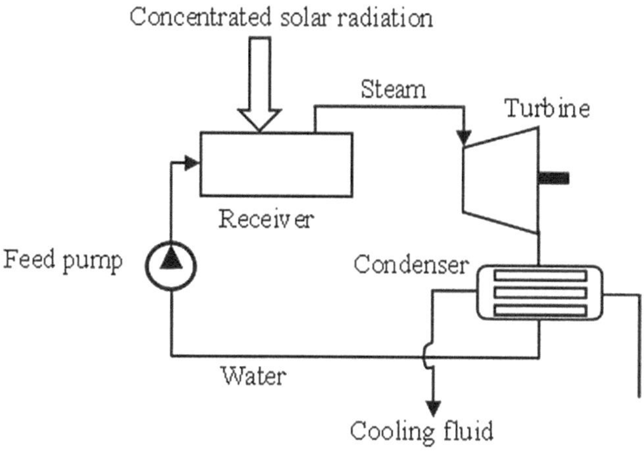

FIGURE 5.2 Schematic representation of direct steam generation.

heat exchanger, (d) easy plant configuration with less capital investment, and (e) lower operation and maintenance costs than for a synthetic oil-based PTC system mainly because there is no need of an auxiliary heating system. The DSG system has some challenges such as: (a) control of the solar field under intermittent solar radiation, (b) possible occurrence of temperature gradients in the receiver pipes, and (c) instability of the binary-phase flow inside a receiver. Nonetheless, the DSG technology has attained commercial stage of development.

Direct steam generation was applied in the first commercial grid-connected CSP plant, Planta Solar 10 (SP 10), that started operating in southern Spain in 2007 (Ho, 2017). The SP 10 plant used saturated steam at T»523−573 K and P = 4.5 MPa for a Rankine cycle that uses wet cooling (refrigeration towers) with net turbine capacities of 11 MW. This plant utilizes a cavity receiver that accommodates tubular panels to heat the water/steam, and it uses pressurized water systems of thermal storage to store heat for off-sunshine operation of up to 1 h. Later, several other DSG-based power plants were constructed and commissioned. As of 2020, the largest CSP power plant in operation in the world was the Ivanpah Solar Electric Generating System (392 MW) that commenced operating in 2014.

Most of the CSP plants are based on the Rankine cycle and utilize the indirect method of steam generation. In the conventional Rankine cycle, water is used as the working fluid. The hot primary heat transfer fluid heats up the working fluid of the thermodynamic cycle via a heat exchanger (Fig. 5.3). The main advantage of this system is the flexibility in choosing heat transfer fluids for a given thermodynamic cycle. Nevertheless, the indirect system has some drawbacks that include: (a) attainment of lower temperature for the

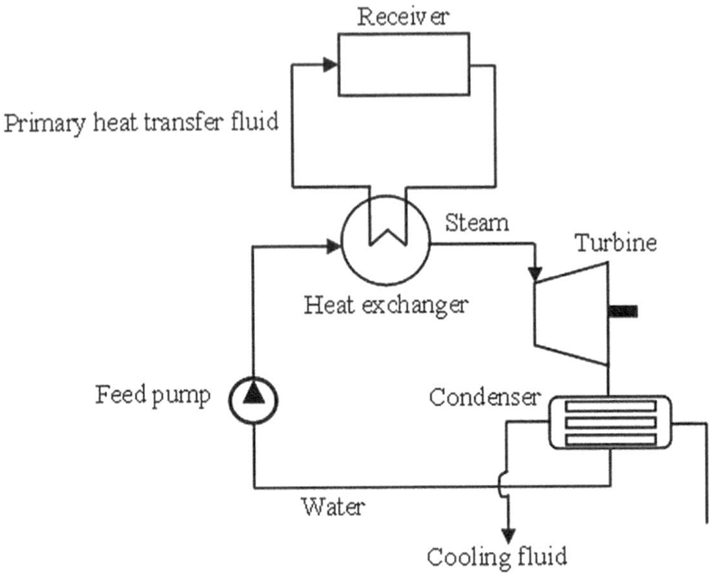

FIGURE 5.3 Schematic representation of indirect steam generation.

thermodynamic cycle than that of the primary heat transfer fluid, (b) environmental risk of fire and leakage, (c) lower overall plant efficiency due to heat exchanging between the two flow loops, (d) complex plant configuration with higher capital investment, and (e) higher operation and maintenance costs than for the DSG system, predominantly because there is need of an auxiliary heating system for the primary heat transfer fluid (such as molten salt and thermal oil).

Even though the indirect steam generator has some disadvantages, most of the existing commercial CSP plants are based on it. The majority of current PTC power plants utilize synthetic oil, while ST plants predominantly use molten salt. The implication of this is that PTC power plants with a storage system may require up to three heat transfer fluids: synthetic oil (primary heat transfer fluid), molten salt (thermal storage fluid), and water (working fluid). Molten salt can be used as a heat transfer fluid as well as storage medium, thereby avoiding the need for heat exchangers between the storage salt and thermal oil. This medium also enables higher operating temperatures than synthetic oil. Thus, the use of molten salt is more cost-effective than that of synthetic oil.

5.2.2 Gases

The rate of heat transfer from a solar receiver increases with the mass flow rate of the heat transfer fluid (see Section 1.5). Nevertheless, the density of gases is generally lower than that of liquids or solids. Consequently, they need compression to raise their density and, therefore, mass flow rate. Thermal conductivity (k) is a critical parameter of heat transportation, and generally, nonmetals (that include gases) have low values of k. A compressed gas can be heated by a receiver and used to directly drive a gas turbine or indirectly propel a Rankine cycle. Part of the hot gas may be used to charge up a storage unit. The gas turbine technology has higher thermodynamic efficiency than the Rankine cycle that is commonly exploited in the electricity sector. Hence, solarization of the gas turbine cycle is a promising pathway for improving the performance of the CSP technology (see Section 1.5.3).

Some advantages of gaseous heat transfer fluids are: (a) low cost, (b) inertness and environmental friendliness, and (c) ability to attain high temperatures above 1273 K (Ho, 2017). Nevertheless, there are some disadvantages of these heat transfer fluids, some of which are: (a) diminished heat capacitance and thermal conductivity, (b) instability of fluid flow, and (c) possibility of costly equipment and materials where additional heat exchanger or high-temperature and high-pressure piping components are compulsory. Air, carbon dioxide, helium, and hydrogen gases are commonly used as heat transfer fluids.

5.2.2.1 Air

Air is predominantly composed of nitrogen, oxygen, argon, and carbon dioxide. It is an abundant resource that is free, friendly to the environment, and easy to handle. Its critical point is at $T = 132.52$ K and $P = 3.766$ MPa. However, air has low density, and it therefore requires a lot of pumping power to achieve the necessary high pressure. A conventional gas turbine utilizes air as the working fluid, and it uses compressors to achieve high operating pressures. In this context, air receivers are currently being designed to heat pressurized air to meet inlet conditions of gas turbine systems. It is possible to heat compressed air by using volumetric receivers up 1773 K (Ho, 2017).

Some air-based projects have been developed for demonstration purposes. The world's first operational solar-gas hybrid power plant (100 kW_e and 170 kW_{th}) was constructed at Kibbutz Samar in southern Israel by AORA Solar and it started operating in 2009. The power plant uses a solar tower with 30 heliostats to heat compressed air to about 1273 K before injecting it into a gas turbine (De Morsella, 2021). The gas turbine can also be fueled by natural gas at night or during periods of low insolation. Compressed air was also used in the SOLGATE project (230 kW_e) that commenced operating in 2001 to demonstrate the performance of a solar-hybrid gas turbine system driven by a solar tower (SOLGATE, 2005). An Allison 250 helicopter engine was used in this project. This engine was modified to allow heat addition from the receiver to drive an electric generator. The receiver could heat compressed air up to 1073 K. Kerosene was utilized as a backup fuel to maintain the stipulated turbine inlet temperature. The demonstration facility was installed at the solar tower test facility at Plataforma Solar de Almería in Spain. In the SOLTREC project, a pressurized volumetric receiver was designed and manufactured with a thermal power output of 1.47 MW_{th}, and it could heat compressed air to 1273 K with solar shares of up to 80% (Del Río et al., 2015). In 2006, another project on the development of solar gas turbines was initiated with a focus on solar-hybrid power and cogeneration (SOLHYCO) plants. This project aimed at developing a hybrid solar gas turbine that was highly efficient, reliable, and economic (SOLHYCO, 2011). The receiver was able to achieve a design outlet air temperature of 1073 K. Tests on a solar hybrid system were again conducted at the PSA test facility and receiver outlet temperatures of up to 1076 K were achieved (Stein & Buck, 2017). Another important milestone in the development of solar gas turbines was achieved through the SOLUGAS project. The aim of this project was to demonstrate the performance of an operational solar hybrid gas turbine system at a precommercial scale (Korzynietz et al., 2016). Even though solar gas turbines were tested in previous studies, SOLUGAS was the first demonstration project at megawatt scale, where compressed air could be preheated by a 3.2 MW_{th} receiver up to 1073 K and returned to the combustor of the 4.6 MW Mercury 50 gas turbine.

5.2.2.2 Carbon dioxide

At present, the Kalina cycle and organic Rankine cycle (ORC) are the most widely studied cycles that can utilize low-grade and waste heat sources (see Section 1.5). A standard Kalina cycle uses a mixture of ammonia and water, while the conventional organic Rankine cycles utilize organic substances. The ammonia–water mixture meets most of the requirements of a heat transfer fluid. Ammonia has a low critical temperature (406.04 K) and is friendly to the environment, but it has a relatively high critical pressure (11.28 MPa). Thus, it requires high operating pressures. Initially, ammonia–water was used as a refrigerant. However, this fluid is also very toxic and corrosive, and therefore, it is classified as a Safety Code Group B2 refrigerant (ASHRAE Standard 34), which means it has a low level of safety (ASHRAE, 2017). Based on the United Nations (UN) globally harmonized system of classification of chemicals, ammonia is also described as corrosive, acutely toxic, and environmentally hazardous (UN, 2011).

Most of the organic fluids exploited in the ORC were originally developed for use in the refrigeration industry. Nevertheless, based on the principle of reversibility of the processes in a thermodynamic cycle, a refrigeration cycle can be reversed to a power cycle. For example, a Carnot refrigeration cycle can be reversed into a Carnot power cycle (Fig. 5.4). In the refrigeration cycle,

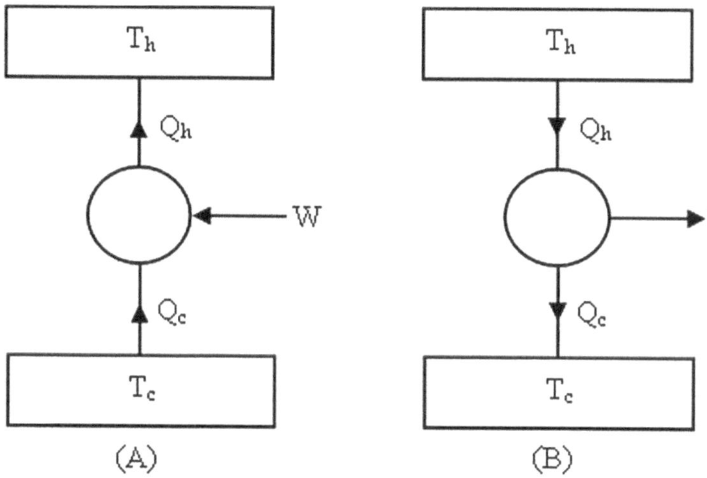

FIGURE 5.4 Schematic representation of: (A) Carnot refrigeration cycle and (B) Carnot power cycle.

work (W) is done on the system to extract heat from a cold reservoir at temperature T_c and reject it to a hot reservoir at a temperature T_h. Work is done by the system when heat is extracted from the hot reservoir and rejected to the cold reservoir. The working fluids in the refrigeration cycle are known as refrigerants, and some of them fall in the category of chlorofluorocarbon substances that contribute to the depletion of the stratospheric ozone layer. For instance, trichlorofluoromethane (R11) is a refrigerant that has high ozone depletion potential even though it is safe to use as a heat transfer fluid (Table 5.1). Other organic fluids such as propane, butane, isobutene, and propylene are highly flammable and risky. Thus, there was a need to continue searching for alternative refrigerants. After studying the properties of CO_2, the refrigeration community started advocating its use as a working fluid (Srinivasan, 2011). This development prompted researchers in the power generation field to investigate the possibility of utilizing CO_2 as a working fluid in power cycles.

Critical temperature and pressure show whether a cycle can be driven as a subcritical cycle, transcritical cycle or supercritical cycle, and the possible onset of condensation and the possible working pressure of the system, respectively. Carbon dioxide attains a critical state at $T = 304.13$ K and $P = 7.38$ MPa. Consequently, it is easier to elevate the state of CO_2 to the supercritical level than that of water. Heating the fluid in the supercritical region helps to eliminate pinching, and the temperature profile of the CO_2 and the heat source can therefore match and glide (Baheta et al., 2018). Ideally, a supercritical fluid has relatively low compressibility, which tends to reduce the required amount of power to compress the fluid.

TABLE 5.1 Properties of selected refrigerants.

Fluid name	ASHRAE number	Critical temperature (K)	Critical pressure (MPa)	ASHRAE level for safety	ODP	GWP_{100}
Trichlorofluoromethane	R11	471.11	4.4076	A1	1	4660
Trifluoromethane	R23	299.44	4.83	A1	0	11,700
Difluoromethane	R32	351.41	5.78	A2	0	580
2.2-dichloro-1.1.1-trifluoroethane	R123	456.98	3.66	B1	0.02	93
Pentafluoroethane	R125	339.32	3.62	A1	0	2800
Hydroflorocarbon	R134a	374.25	4.07	A1	0	1300
1.1.1-trifluoroethane	R143a	346.01	3.76	A2	0	3900
1.1-difluoroethane	R152a	386.56	4.52	A2	0	140
Octafluoropropane	R218	345.17	2.64	A1	0	7000
Propane	R290	642.95	4.247	A3	0	<10
Butane	R600	425.15	3.79	A3	0	<10
Isobutane	R600a	407.85	3.64	A3	0	<10
Ammonia	R717	406.04	11.28	B2	0	0
Water	R718	647.04	22.1	A1	0	0
Carbon dioxide	R744	304.13	7.38	A1	0	1
Propylene	R1270	365.72	4.66	A3	0	0

GWP_{100}, global warming potential over an integration time horizon of 100 year; *ODP*, ozone depletion potential.

The performance of CO_2 cycles has been investigated in a wide range of temperatures covering the subcritical, transcritical, and supercritical regions. It should be noted that the distinction among these three cycles is the operating pressure (see Section 1.5.3). (Garg et al., 2013) studied the performance of subcritical (sn-CO_2), transcritical (tn−CO_2), and supercritical (s-CO_2) carbon dioxide cycles in the range 308−1000 K. As expected, they found that the s-CO_2 cycle was thermodynamically most efficient. Thus, there is great potential of exploiting CO_2 as a working fluid in power cycles driven by a wide range of heat sources, from low-to high-grade heat sources including concentrated solar radiation.

5.2.2.3 Helium

Helium is an inert and nontoxic gas that has recently found potential use as a heat transfer fluid. It is an ideal working fluid because it possesses some good heat transfer properties. For instance, helium has higher thermal conductivity than air, and its critical point is at $T = 2$ K and $P = 0.229$ MPa. It is chemically more inert and has higher specific heat capacity than air. For example, the composition of air includes the element oxygen that is very reactive. At a given temperature and pressure, the specific heat capacity of helium is remarkably higher than that of air.

Helium has been applied in the gas turbine technology. For the first time, helium was used as a working fluid in the development of cryogenic processes in gas turbines. In this context, the first two small fossil-fired helium gas turbines were for air liquefaction and nitrogen production facilities in the United States of America. In 1970s, the Oberhausen II 50 MW_e helium turbine power plant (EVO) and the Jülich 300 MW_e high-temperature helium test facility (HHV) were constructed and operated in Germany. The Oberhausen helium turbine plant used a coke oven gas-fired heater to achieve a turbine inlet temperature of 1023 K (McDonald et al., 1994). The plant provided electricity and district heating to the city of Oberhausen. The Jülich facility used a nuclear reactor to achieve a high temperature of 1123 K (Hebel & Weiskopf, 1982). The aim of these two facilities was to establish a technology foundation for the advancement of future concepts of large nuclear gas turbine power plants based on closed power cycles.

5.2.2.4 Hydrogen

Hydrogen (H_2) is the lightest gas and one of the most abundant elements on earth. This gas is a clean energy carrier for heat and electricity. Advantages of hydrogen include: (a) large storage capacity: H_2 is a fuel with very high energy density, and it is considered as the fuel with the highest energy content per unit mass because its higher heating value (HHV) is about 39.42 kWh kg^{-1}, which is approximately three times higher than that of methane and gasoline (Rajeshwar et al., 2008; Ursúa et al., 2012), (b) high efficiency of energy conversion, (c) cleanliness and environmental friendliness, (d) renewable production, (e) wide sources, (f) reliability, and (g) simple means of storage and regeneration. In addition, hydrogen is an important chemical raw material in large-scale industrial processes such as petroleum refining, water-gas shift reaction, and ammonia synthesis (Zhang et al., 2021). The major drawback of this gas is that it does not naturally occur as an element, and therefore, it has to be produced from chemical compounds, which requires a significant input of energy and raises the production cost.

Hydrogen can be produced through thermochemical, direct solar water splitting, electrolytic, and biological processes. Thermochemical processes utilize heat from various sources such as fossil fuels, renewable energy, and others to produce hydrogen. Steam-methane reforming is the most matured thermochemical pathway for H_2 production, that typically proceeds at temperatures of 1023–1173 K and pressure $P = 0.15–0.40$ MPa (Karatza et al., 2021). This process yields hydrogen and carbon monoxide. In the direct solar water splitting processes, light is used to split water into hydrogen and oxygen gases. These processes are still at infant stage of research and development. Electricity is used through electrolyzers to split water into hydrogen and oxygen in the electrolytic processes. At present, there is a growing interest to produce hydrogen during periods of excess electricity supply and low electricity prices (Tang et al., 2021). Similarly, electricity produced from intermittent renewable energy sources (such as solar and wind) can be stored via hydrogen. The stored hydrogen can be converted back to electricity by using suitable generation technologies. Biological processes involve use of microorganisms such as bacteria and microalgae to produce hydrogen through biological reactions. These processes are also at an early stage of development.

Hydrogen has also found application as a heat transfer fluid in solar receiver technology. It is the most efficient heat transfer fluid for improving the output power of a Stirling engine compared to helium and air (Musmar & Tlili, 2015). Air exhibits the least efficiency of this heat engine. (Hafez et al., 2016) made a similar observation on the superiority of hydrogen over air and helium gases as heat transfer fluids. Hydrogen was used as a heat transfer fluid in solar dish-Stirling systems (Nepveu et al., 2009), but it has risks of inflammation unless appropriate technical solutions are found.

5.2.3 Molten salts

A molten salt is a synthetic mixture of salts composed mainly of nitrates, which can be used as a thermal storage medium as well as a heat transfer fluid. Molten salts have various required qualities of heat transfer at elevated temperature. They have high density, specific heat capacity and thermal stability, and low vapor pressure (Benoit et al., 2016). These salts are also cheaper and more friendly to the environment than synthetic oil. Nevertheless, the main drawback of molten salts is their relatively high freezing point, which contributes to an increase in operation and maintenance costs through the implementation of freeze protection. For instance, current binary (two-component) and ternary (three-component) molten salts freeze in the temperature range 393–493 K, whereas the freezing point of synthetic oil is about 288 K. Unfortunately, salts with high freezing points are associated with higher viscosities at elevated temperatures, thereby raising the costs of pumping power (Coscia et al., 2013).

Three of the commercially available molten salts are: solar salt, Hitec, and HitecXL. Solar salt is a common binary mixture of 60% sodium nitrate ($NaNO_3$) and 40% potassium nitrate (KNO_3) (Zhang et al., 2013). This mixture melts at 493 K, and it is stable with a diminished vapor pressure and can be used in a temperature range of 533–894 K. Nevertheless, the salt commences to crystallize at 511 K and solidifies at 494 K. Solar salt has a high melting point, and therefore, it is not a satisfactory heat transfer fluid, though in use. Hitec salt is a ternary mixture of sodium nitrite ($NaNO_2$), sodium nitrate ($NaNO_3$), and potassium nitrate (KNO_3). It melts at 415 K and can be applied up to a maximum temperature of 811 K. Hitec XL salt is a ternary mixture of potassium nitrate (KNO_3), sodium nitrite ($NaNO_2$), and calcium nitrate ($Ca(NaNO_3)_2$). This salt melts at a temperature of 393 K, and it is applicable up to a maximum temperature of 773 K. Hitec and Hitec XL salts are promising to replace Solar Salt because they have lower melting points. Based on the Global Harmonized System of Classification of Chemicals, some of the salt components are hazardous and unsafe. For example, sodium nitrate is toxic, an oxidizing solid, and an irritant, while potassium nitrate is an irritant and oxidizing solid. Sodium nitrite is acutely toxic and an irritant and oxidizing solid. Other molten salts have been developed, and they include chloride-based salts that are promising to have higher temperature limits of stability (Table 5.2).

5.2.4 Thermal oil

Originally, CSP plants started using synthetic oil as a heat transfer fluid, commonly known under the brand names Therminol VP-1 or Dowtherm A. Both Therminol VP-1 and Dowtherm A are eutectic mixtures of diphenyl oxide ($C_{12}H_{10}O$) and biphenyl ($C_{12}H_{10}$). A eutectic mixture is a composition of two or more solid components that produces a change of phase to liquid at a certain temperature. This temperature is known as the eutectic point temperature and it corresponds to the minimum melting temperature of the various possible compositions. Therminol VP-1 and Dowtherm A have the same eutectic mixture. This mixture comprises 73.5% diphenyl oxide and 26.5% biphenyl (Lang & Lee, 2015).

The use of thermal oil avoids the high-pressure requirement and phase change encountered when using water. This heat transfer fluid can be used at temperatures as high as 673 K (Wang et al., 2013). However, above this temperature limit, the hydrocarbons of the fluid breakdown fast, resulting in the production of hydrogen. Chemical deterioration can augment makeup fluid requirements, decrease overall fluid lifetime, and cause accumulation of sludge or other undesirable by-products that reduce the heat transfer efficiency of a system and raise the costs of maintenance. Moreover, based on the United Nations system of classification and labeling of chemicals, diphenyl oxide and biphenyl are very toxic to aquatic life with long-lasting effects (UN, 2011). Thus, they are not friendly to the environment.

5.3 Advanced heat transfer media

As seen in Section 5.2, conventional heat transfer fluids have some deficiencies in desirable characteristics. This has led to the search for materials that can meet most of the requirements for heat transfer fluids. Among the recent innovations are the supercritical cycles, liquid metals, nanofluids, and suspended solid particles.

5.3.1 Supercritical cycles

Every working fluid has a critical point, which is specified by temperature and pressure. A supercritical cycle is a closed cycle heat engine that completely operates above the critical pressure of the working fluid. It exhibits high thermal efficiency and compactness of the machinery. The Rankine and Brayton (gas turbine) cycles are the most widely exploited heat engine cycles in the power sector. The former cycle is inherently a closed cycle heat engine, while the latter cycle has open, closed, and semi-closed configurations (see Section 1.5). Thus, supercritical fluids can be used as heat transfer media in the Rankine cycle and closed cycle gas turbine. In this vein, supercritical steam and supercritical carbon dioxide are the most promising materials for exploitation as heat transfer fluids in advanced power cycles.

5.3.1.1 Supercritical steam

Most of the thermal power plants operating in the world use water (that has a critical point at $T = 647.04$ K and $P = 22.1$ MPa) as the working fluid in the Rankine cycle. These power plants can be operated in the subcritical region (upper side pressure <22.1 MPa), transcritical region (lower side pressure$<$ 22.1 MPa and upper side pressure >22.1 MPa), or supercritical region (lower side pressure >22.1 MPa). Operation of steam plants in the subcritical range (sn-H_2O) has relatively low efficiency. On the other hand, the supercritical steam cycle (s-H_2O) yields higher efficiency and therefore more attractive.

For decades, the supercritical water cycle has been used in conventional thermal power plants, such as fossil-fueled plants. Nevertheless, recent research efforts are directed toward the development of ultrasupercritical steam power plants. The parameters of the ultrasupercritical steam are

TABLE 5.2 Selected molten salts.

Molten salt	Composition	Melting point (K)	Stability limit (K)
Solar salt	60% $NaNO_3$, 40% KNO_3	513	838–803
Hitec	7% $NaNO_2$, 53% $NaNO_3$, 40% KNO_3	415	813–723
Hitec XL	45% KNO_3, 7% $NaNO_2$, 48% $Ca(NaNO_3)_2$	393	773
$K_2CO_3/Li_2CO_3/Na_2CO_3$	32% K_2CO_3, 35% Li_2CO_3, 33% Na_2CO_3	670	>923
KF/LiF/NaF	59% KF, 29% LiF, 12% NF	727	>973
$KCl/NaCl/ZnCl_2$	23.9% KCl, 7.5% NaCl, 68.6% $ZnCl_2$	477	1123
$KCl/MgCl_2/NaCl$	17.8% KCl, 68.2% $MgCl_2$, 14.0% NaCl	653	>1073

$T = 893$ K and $P = 31.0$ MPa (Boss et al., 2007). One of the main bottlenecks to the evolution of ultrasupercritical steam power plants is material limitation. In this context, some relevant materials have been developed and tested at laboratory and pilot scales. Suitable metallic alloy materials and fabrication processes have been successfully developed for power generation applications with metal temperatures of up to about 1073 K (Purgert et al., 2015). Nevertheless, these advanced materials and components need to be tested at a larger demonstration scale before implementation in commercial power plants to reduce financial risks. At the time of writing this book, there was no CSP plant operating with s-H_2O as a heat transfer fluid. For solar tower plants, steam turbines will require down-scaling to operate on the s-H_2O cycle (Zhang et al., 2016).

5.3.1.2 Supercritical carbon dioxide

Carbon dioxide has a critical point at $T = 304.5$ K and $P = 7.38$ MPa. In the supercritical region ($P > 7.38$ MPa), carbon dioxide is in the supercritical state (s-CO_2), and it has favorable heat transfer properties that allow the design of innovative conversion systems. The Rankine and Brayton cycles are the most promising power cycles where s-CO_2 can be utilized. Due to the low critical temperature of CO_2, it is possible to power an s-CO_2 cycle by using a heat source of low grade. In this regard, the s-CO_2 Rankine cycle performs better than both steam and organic Rankine cycles (Sarkar, 2015). Other exceptional advantages of the s-CO_2 Rankine cycle are: (a) system compactness and

(b) environmental benefits. The s-CO_2 gas turbine can be operated in a wide range of temperatures: low to high. To obtain high efficiency, it is necessary to drive the cycle at high temperature (Dutta, 2017). High-temperature power cycles also entail high pressures that require solar receiver and thermal storage systems with suitable designs. It is perceived that power plants based on the high-temperature s-CO_2 gas turbine can achieve efficiencies of up to about 50% when the cycle is operated at a turbine inlet temperature of about 973 K.

Different researchers have investigated means of improving the thermodynamic performance of the Brayton cycle. There is a general consensus that the combined cycle solar gas turbine exhibits a higher thermal efficiency than single cycles. The Brayton-Rankine cycle is often referred to as the conventional combined cycle, with reported thermal efficiencies of up to 60% (Poullikkas, 2005). Moreover, the s-CO_2 cycle can be driven by a wide range of grades of heat sources. Thus, the s-CO_2 cycle can be exploited as a topping or bottoming cycle of a combined cycle system. The s-CO_2 cycle has been investigated as a topping cycle in the combined cycle with bottoming Rankine and organic Rankine cycles. It was concluded that the CO_2 Brayton cycles have the potential to compete in terms of cost and efficiency with the conventional cycles (Chacartegui et al., 2011). A thermo-economic analysis was performed on a combined cycle power plant that comprised a topping s-CO_2 recompression Brayton cycle and a bottoming organic Rankine cycle using different organic fluids (Akbari & Mahmoudi, 2014). It was found that the second law efficiency of the combined cycle was higher than the supercritical carbon dioxide recompression Brayton cycle.

Some notable milestones have been achieved in the development of the s-CO_2 gas turbine. A 150 kWe s-CO_2 test loop for a small terrestrial nuclear reactor was developed by Hoffman and Feher in 1970 (Hoffmann & Feher, 1971). In 1997, the Tokyo Institute of Technology developed a test loop for corrosion studies on materials for the s-CO_2 cycle (Dostal et al., 2001; Kato et al., 2001). The final design and installation of a megawatt class s-CO_2 recompression gas turbine test unit was accomplished by Sandia National Laboratory (SNL) contractor Barber-Nichols Inc. in 2012 (Pasch et al., 2012). Later in December 2014, an 8 MW closed power cycle utilizing s-CO_2 as a working fluid was commissioned for commercial demonstration by Echogen Power Systems (Echogen Power Systems, 2021; Olumayegun et al., 2016). This demonstration plant was powered by waste heat, and it was the first commercial power plant based on the s-CO_2 cycle. Thus, there is great potential of applying the advanced s-CO_2 power cycles for concentrated solar energy conversion.

Integration of CSP with s-CO_2 requires a suitable receiver for collecting concentrated solar radiation and heating the s-CO_2. Solar receivers for s-CO_2 cycles ought to withstand the high pressure, high temperature, and high solar flux (Teng & Xuan, 2019). Meeting these requirements is an enormous challenge for receiver designers. It has been observed that achievement of efficiencies higher than 50% is possible for turbine inlet temperatures above 973 K with cycle configurations such as recompression cycles combined with intercooling and/or turbine reheating (Benoit et al., 2016). Recompression of the working fluid raises the total compressive work, but the net output work increases. For example, the performance of five s-CO_2 Brayton cycles (simple, regenerative, recompression, precompression, and split Brayton cycles) driven by a solar tower in Saudi Arabia was investigated by Al-Sulaiman and Atif (2015). They found that recompression yielded the highest efficiency (52% observed in summer). This level of thermodynamic performance clearly demonstrates the potential of the s-CO_2 cycle to reduce the generation costs of CSP-based electricity.

5.3.2 Liquid metals

Naturally, most metals occur in solid form at or around room temperature. However, a few metals remain in the liquid state at or close to room temperature. Liquid metals have generally low viscosity, low vapor pressure, high thermal conductivity, and a wide temperature range between melting and boiling points. Consequently, they are attractive for applications at high temperatures and can significantly contribute to the development of CSP. However, some disadvantages of liquid metals are low specific heat capacity and corrosive properties. Low specific heat capacity deters the use of liquid metals as storage media. A heat transfer fluid with high heat capacity can be utilized as a storage medium too, thereby removing the need for another storage medium and reducing the costs of the system. Corrosive properties of metals also discourage the exploitation of liquid metals as heat transfer fluids. Nevertheless, some of the liquid metals, such as sodium (Na), lead-bismuth eutectic (PbBi), and gallium (Ga) are promising candidates of primary heat transfer fluids.

5.3.2.1 Liquid sodium

Sodium (Na) is a common alkali metal with melting and boiling points of 371 and 1155 K at normal pressure, respectively. Thus, it has a very wide range of the operational temperature. Liquid sodium is chemically stable up to near 1173 K with high thermal conductivity (Liu et al., 2019; Pacio et al., 2013). When alloyed with potassium (K), a eutectic mixture is obtained with melting and boiling points at 260 and 1058 K at normal pressure, respectively (Kotzé et al., 2014). It is therefore not surprising that this liquid metal is perceived as a potential heat transfer fluid in the next-generation CSP systems.

In the early work, in 1980s, liquid sodium was used as a heat transfer fluid in external and cavity receivers (Wetzel et al., 2014). With liquid sodium as a working fluid, receiver efficiencies of 88%–96% were achieved at that time. These experiences revealed some shortcomings of using liquid sodium as a heat transfer fluid. There was lack of suitable standards for monitoring, handling, and technological benchmarking. In addition, sodium is chemically very reactive in water and air, but provisions were not available relating to the safety and hazards of using this material as a heat transfer fluid. Since then, there has been significant progress in the development of liquid metals as heat transfer fluids.

Boerema et al. (2012) theoretically compared the performance of Hitec (53% KNO_3, 40% $NaNO_2$, 7% $NaNO_3$) and liquid sodium (Na). Liquid sodium exhibited a significantly higher coefficient of heat transfer than Hitec. So, this liquid metal has great potential to replace molten salts in next-generation CSP technology. One of the advantages of using liquid metals is the chemical stability at high temperatures. Nevertheless, it is a big challenge to collect, transport, and store liquid metal at extreme temperatures exceeding 1300 K. Recently, a ceramic, mechanical pump that has ability to continuously distribute liquid tin (Sn) at temperatures of about 1473−1673 K has been demonstrated (Amy et al., 2017).

5.3.2.2 *Lead-bismuth eutectic*

A lead-bismuth eutectic is composed of 44.5% lead (Pb) and 55.5% bismuth (Bi). This eutectic has melting and boiling points of 397 and 1943 K at normal pressure, respectively. The Pb−Bi eutectic has a very wide range of operating temperature, and it is proposed as a potential coolant in the IV-generation (Gen IV) fast reactor.

Gen IV lead-cooled reactors are expected to operate at temperatures of 753−843 K, and industrial deployment of this technology is expected to commence in 2025 (WNA, 2021). Sobolev (2007) reviewed data on the characteristic temperatures, density, heat capacity, viscosity, thermal conductivity, latent heats of melting and evaporation, surface tension, and electric conductivity of lead and lead-bismuth eutectic. Correlations for calculating these thermophysical properties were also established. Most of the studies on Pb−Bi eutectic relate to applications in the nuclear power industry. However, the knowledge and experience gained from these works can be applied in the development of the next-generation CSP technology.

5.3.2.3 *Gallium*

Gallium-based liquid metals have low melting point, high thermal conductivity, nonflammable, and nontoxic characteristic (Deng et al., 2021). The melting point of gallium is 302.85 K, and it boils at 2676 K (Bo et al., 2018). Thus, it remains in the liquid state over a very wide range of temperature. Moreover, the melting point of gallium can be reduced through alloying it with other metal elements. For example, the melting point of a eutectic gallium-indium alloy is 288 K. The introduction of tin to the gallium-indium alloy produces a eutectic gallium-indium-tin alloy with a melting point of 254 K.

Thus, gallium and gallium-based alloys are promising candidates for exploitation as heat transfer fluids in the next-generation of CSP plants.

5.3.3 Nanofluids

The rate of heat transfer increases with the thermal conductivity of a fluid. Consequently, enhancement of thermal conductivity is one of the most important strategies of augmenting the performance of solar thermal systems. This may be achieved by replacing working fluids with nanofluids as advanced heat transfer fluids. The concept of adding metallic nanoparticles to conventional heat transfer fluids to obtain a nanofluid was first proposed by Choi and Eastman (1995). A nanofluid is a fluid that contains a base fluid and nanoparticles that are dispersed in the base fluid. Common base fluids are water, ethylene glycol, and synthetic oil. A nanoparticle is a small particle with a size that falls in the nanometre range. Various ranges of nanoparticle size have been reported in literature but the International Organization for Standardization (2015) specifies a range of 1-100 nm. Nanoparticles raise the thermophysical properties of base fluids, and they fall into two main categories: metallic and nonmetallic nanoparticles. Metallic nanoparticles include copper (Cu), iron (Fe), aluminum (Al), gold (Au), nickel (Ni), and silver (Ag). Aluminum oxide (Al_2O_3), copper oxide (CuO), silicon (Si), silicon carbide (SiC), zinc oxide (ZnO), carbon nanotubes (CNTs), and boron nitrate nanotubes are examples of nonmetallic nanoparticles.

Nanofluids are two-phase systems with solid and liquid phases and possess boosted thermophysical properties compared to those of base fluids. In some formulations, the nanoparticles are hybridized when preparing a nanofluid. For example, copper (Cu) nanoparticles and CNTs can be dispersed in water to obtain a hybrid Cu-CNT nanofluid. The shape and size of nanoparticles have an influence on the heat transfer characteristics of the nanofluid. Advantages of using a nanofluid as a heat transfer medium in solar energy systems are (Elsheikh et al., 2018): (a) Higher thermal conductivity of the nanofluids than the base fluid because of the presence of nanoparticles. (b) Increase in solar absorption and heat capacity because nanoparticles are very small with a large surface area. (c) High stability of nanofluids in a wide range of temperature gradients combined with high solar absorption coefficient. (d) Better optical characteristics of nanofluids than those of the base fluid. (e) The use of nanofluids prevents the sedimentation, clogging, and fouling of pipes and pumps because nanoparticles are extremely smaller than micro- or millimeter-sized particles, which is a useful property in many solar applications. (f) Nanofluids decrease the required heat transfer area of a thermal device and, therefore, the total cost of the device. (g) Increased thermal efficiency of devices because of the generally high density and convective heat transfer coefficient of nanofluids.

5.3.3.1 Preparation of nanofluids

Nanofluids are formed by suspending nanoparticles in a base fluid, and two main methods of producing nanofluids are single-step and two-step methods (Yu & Xie, 2012). In the single-step method, nanoparticles are concurrently produced and dispersed in the base fluid. The processes of drying, storage, transportation, and dispersion of nanoparticles are avoided in this method, and so the aggregation of nanoparticles is minimized, thereby raising the stability of the fluids.

One-step physical technique is not able to synthesize nanofluids on large scale, and the product cost is also high. Consequently, the one-step chemical method is developing fast. In the two-step method, nanoparticles are produced in powder form by using physical or chemical techniques, and then the powder is dispersed into the base fluid in the second step with the assistance of intensive magnetic force agitation, ultrasonic agitation, high-shear mixing, homogenizing, and ball milling. This is the most economic technique of producing nanofluids on large scale because synthesis techniques for nano powders are already matured at the industrial scale. Nanoparticles tend to agglomerate because of the high surface area and surface activity. In view of this, surfactants are employed to enhance the stability of the nanofluid, but the functionality of the surfactants under elevated temperature also causes a big concern.

5.3.3.2 Application of nanofluids in solar receivers

Nanoparticles thermally and optically affect the performance of CSP systems. In a conventional solar receiver, concentrated solar radiation is absorbed and converted to heat by an opaque absorber surface (Fig. 5.5). Then, this heat is transmitted from the hot surface to the heat transfer fluid through convection. To achieve good performance of the receiver, it is necessary for: (a) the absorber surface to have high absorptance of solar radiation and low emittance

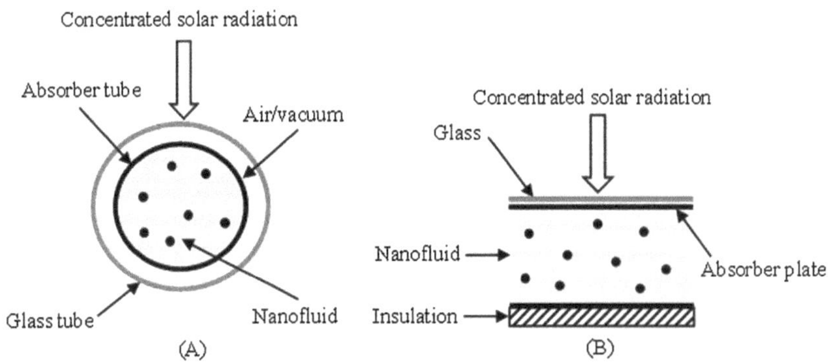

FIGURE 5.5 Schematic representation of surface-based receivers with nanofluid: (A) tubular absorber and (B) flat plate absorber.

of infrared radiation (see Chapter 4) and (b) heat transfer fluid to possess desirable thermophysical properties. The use of nanofluids in surface-based absorbers enhances the coefficient of heat transfer from the surface to the fluid.

Surface-based receivers are efficient at conversion of solar radiation to heat, but they are not well suited for transmitting heat to a fluid (Lenert & Wang, 2012). High levels of solar concentration results in a large temperature gradient between the absorber and the fluid. So, the temperature of the absorber surface is significantly higher than that of the fluid, thereby augmenting emissive losses from the surface. In contrast, direct absorption of solar radiation by the fluid promotes uniform distribution of the heat in the surrounding fluid, which reduces the temperature differentials between the absorber and the fluid. Consequently, there has been increasing interest to develop receivers in which concentrated solar radiation is directly absorbed by the heat transfer fluid. Unfortunately, most of the heat transfer fluids weakly absorb solar radiation. This drawback has led to the development of the concept of nanofluid volume absorbers of solar radiation. Optically, nanofluids can be engineered to exhibit very high absorptance by using a small volume fraction of suitable nanoparticles. The concentrated solar radiation is directly absorbed by the bulk of the nanofluid. Figs. 5.6 and 5.7 show surface- and volume-based absorption of solar radiation.

A nanofluid comprises nanoparticles and a base fluid, and so it has composite thermophysical properties. The specific heat capacity ($C_{p,nf}$), density (ρ_{nf}), thermal conductivity (k_{nf}), and viscosity (μ_{nf}) of the fluid can be computed from following expressions (Mahmoudi et al., 2020):

$$C_{p,nf} = \frac{(1-\phi)\rho_{bf}C_{p,bf} + \phi\rho_{np}C_{p,np}}{\rho_{nf}} \tag{5.14}$$

Where $C_{p,bf}$ is the specific heat capacity of the base fluid, $C_{p,np}$ is the specific heat capacity of the nanoparticles, ρ_{bf} is the density of the base fluid, ρ_{nf} is the

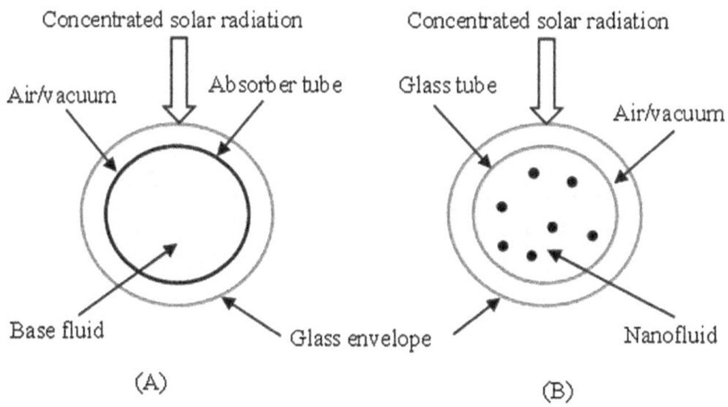

FIGURE 5.6 Schematic representation of: (A) tubular surface absorber and (B) nanofluid-based tubular volume absorber.

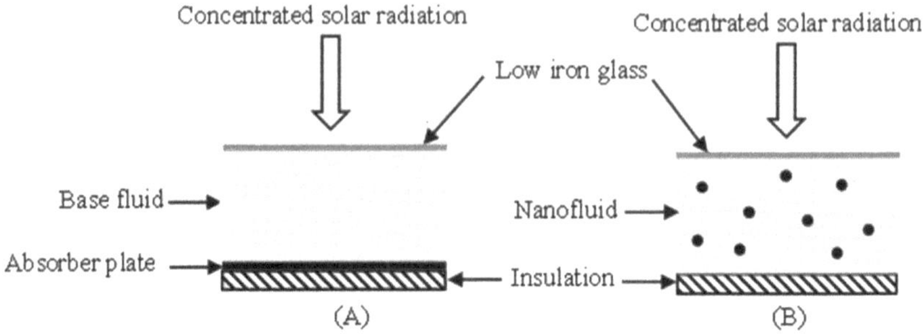

FIGURE 5.7 Schematic representation of (A) flat plate surface absorber and (B) nanofluid-based flat plate volume absorber.

density of the nanofluid, ρ_{np} is the density of the nanoparticles, and ϕ is the volume fraction of nanoparticles in the nanofluid.

The density of the nanofluid (ρ_{nf}) can be given by:

$$\rho_{nf} = (1-\phi)\rho_{bf} + \phi\rho_{np} \tag{5.15}$$

There are different models for calculating the thermal conductivity (k) of the nanofluid. These models have different accuracies and requirements (such as shape of nanoparticles) of input values. The Maxwell correlation can be used to compute k (Wang & Xu, 1999):

$$k_{nf} = k_{bf} + \frac{3\phi\left(k_{np} - k_{bf}\right)}{\frac{k_{np} + 2k_{bf}}{k_{bf}} - \phi\left(\frac{k_{np} - k_{bf}}{k_{bf}}\right)} \tag{5.16}$$

The viscosity of the nanofluid (μ_{nf}) is also computed from:

$$\mu_{nf} = \mu_{bf}\left(1 + 2.5\phi + 6.5\phi^2\right) \tag{5.17}$$

At present, most of the operating CSP power plants use thermal oils as HTFs (see Section 5.1). Nonetheless, the thermal conductivity of these fluids is generally very diminished (in the order of 0.1 W $m^{-1}K^{-1}$), and it decreases with increasing temperature (Wang et al., 2013). Therefore, the addition of nanoparticles to these oils has potential to improve the optical and thermal properties of receivers, and the overall performance of the CSP system.

Researchers have examined the possibility of replacing thermal oil (and other base fluids) with nanofluids. Wang et al. (2013) prepared gold-Therminol VP-1 nanofluids and characterized them for their thermophysical properties and stability. They found that the mass of gold (Au) significantly influences the stability of the Au-Therminol VP-1 nanofluids, while the viscosity is marginally affected by the increase in the concentration of nanoparticle within the investigated range of concentration (0.005%–0.05% by mass). Gold nanoparticles improved the thermal conductivity of the fluids by up to 6.5%.

Nevertheless, gold is an expensive metal. Consequently, it is important to search for cheaper alternatives. Lenert and Wang (2012) used 28 nm carbon-coated cobalt nanoparticles in Therminol VP-1 for varying nanofluid optical thicknesses. Many other studies have been conducted using different types of nanofluids such as aluminum oxide-water, copper-water, nickel nanoparticle-alkyl naphthalene, zinc oxide-water, titanium dioxide-water, and others. Generally, there is improvement to the heat transfer capability of solar collectors when nanofluids replace base fluids in indirect absorption. Thermal efficiency improvements of more than 50% have been reported in literature (Said et al., 2015, 2016), which is very significant. At the same time, some nanofluids have yielded insignificant improvements (for instance, Bellos et al., 2016; Coccia et al., 2016). So, there is wide variation in the performance of nanofluid-based HTFs.

For direct solar absorption, the absorption coefficient generally increases with the volume fraction, nanoparticle size, and path length. Nanoparticles of some metals and metal oxides have excellent optical absorption properties that can be tuned by altering their size and geometry. The performance of nanofluids has been investigated in a wide range of temperatures up to 1200 K in case of volumetric receivers (Lenert & Wang, 2012). Nanofluid-based volumetric receivers achieve higher efficiency than that of surface-based receivers (Mahmoudi et al., 2020). It is evident that nanofluids have great potential of improving the optical and thermal performance of receivers. However, the following major challenges need to be addressed before their wide exploitation (Elsheikh et al., 2018):

(a) Difficulties in the production process: nanofluids are commonly produced by using single step and two-step methods. In the latter method, residual reactants are left in the nanofluids because of incomplete reaction or stabilization. Both methods utilize processes that involve reduction or ion-exchange reactions. In addition, base fluids comprise reaction products and ions that are hard to separate from the base fluids. These undesirable chemical constituents negatively affect the performance of nanofluids.

(b) High production cost: The two aforementioned methods for producing nanofluids require sophisticated and advanced equipment that is expensive, thereby raising the production cost. The high production cost is the main barrier to the industrial application of nanofluids. Through research and development, it is possible that the cost will be reduced in future.

(c) Instability and agglomerating: When suspended in a base fluid, nanoparticles tend to agglomerate because of the strong van der Waals interactions. This aggregation process affects the homogeneity of the nanofluid, and the problem of agglomeration becomes worse at elevated temperature gradients. Some remedies to this challenge are surface modification of the suspended nanoparticles and addition of surfactants to augment the stability of the nanofluids. In addition, raising the stability of nanofluids results in the enhancement of the thermal conductivity of nanofluids.

(d) Elevated pumping power: The viscosity of nanofluids is higher than that of their corresponding base fluids. Consequently, the use of nanofluids increases the pressure drop and the required power for pumping. For example, it has been found that the pressure drop increases with the volume fraction of titanium dioxide-water nanofluid.

5.3.4 Suspended solid particles

Recently, the concept of using suspended solid particles as a heat transfer medium has been suggested. A particle suspension is formed by suspending solid particles in a gas (such as air), and the suspension is directly or indirectly heated by concentrated solar radiation. In the direct-absorption system, solid particles absorb concentrated solar radiation and convert it to heat in a receiver. Thus, it is necessary for the particles to have high absorbance in the solar spectrum and low emittance in the infrared region. In the indirect-absorption system, a flow of solid particles is created inside opaque tubes of the receiver that are irradiated with concentrated solar radiation. Heat is then transferred from the walls of the hot tubes to the particle suspension. The apparent density of the suspension can be as high as 1000 kg m^{-3}, which is approximately 1000 times higher than that of air at atmospheric pressure (Benoit et al., 2016). Consequently, most of the thermal energy is transported by the solid particles.

For metallic tubes, it is possible to reach a particle temperature of up to 1023 K, while ceramic tubes can withstand temperatures above 1273 K. The hot particle suspension can heat the working fluid of a thermodynamic cycle (such as the Rankine, Brayton, supercritical steam, and supercritical carbon dioxide) via a heat exchanger. Advantages of the direct absorption system are: (a) avoidance of the need for a window and (b) acceptance of very high solar flux density. The main drawbacks of the system are: (a) difficulty of controlling particle flow stability and (b) possibility of high convection losses. The indirect-absorption system is attractive because it offers: (a) improved control of particle circulation within the receiver and (b) possible management of the operating pressure and atmosphere composition. Nevertheless, this system accepts concentrated solar radiation of lower intensity (200−400 kW m^{-2}). Fig. 5.8 is an illustration of a direct-absorption particle receiver that drives a conventional Rankine cycle. Advantages of utilizing solid particles as heat transfer media are (Jiang et al., 2019): (a) solid particles can function as both heat transfer and thermal storage media, (b) low cost of the materials, (c) the medium is very stable under high temperatures (as high as 2773 K), and (d) ability to drive advanced thermodynamic cycles such as the supercritical Rankine cycle and supercritical carbon dioxide Brayton cycle that can improve the power-generation efficiency to 40%−48%.

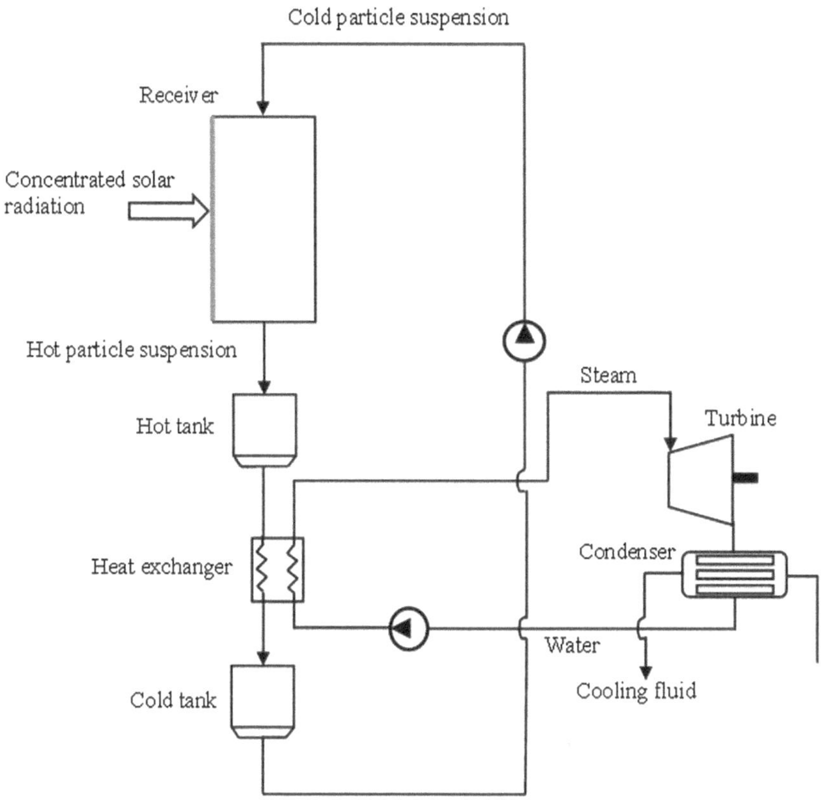

FIGURE 5.8 Schematic representation of a conventional Rankine cycle driven by a particle receiver.

There are variations in the fluidity and heat transfer properties of solid particles. Consequently, it is necessary to have some criteria for selecting suitable materials. The following criteria are recommended (Jiang et al., 2019): (a) Suitable thermophysical properties (high solid density, thermal conductivity, and specific heat capacity) in order to achieve greater heat transfer and heat storage capacities within a certain temperature gradient, (b) high resistance to heat such that the particles do not sinter and melt when the operational temperature exceeds 1273 K, and low content of impurities to make sure that eutectic melting does not take place at high temperatures, (c) high resistance to cracking to decrease the wear and tear of particles during elevated-temperature cycles and to enhance the stability of the performance of heat transfer, (d) low hazards of health and safety to protect operators and other people, (e) low cost and high storage capacity of solid particles, and (f) low impact on the environment, particularly when soft particles are selected because they are susceptible to wear, thereby producing dust pollution.

A wide variety of solid materials meet these criteria, and they include sand, quartz, cristobalite, alumina, bauxite, silicon carbide, calcined flint clay (CFC), and olivine. Solid particles of these materials can be used as heat transfer and thermal storage media. Selected properties of these materials are presented in Table 5.3.

A particle suspension is made up of solid particles and often air, and it has composite thermophysical properties with two mass flow rates: particle and gas. As the suspension flows through the receiver, concentrated solar radiation is absorbed by the solid particles. Heat is then transferred from the hot particles to the air. The heat collected by the particles from the solar receiver (Q_{sp}) can be expressed as (Reyes-Belmont et al., 2019):

$$Q_{sp} = \dot{m}_{sp} C_{p,sp} \left(T_{sp,out} - T_{sp,in} \right) \tag{5.18}$$

Where $C_{p,sp}$ is the specific heat capacity of the solid particles at constant pressure, $\dot{m}_{sp}$ is the mass flow rate of the solid particles, $T_{sp,in}$ is the inlet temperature of the solid particles, and $T_{sp,out}$ is the outlet temperature of the solid particles.

The mass flow rate of the gas $\left(\dot{m}_{ga}\right)$ drives the particles through the receiver. For a fluidized bed receiver, mass flow rate can be computed based on the volume fraction the solid particles (ϕ_{sp}), particle velocity (v_{sp}), gas velocity gas (v_{ga}), and the characteristics of the particle suspension (Eq. 5.19). The gas absorbs some heat while passing through the receiver, but its thermal energy is not utilized, it forms part of the thermal losses. More details about the concept of solid particle receivers are presented in Section 3.3.3.

$$\dot{m}_{ga} = \frac{\left(1 - \phi_{sp}\right) \rho_{ga} v_{ga}}{\phi_{sp} \rho_{sp} \left(v_{ga} - v_{mf} \right)} \dot{m}_{sp} \tag{5.19}$$

Where v_{mf} is the minimum fluidization velocity.

Example 5.2

A fluidized bed particle receiver uses olivine particles as a heat transfer and storage medium. The temperatures of the particles at the inlet and outlet points are 300 and 1023 K. If the heat extracted by the particles is 20 kW, calculate the total mass flowrate of the suspension. Thermophysical properties given in Table 5.3 at the mean temperature (density of gas $\rho_{ga} = 1.05$ kg m^{-3}, velocity of gas $v_{ga} = 0.1$ ms^{-1}, minimum velocity of fluidization $v_{mf} = 0.02$ ms^{-1}, and volume fraction $\phi_{sp} = 0.25$) are assumed.

Solution

Total mass flow, $\dot{m}_{total} = \dot{m}_{sp} + \dot{m}_{ga}$

TABLE 5.3 Characteristics of selected solid particles.

Parameter	Quartz	Sand	Cristobalite	Bauxite	Alumina	Silicon carbide	CFC	Olivine
C_p (kJ kg^{-1}K^{-1})	0.95	0.95	1.05	1.05	1.05	1.05	1.05	1.25
r (kg m^{-3})	2610	2610	2330	3600	3960	3210	2600	3400
E (kJ m^{-3})	2480	2480	2447	4095	4198	3371	2730	4284
k (Wm^{-1}K^{-1})	0.4	0.4	0.48	0.65	0.65	0.67	0.55	0.56
T_{sint} (K)	1600	750	1714	1550	2072	1800	1200	1450
Hardness (Mohs scale)	6	6	7	9	9	9	8	8
Cost (USD/ton)	150	75	200	400	300	1900	180	175

CFC, calcined flint clay.

From Eq. (5.18),

$$\dot{m}_{sp} = \frac{C_{p,sp}\left(T_{sp,out} - T_{sp,in}\right)}{Q_{sp}} = \frac{1250 * (1023 - 300)}{20000} = 23.313 \text{ kg s}^{-1}$$

$$\dot{m}_{ga} = \frac{\left(1 - \phi_{sp}\right)\rho_{ga}v_{ga}}{\phi_{sp}\rho_{sp}\left(v_{ga} - v_{mf}\right)}\dot{m}_{sp} = \frac{(1 - 0.25) * 1.05 * 0.1 * 23.313}{0.25 * 3400 * (0.1 - 0.02)} = 0.027 \text{ kg s}^{-1}$$

$$\dot{m}_{total} = (23.313 + 0.027)\text{kg s}^{-1} = 23.340 \text{ kg s}^{-1}$$

Nomenclature

C_p specific heat capacity at constant pressure (J $kg^{-1}K^{-1}$)
D diameter (m)
g acceleration due to gravity (m s^{-2})
Gr Grashof number (dimensionless)
h coefficient of heat transfer ($Wm^{-2}K^{-1}$)
k thermal conductivity ($Wm^{-1}K^{-1}$)
Nu Nusselt number (dimensionless)
P pressure (Pa)
Pr Prandtl number (dimensionless)
Q heat rate (W)
Ra Rayleigh number (dimensionless)
Re Reynolds number (dimensionless)
S characteristic length (m)
T temperature (K)
$\dot{m}$ mass flowrate (kg s^{-1})

Greek symbols

β coefficient of thermal expansion (K^{-1})
ρ density (kg m^{-3})
ξ thermal diffusivity (m^2 s^{-1})
μ dynamic viscosity (Pa s)
ϑ friction factor (dimensionless)
ν kinematic viscosity (m^2 s^{-1})

References

Akbari, A. D., & Mahmoudi, S. M. S. (2014). Thermoeconomic analysis & optimization of the combined supercritical CO_2 (carbon dioxide) recompression Brayton/organic Rankine cycle. *Energy, 78*, 501–512. https://doi.org/10.1016/j.energy.2014.10.037

Al-Sulaiman, F. A., & Atif, M. (2015). Performance comparison of different supercritical carbon dioxide Brayton cycles integrated with a solar power tower. *Energy, 82*, 61–71. https://doi.org/10.1016/j.energy.2014.12.070

Amy, C., Budenstein, D., Bagepalli, M., England, D., Deangelis, F., Wilk, G., Jarrett, C., Kelsall, C., Hirschey, J., Wen, H., Chavan, A., Gilleland, B., Yuan, C., Chueh, W. C., Sandhage, K. H., Kawajiri, Y., & Henry, A. (2017). Pumping liquid metal at high temperatures up to 1,673 kelvin. *Nature, 550*(7675), 199–203. https://doi.org/10.1038/nature24054

ASHRAE. (2017). *ASHRAE fundamentals handbook*. American Society of Heating, Refrigerating and Air-Conditioning Engineers (ASHRAE).

Baheta, A. T., Hailegiorgis, S. M., Oumer, A. N., Sulaiman, S. A. B., Abdul Karim, S. A., Zainuddin, N., Yusof, M. H., & Saad, N. (2018). Performance analysis of transcritical carbon dioxide rankine cycle with regenerator. *MATEC Web of Conferences, 225*, 05020. https://doi.org/10.1051/matecconf/201822505020

Bellos, E., Tzivanidis, C., & Antonopoulos, K. A. (2017). Parametric analysis and optimization of a solar assisted gas turbine. *Energy Conversion and Management, 139*, 151–165. https://doi.org/10.1016/j.enconman.2017.02.042

Bellos, E., Tzivanidis, C., Antonopoulos, K. A., & Gkinis, G. (2016). Thermal enhancement of solar parabolic trough collectors by using nanofluids and converging-diverging absorber tube. *Renewable Energy, 94*, 213–222. https://doi.org/10.1016/j.renene.2016.03.062

Benoit, H., Spreafico, L., Gauthier, D., & Flamant, G. (2016). Review of heat transfer fluids in tube-receivers used in concentrating solar thermal systems: Properties and heat transfer coefficients. *Renewable and Sustainable Energy Reviews, 55*, 298–315. https://doi.org/10.1016/j.rser.2015.10.059

Bignon, M. J. (1980). The influence of the heat transfer fluid on the receiver design. *Electric Power Systems Research, 3*(1–2), 99–109. https://doi.org/10.1016/0378-7796(80)90026-7

Boerema, N., Morrison, G., Taylor, R., & Rosengarten, G. (2012). Liquid sodium versus Hitec as a heat transfer fluid in solar thermal central receiver systems. *Solar Energy, 86*(9), 2293–2305. https://doi.org/10.1016/j.solener.2012.05.001

Bo, G., Ren, L., Xu, X., Du, Y., & Dou, S. (2018). Recent progress on liquid metals and their applications. *Advances in Physics: X, 3*(1), 1446359. https://doi.org/10.1080/23746149.2018.1446359

Boss, M., Gadoury, T., Feeny, S., Montgomery, M., Nah, U. H., Cho, S. I., & Yang, S. H. (2007). *Recent advances in ultra super critical steam turbine technology*. GE Energy.

Boyd, C. E. (2015). Physical properties of water. In *Water quality* (pp. 1–19). Cham: Springer. https://doi.org/10.1007/978-3-319-17446-4_1

Chacartegui, R., Muñoz De Escalona, J. M., Sánchez, D., Monje, B., & Sánchez, T. (2011). Alternative cycles based on carbon dioxide for central receiver solar power plants. *Applied Thermal Engineering, 31*(5), 872–879. https://doi.org/10.1016/j.applthermaleng.2010.11.008

Choi, S.U.S. & Eastman, J.A. (1995). Enhancing thermal conductivity of fluids with nanoparticles. http://www.osti.gov/scitech/biblio/196525.

Clauser, C., & Ewert, M. (2018). The renewables cost challenge: Levelized cost of geothermal electric energy compared to other sources of primary energy – review and case study. *Renewable and Sustainable Energy Reviews, 82*, 3683–3693. https://doi.org/10.1016/j.rser.2017.10.095

Coccia, G., Di Nicola, G., Colla, L., Fedele, L., & Scattolini, M. (2016). Adoption of nanofluids in low-enthalpy parabolic trough solar collectors: Numerical simulation of the yearly yield. *Energy Conversion and Management, 118*, 306–319. https://doi.org/10.1016/j.enconman.2016.04.013

Coscia, K., Nelle, S., Elliott, T., Mohapatra, S., Oztekin, A., & Neti, S. (2013). Thermophysical properties of $LiNO_3-NaNO_3-KNO_3$ mixtures for use in concentrated solar power. *Journal of Solar Energy Engineering, 135*(3). https://doi.org/10.1115/1.4024069

De Morsella, C. (2021). *World's first solar gas hybrid power plant launched in Israel.* http://greeneconomypost.com/solar-gas-hybrid-power-plant-3368.htm.

Del Río, A., Korzynietz, R., Brioso, J. A., Gallas, M., Ordóñez, I., Quero, M., & Díaz, C. (2015). Soltrec - pressurized volumetric solar air receiver technology. In *Energy procedia* (Vol. 69, pp. 360–368). Elsevier Ltd. https://doi.org/10.1016/j.egypro.2015.03.042

Deng, Y., Jiang, Y., & Liu, J. (2021). Low-melting-point liquid metal convective heat transfer: A review. *Applied Thermal Engineering, 193*, 117021. https://doi.org/10.1016/j.applthermaleng.2021.117021

Dostal, V., Todreas, N., Hejzlar, P., & Kazimi, M. (2001). *Power conversion cycle selection for the LBE cooled reactor with forced circulation.* Massachusetts Institute of Technology.

Dutta, P. (2017). High temperature solar receiver and thermal storage systems. *Applied Thermal Engineering, 124*, 624–632. https://doi.org/10.1016/j.applthermaleng.2017.06.028

Echogen Power Systems. (2021). *Echogen power systems' waste heat recovery system Available as turnkey solution.* https://www.echogen.com.

Elsheikh, A. H., Sharshir, S. W., Mostafa, M. E., Essa, F. A., & Ahmed Ali, M. K. (2018). Applications of nanofluids in solar energy: A review of recent advances. *Renewable and Sustainable Energy Reviews, 82*, 3483–3502. https://doi.org/10.1016/j.rser.2017.10.108

Fernández-García, A., Zarza, E., Valenzuela, L., & Pérez, M. (2010). Parabolic-trough solar collectors and their applications. *Renewable and Sustainable Energy Reviews, 14*, 1695–1721. https://doi.org/10.1016/j.rser.2010.03.012

Garg, P., Kumar, P., & Srinivasan, K. (2013). Supercritical carbon dioxide Brayton cycle for concentrated solar power. *Journal of Supercritical Fluids, 76*, 54–60. https://doi.org/10.1016/j.supflu.2013.01.010

Hachicha, A. A. (2017). Thermo-hydraulic modelling for direct steam generation. In *Energy procedia* (Vol. 143, pp. 705–712). Elsevier Ltd. https://doi.org/10.1016/j.egypro.2017.12.750

Hafez, A. Z., Soliman, A., El-Metwally, K. A., & Ismail, I. M. (2016). Solar parabolic dish Stirling engine system design, simulation, and thermal analysis. *Energy Conversion and Management, 126*, 60–75. https://doi.org/10.1016/j.enconman.2016.07.067

Hebel, G., & Weiskopf, H. (1982). High-temperature Helium test facility - experience in construction and commissioning. *Atomwirtschaft-Atomtechnik, 27*(5), 258–262.

Ho, C. K. (2017). Advances in central receivers for concentrating solar applications. *Solar Energy, 152*, 38–56. https://doi.org/10.1016/j.solener.2017.03.048

Hoffmann, J. R., & Feher, E. G. (1971). 150 kwe supercritical closed cycle system. *Journal of Engineering for Gas Turbines and Power, 93*(1), 70–80. https://doi.org/10.1115/1.3445409

Ho, C. K., & Iverson, B. D. (2014). Review of high-temperature central receiver designs for concentrating solar power. *Renewable and Sustainable Energy Reviews, 29*, 835–846. https://doi.org/10.1016/j.rser.2013.08.099

International Organization for Standardization. (2015). *ISO/TS 8004 – 1: 2015: Nanotechnologies – Vocabulary - Part 1: Core terms.* Geneva: International Organization (ISO) for Standardization.

IRENA. (2013). *Southern African power pool: Planning and prospects for renewable energy.* International Renewable Energy Agency (IRENA).

IRENA. (2020). *Power generation costs in 2019.* International Renewable Energy Agency (IRENA).

Islam, M. T., Huda, N., Abdullah, A. B., & Saidur, R. (2018). A comprehensive review of state-of-the-art concentrating solar power (CSP) technologies: Current status and research trends. *Renewable and Sustainable Energy Reviews, 91*, 987–1018. https://doi.org/10.1016/j.rser.2018.04.097

Jiang, K., Du, X., Kong, Y., Xu, C., & Ju, X. (2019). A comprehensive review on solid particle receivers of concentrated solar power. *Renewable and Sustainable Energy Reviews, 116*, 109463. https://doi.org/10.1016/j.rser.2019.109463

Karatza, D., Konstantopoulos, C., Chianese, S., Diplas, S., Svec, P., Hristoforou, E., & Musmarra, D. (2021). Hydrogen production through water splitting at low temperature over Fe_3O_4 pellet: Effects of electric power, magnetic field, and temperature. *Fuel Processing Technology, 211*, 106606. https://doi.org/10.1016/j.fuproc.2020.106606

Kato, Y., Nitawaki, T., & Yoshizawa, Y. (2001). A carbon dioxide partial condensation direct cycle for advanced gas cooled fast and thermal reactors. In *Global 2001 international conference on: "back-end of the fuel cycle: From research to solutions", Paris (France), 9–13 September 2001, Paris*. https://www.osti.gov/etdeweb/biblio/20264259.

Korzynietz, R., Brioso, J. A., Del Río, A., Quero, M., Gallas, M., Uhlig, R., Ebert, M., Buck, R., & Teraji, D. (2016). Solugas - comprehensive analysis of the solar hybrid Brayton plant. *Solar Energy, 135*, 578–589. https://doi.org/10.1016/j.solener.2016.06.020

Kotzé, J. P., Backström, T. W., & Erens, P. J. (2014). NaK as a primary heat transfer fluid in thermal solar power installations. In *Proceedings of the SolarPACES 2012 international conference* (pp. 27–29).

Lang, C., & Lee, B. (2015). Heat transfer fluid life time analysis of diphenyl oxide/biphenyl grades for concentrated solar power plants. In *Energy procedia* (Vol. 69, pp. 672–680). Elsevier Ltd. https://doi.org/10.1016/j.egypro.2015.03.077

Larsson, S., Fantazzini, D., Davidsson, S., Kullander, S., & Höök, M. (2014). Reviewing electricity production cost assessments. *Renewable and Sustainable Energy Reviews, 30*, 170–183. https://doi.org/10.1016/j.rser.2013.09.028

Lenert, A., & Wang, E. N. (2012). Optimization of nanofluid volumetric receivers for solar thermal energy conversion. *Solar Energy, 86*(1), 253–265. https://doi.org/10.1016/j.solener.2011.09.029

Liu, J., He, Y., & Lei, X. (2019). Heat-transfer characteristics of liquid sodium in a solar receiver tube with a nonuniform heat flux. *Energies, 12*(8), 1432. https://doi.org/10.3390/en12081432

Mahmoudi, A., Fazli, M., Morad, M. R., & Gholamalizadeh, E. (2020). Thermo-hydraulic performance enhancement of nanofluid-based linear solar receiver tubes with forward perforated ring steps and triangular cross section; a numerical investigation. *Applied Thermal Engineering, 169*. https://doi.org/10.1016/j.applthermaleng.2020.114909

McDonald C.F., Orlando R.J., & Cotzas G.M. (1994). Helium turbomachine design for GT-MHR power plant. International Joint Power Generation Conference, October 8–13, 1994, Phoenix, Arizona.

Musmar, S. A., & Tlili, I. (2015). Numerical investigation of working fluid effect on stirling engine performance. *International Journal of Thermal and Environmental Engineering, 10*(1), 31–36.

Nepveu, F., Ferriere, A., & Bataille, F. (2009). Thermal model of a dish/Stirling systems. *Solar Energy, 83*(1), 81–89. https://doi.org/10.1016/j.solener.2008.07.008

NREL (2021). Concentrating Solar Power Projects. National Renewable Energy Laboratory (NREL). https://solarpaces.nrel.gov

Olumayegun, O., Wang, M., & Kelsall, G. (2016). Closed-cycle gas turbine for power generation: A state-of-the-art review. *Fuel, 180*, 694–717. https://doi.org/10.1016/j.fuel.2016.04.074

Pacio, J., Singer, C., Wetzel, T., & Uhlig, R. (2013). Thermodynamic evaluation of liquid metals as heat transfer fluids in concentrated solar power plants. *Applied Thermal Engineering, 60*(1–2), 295–302. https://doi.org/10.1016/j.applthermaleng.2013.07.010

Pasch, J., Conboy Fleming, D., & Rochau, G. (2012). *Supercritical CO_2 recompression Brayton cycle: Completed assembly description.* Sandia National Laboratories (Sandia Report).

Poullikkas, A. (2005). An overview of current and future sustainable gas turbine technologies. *Renewable and Sustainable Energy Reviews, 9*(5), 409–443. https://doi.org/10.1016/j.rser.2004.05.009

Purgert, R., Phillips, J., Hendrix, H., Shingledecker, J., & Tanzosh, J. (2015). *Materials for advanced ultrasupercritical steam turbines.* Final Technical Report. U.S.A. Department of Energy. https://doi.org/10.2172/1243058

Rajeshwar, K., McConnell, R., & Licht, S. (2008). *Solar hydrogen generation. Toward A renewable energy future.* Springer-Verlag.

Reyes-Belmonte, M. A., Sebastián, A., Spelling, J., Romero, M., & González-Aguilar, J. (2019). Annual performance of subcritical Rankine cycle coupled to an innovative particle receiver solar power plant. *Renewable Energy, 130*, 786–795. https://doi.org/10.1016/j.renene.2018.06.109

Said, Z., Sabiha, M. A., Saidur, R., Hepbasli, A., Rahim, N. A., Mekhilef, S., & Ward, T. A. (2015). Performance enhancement of a flat plate solar collector using titanium dioxide nanofluid and polyethylene glycol dispersant. *Journal of Cleaner Production, 92*, 343–353. https://doi.org/10.1016/j.jclepro.2015.01.007

Said, Z., Saidur, R., & Rahim, N. A. (2016). Energy and exergy analysis of a flat plate solar collector using different sizes of aluminium oxide based nanofluid. *Journal of Cleaner Production, 133*, 518–530. https://doi.org/10.1016/j.jclepro.2016.05.178

Sarkar, J. (2015). Review and future trends of supercritical CO_2 Rankine cycle for low-grade heat conversion. *Renewable and Sustainable Energy Reviews, 48*, 434–451. https://doi.org/10.1016/j.rser.2015.04.039

Sobolev, V. (2007). Thermophysical properties of lead and lead-bismuth eutectic. *Journal of Nuclear Materials, 362*(2–3), 235–247. https://doi.org/10.1016/j.jnucmat.2007.01.144

Sohal, M. S., Ebner, M. A., Sabharwall, P., & Sharpe, P. (2013). *Engineering database of liquid salt thermophysical and thermochemical properties.* Idaho National Laboratory.

SOLGATE. (2005). *Solar hybrid gas turbine electric power system: Final publishable report.* European Communities: European Communities.

SOLHYCO. (2011). *Solar-hybrid power and cogeneration plants: Final public report.* German Aerospace Centre (Deutsches Zentrum für Luft- und Raumfahrt, DLR).

Srinivasan, K. (2011). Identification of optimum inter-stage pressure for two-stage transcritical carbon dioxide refrigeration cycles. *Journal of Supercritical Fluids, 58*, 26–30. https://doi.org/10.1016/j.supflu.2011.04.015

Stein, W. H., & Buck, R. (2017). Advanced power cycles for concentrated solar power. *Solar Energy, 152*, 91–105. https://doi.org/10.1016/j.solener.2017.04.054

Tang, O., Rehme, J., Cerin, P., & Huisingh, D. (2021). Hydrogen production in the Swedish power sector: Considering operational volatilities and long-term uncertainties. *Energy Policy, 148*, 111990. https://doi.org/10.1016/j.enpol.2020.111990

Teng, L., & Xuan, Y. (2019). Design of a composite receiver for solar-driven supercritical CO_2 Brayton cycle. *Journal of CO_2 Utilization, 32*, 290–298. https://doi.org/10.1016/j.jcou.2019.05.006

UN. (2011). *Globally harmonized system of classification and labelling of chemicals (GHS)* (4th ed.) United Nations (UN).

Ursúa, A., Gandía, L. M., & Sanchis, P. (2012). Hydrogen production from water electrolysis: Current status and future trends. In *Proceedings of the IEEE* (Vol. 100, pp. 410–426). Institute of Electrical and Electronics Engineers Inc. https://doi.org/10.1109/JPROC.2011.2156750. Issue 2.

Wang, X., Xu, X., & Choi, S. U. S. (1999). Thermal conductivity of nanoparticle-fluid mixture. *Journal of Thermophysics and Heat Transfer, 13*(4), 474–480. https://doi.org/10.2514/2.6486

Wang, C., Yang, J., & Ding, Y. (2013). Phase transfer based synthesis and thermophysical properties of Au/Therminol VP-1 nanofluids. *Progress in Natural Science: Materials International, 23*(3), 338–342. https://doi.org/10.1016/j.pnsc.2013.04.006

Wetzel, T., Pacio, J., Marocco, L., Weisenburger, A., Heinzel, A., Hering, W., Schroer, C., Muller, G., Konys, J., Stieglitz, R., Fuchs, J., Knebel, J., Fazio, C., Daubner, M., & Fellmoser, F. (2014). Liquid metal technology for concentrated solar power systems: Contributions by the German research program. *AIMS Energy*, 89–98. https://doi.org/10.3934/energy.2014.1.89

WNA. (2021). *Generation IV nuclear reactors*. World Nuclear Association (WNA).

Yu, W., & Xie, H. (2012). A review on nanofluids: preparation, stability mechanisms, and applications. *Journal of Nanomaterials*, 1–17. https://doi.org/10.1155/2012/435873, 435873.

Zhang, H. L., Baeyens, J., Degrève, J., & Cacères, G. (2013). Concentrated solar power plants: Review and design methodology. *Renewable and Sustainable Energy Reviews, 22*, 466–481. https://doi.org/10.1016/j.rser.2013.01.032

Zhang, H., Benoit, H., Gauthier, D., Degrève, J., Baeyens, J., López, I. P., Hemati, M., & Flamant, G. (2016). Particle circulation loops in solar energy capture and storage: Gas-solid flow and heat transfer considerations. *Applied Energy, 161*, 206–224. https://doi.org/10.1016/j.apenergy.2015.10.005

Zhang, B., Zhang, S.-X., Yao, R., Wu, Y.-H., & Qiu, J.-S. (2021). Progress and prospects of hydrogen production: Opportunities and challenges. *Journal of Electronic Science and Technology, 19*(2), 100080. https://doi.org/10.1016/j.jnlest.2021.100080

Chapter 6

Concepts of thermal energy storage and solar receivers

6.1 Introduction

In a concentrating solar power (CSP) plant, solar radiation is used to directly or indirectly heat a working fluid that drives a steam or gas turbine. In this vein, turbine inlet temperature (TIT) is an important parameter in the performance of the plant. It is desirable that this temperature should be steady for good performance of the CSP system. Some systems (such as the micro gas turbine) may even be damaged in case of dramatic changes in the TIT (Bashir et al., 2019). A solar receiver provides heat to the working fluid of a thermodynamic cycle so that the fluid attains the specified TIT. However, most of the receivers have low thermal inertia. Unfortunately, solar radiation is intermittent, which implies that fluctuations in solar radiation can drastically alter the outlet temperature of the working fluid, thereby causing instability in the system. Solutions to this challenge are to (a): operate the system in hybrid mode (solar and conventional fuels such as fossil fuel and nuclear resources) and (b) store thermal energy for later use. On a global scale, most of the electricity is based on fossil fuels. Nevertheless, exploitation of fossil fuels is associated with environmental pollution, especially climate change. Moreover, these fuels occur in limited reserves and therefore not sustainable. Thus, storage of solar thermal energy is a promising pathway for sustainable energy production. Different concepts of thermal storage have evolved over the time.

6.1.1 Classification of thermal energy storage concepts

Thermal energy storage (TES) is the temporary accumulation of heat for later use. In this process, a storage medium is charged with heat during the period of excess thermal energy, and interchangeably discharged when there is inadequate heat supply. The two charging and discharging processes constitute a thermal cycle.

TES concepts can be classified based on: (a) type of stored heat, (b) number of media involved in the heat transfer and storage processes, and (c) energy conversion. Based on the type of stored heat, three commonest

Solar Receivers for Thermal Power Generation. https://doi.org/10.1016/B978-0-323-85271-5.00006-9

types are: sensible heat, latent heat, and thermochemical heat storage (Dutta, 2017; Elmaazouzi et al., 2020), as shown in Fig. 6.1. The three types of thermal energy storage are presented and discussed in detail in Sections 6.2–6.4.

Storage systems can also be classified into direct and indirect types (Bauer et al., 2012). In the direct systems, a single fluid is used as a heat transfer fluid (HTF) and storage medium. Some fluids (such as water) are also used as working fluids. An extra volume of the hot working fluid is used to store heat in simple direct storage systems. The direct-storage system is generally more economical than the indirect-storage system, as long as the medium has a low partial pressure to avert costly pressure vessels, and the medium itself is cheap. Water possesses these characteristics, and so it is used in the direct thermal storage systems.

Nevertheless, many working fluids are not amenable to the process of direct thermal energy storage. The thermal energy must be transmitted from the HTF to another storage medium. In such systems, two distinct media are required to transfer and store heat. Hence, this concept is often referred to as the indirect storage concept of TES. The HTF and storage medium may be in direct contact or indirect contact, depending on the economic feasibility of the configuration. In the direct system, the HTF is in direct contact with the storage medium. For instance, hot air from a solar receiver can be passed through a bed of dry pebbles that are used as storage medium. Generally, direct contact helps to avoid the need for heat exchangers and to reduce the cost of the system. However, some cases of the direct contact between the heat carrier and storage material are not feasible or economically viable. For example, a pressurized gas such as steam is not compatible with a pebble storage medium. In view of this, indirect contact of HTF and storage medium is the only option. Often, indirect-contact systems require exchangers or tube registers (bundles) within the storage medium.

Based on the type of energy conversion process, TES can be classified into three main categories: (a) heat to heat, (b) chemical energy to heat, and (c) electricity to heat. Reversible heat-to-heat processes are taken into account in the heat-to-heat TES systems. Fossil and nuclear fuels constitute a special class of thermal storage media (chemical energy to heat), and they are characterized by a very high density of energy. Their application involves irreversible processes (for example, the combustion of fossil fuels is an irreversible process). The maximum ideal efficiency of a thermal system is attained when there are no irreversibilities. Consequently, the reversibility of processes in a TES system is necessary to enhance the efficiency of the system. The concept of irreversible storage processes and their materials falls beyond the scope of this book. The electricity-to-heat process is commonly exploited by night storage heaters.

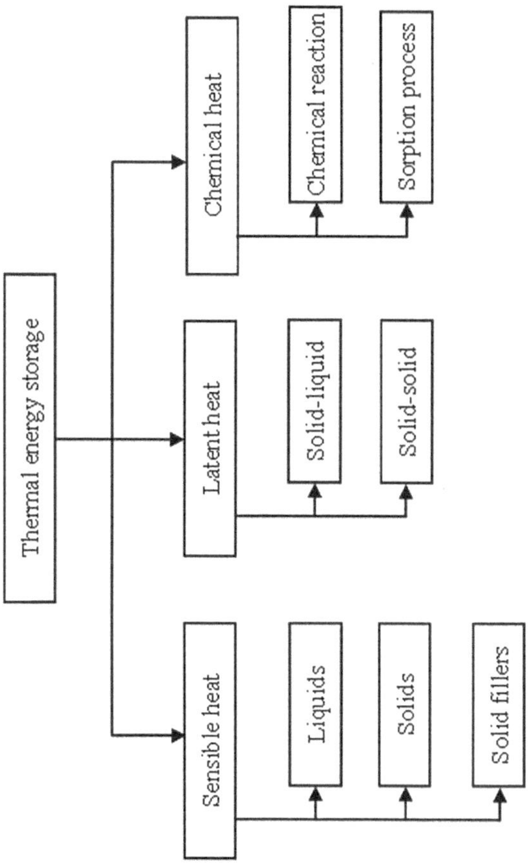

FIGURE 6.1 Classification of the concepts of thermal energy storage based on the type of stored heat.

6.1.2 Characteristics of thermal energy storage media and systems

6.1.2.1 Thermophysical properties

Every storage concept requires specific materials that may exist in different physical phases (solids, liquids, or gas). These materials need to have desirable thermophysical properties and other characteristics. Thermophysical properties are classified into transport and thermodynamic categories (Bergman et al., 2011). Included in the transport properties are diffusion rate coefficients such as the thermal conductivity (k) and the kinematic viscosity (ν).

Thermodynamic properties relate to the equilibrium state of a given system, and they include density (ρ) and specific heat capacity (C_p) that are widely applied in thermodynamic analysis. The ability of a material to store thermal energy is measured by the product ρC_p, which is commonly known as the volumetric heat capacity. Generally, substances with large density are characterized by low specific heats. Due to the large density, these substances have relatively high volumetric heat capacity. Many solids and liquids fall in this category, and they are good energy storage media with comparable heat capacities ($\rho C_p > 1\ \mathrm{MJ\ m^{-3}\ K^{-1}}$). On the other hand, gases have very small densities with relatively low volumetric heat capacity ($\rho C_p \approx 1\ \mathrm{kJ\ m^{-3} K^{-1}}$). Moreover, the gas phase entails large variations in volume or pressure when the temperature of the medium changes. Consequently, gaseous materials are weakly suitable for TES. The capacity of the storage system is influenced by the volumetric and gravimetric energy densities of the materials. A high volumetric heat capacity results in a compact sensible heat storage system. The ratio of the thermal conductivity to the heat capacity is known as the thermal diffusivity (see Section 5.1.3). It is a very important property that measures the ability of a substance to conduct heat relative to its ability to store heat. A high thermal diffusivity enhances transient heat transfer and therefore, reduces the time for charging and discharging processes. Effusivity (e) is another property that is considered in the selection of materials for utilization as storage media. It is defined as the square root of the product of the thermal conductivity and volumetric heat capacity of the material (Eq. 6.1). Metals and graphite are the most suitable materials for fast charging and discharging (high thermal diffusivity) and for storage of a large amount of heat in a given time (high thermal effusivity). Other solid materials with low levels of thermal diffusivity and effusivity, such as stones, are technically less advantageous.

$$e = \sqrt{k \rho c_p} \tag{6.1}$$

6.1.2.2 Economic and environmental characteristics

Materials may have merit from an economic and environmental perspectives. In this regard, economic and environmental characteristics of a material also influence its potential for utilization as a storage medium. The price of a storage medium and the costs associated with its utilization constitute a decisive criterion for material selection. Recently, there has been growing concern about environmental degradation arising from anthropogenic activities. In this context, materials that are friendly to the environment are desirable.

Two main parameters for environmental soundness are global warming potential and ozone depletion potential (ASHRAE, 2017). It has been observed that the production and consumption of some materials contribute to climate change through the emission of carbon dioxide and other greenhouse gases. Climate change is one of the global challenges of the 21st century. In addition, emission of chlorofluorocarbons into the atmosphere is contributing to the depletion of the ozone layer. Consequently, national and international efforts have been pooled together to address this challenge through various interventions, including the evolution of pertinent legislation to control pollution. In this connection, the utilization of various materials must be in compliance with the relevant environmental protection laws.

6.1.2.3 Safety and health hazard

It is also necessary to consider the aspect of safety and health hazards of materials that are used in storage systems. The toxicity and flammability characteristics of a material influence the hazard involved in its use. Toxicity is the degree to which a substance can cause harm to humans or animals. It is dependent on the quantity and concentration of the material used, frequency of use, interactions of the organism receiving the substance, and individual reaction of the organism. The effects of toxicity may be acute, subacute, subchronic, or chronic (Hidayat et al., 2022; Yao et al., 2021).

Flammability is the easiness with which a substance is ignited, the intensity with which it burns and releases heat once ignited, its potential to spread fire, and the rate at which it generates smoke and toxic combustion products during gasification and burning. Substances are classified according to the level of flammability. Some materials (such as oil) are flammable. In view of the foregoing discussion, the classification and labelling of the hazard involved in the exploitation of chemicals are based on the toxicity and flammability characteristics (ASHRAE, 2017; UN, 2011), and it is a requirement that TES materials must have low values of these parameters (see Section 5.1.1.4).

6.1.2.4 Summary of required characteristics of thermal energy storage materials

Heat storage materials should meet the following requirements: (a) small density variations during temperature changes to reduce thermomechanical stress, (b) large gravimetric energy storage capacity (high heat capacity, latent heat, or heat of reaction), (c) high levels of thermal diffusivity and effusivity, (d) large volumetric storage capacity (high density, high heat capacity, latent heat, or heat of reaction), (e) long span of service life, nontoxic, nonflammable, absence of explosive phases, and simple in handling, (f) noncorrosive to the containment, heat exchanger, and heat transfer enhancement structures (such as inserts and fins), (g) ability to maintain the levels of performance and storage capacity over many charging–discharging cycles, (h) high availability and affordability of materials, (i) low carbon footprint, and (j) low ozone depletion potential. It should be noted that not all required thermophysical properties are always available for a given material, and their values may vary among sources of data. Some of the thermophysical properties strongly depend on temperature, and the presence of impurities can also significantly alter these properties.

Usually, heat transfer in sensible and latent thermal storage systems takes place without chemical reactions in the storage medium. On the other hand, chemical TES systems use reversible thermochemical reactions for heat storage. Integrally, thermochemical energy storage systems are more complex than sensible and latent heat systems because thermochemical processes involve heat and mass transfer as well as the kinetics of the chemical reactions (Bauer et al., 2012). TES systems are diverse regarding temperature, power level, and HTFs, and, in addition, each application is characterized by its specific operation parameters. Consequently, it is necessary to have a good understanding of the wide range of storage designs, media, and methods.

Thermodynamic cycles for CSP systems operate in the medium- and high-temperature ranges (>373 K) with different requirements of the grades of heat. For example, the conventional Rankine cycle operates at lower temperature than advanced cycles (see Section 1.5). In this case, the TES system must supply heat of appropriate grade. Sensible heat (using solids) and thermochemical reaction storage mechanisms have a wide range of operational temperature (low to high) (Fig. 6.2). They can store heat of very high grade (up to 1273 K) and therefore, suitable for integration with conventional and advanced thermodynamic cycles. Energy density is another important parameter of TES systems. Generally, sensible heat storage systems have relatively low energy density. In contrast, thermochemical energy storage systems have high energy density. TES systems also utilize different HTFs, some of which (such as air, carbon dioxide, thermal oil) are utilized in a single phase while others are used in two phases with condensation and evaporation processes in the system. Desirable characteristics of HTFs have been presented and discussed in Chapter 5. Some fluids can be used both as TES media and HTFs. In this system, part of the HTF from the solar receiver is stored in a tank, while the remainder is used to drive a thermodynamic cycle directly or indirectly. At present, sensible heat mechanisms have attained the highest level of technological maturity.

6.1.3 Benefits of integrating concentrating solar power with thermal energy storage

Worldwide, most of the electricity is currently produced by conventional thermal power plants, predominantly using the Rankine and Brayton cycles. Power plants based on the Rankine cycle are often exploited to supply base loads, with high-capacity factors. A lot of experience has been gained over decades of operating these plants, and it is currently being used to develop the

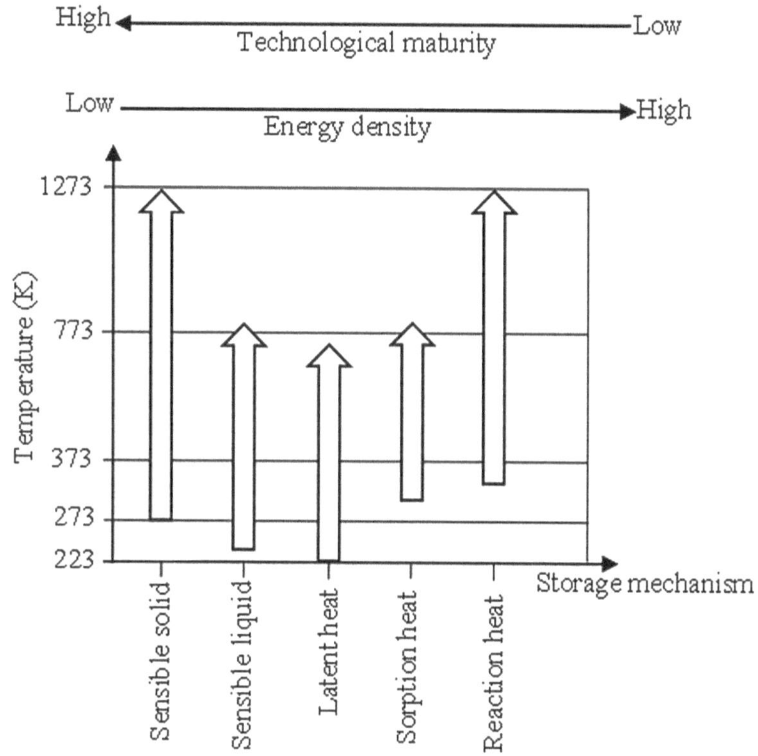

FIGURE 6.2 Variation of working temperature, energy density, and technological maturity of various thermal energy storage mechanisms. *Adapted from Bauer et al. (2012).*

CSP technology. The majority of the CSP plants in the world are based on the Rankine cycle at present. Thus, the CSP technology has many common features with conventional thermal power plants except for the source of heat. Nevertheless, the capacity factor of the CSP technology is lower than that of conventional power plants due to the intermittency of the solar resource. In addition, the cost of CSP-based electricity is currently higher than that of the largest carbon emitter (coal power) (see Section 5.1.2). Integration of CSP with TES enhances the capacity factor and reliability of the CSP technology. It is therefore imperative to exploit cost-effective thermal storage systems to achieve the declination of the generation costs.

TES gives the CSP technology an advantage over other energy systems that are limited to the deployment of batteries to store energy (Kuravi et al., 2013). The benefits of TES are (Bauer et al., 2012): (a) increase in the solar contribution and operation of the system during times of low insolation or at night, (b) compensation of abrupt flux variations, which prevents high gradients for connected components (such as pipes, heat exchangers, boilers, and turbines), thereby augmenting the life time of components, (c) utilization of excess energy, thereby avoiding the need for dumping, (d) possible reduction of the size of subsequent components such as evaporator, condenser, boiler, and turbines, (e) improvement in the thermal management of the solar system (such as increased start-up time and accurate preheating of solar steam cycle), and (f) coverage of peak demand (integration of CSP with TES provides the potential of full dispatchability of bulk power). The dispatchability of electricity allows power producers to sell electricity when tariffs are maximum, thereby improving the competitiveness of the CSP technology on the electricity market.

6.2 Sensible thermal energy storage concepts

In sensible heat storage, the gain or loss of heat causes a temperature change in the storage medium without phase change. Sensible heat storage results in a temperature rise of the storage medium during the charging process, with a reversed trend of the temperature when the storage medium is discharged. There are different materials for sensible heat storage, and they fall into two groups: liquids and solids.

6.2.1 Liquid thermal energy storage

A liquid is a fluid that is almost incompressible and takes the shape of its container but maintains a nearly fixed volume independent of its pressure. Very high pressures are required to change the volume of a liquid by a small margin. So, the bulk modulus of liquids (at constant temperature) is significantly higher than that of gases. Major characteristics of selected liquid media are presented in Table 6.1.

One main advantage of liquids is the possibility of using them as both storage medium and HTF. In this vein, water, thermal oil and molten salt are commonly used as thermal storage media. Liquid-based TES systems can utilize a single-tank or a two-tank concept. Thermal stratification is required in a single-tank system to retain high-temperature heat in one part of the tank, and low-temperature fluid that is fully mixed or stratified in another part of the tank. The concept of two tanks comprises two distinct tanks that are maintained at different temperatures and fluid levels. Solid filler materials can be utilized to curb natural convection within the liquid, and therefore enhance thermal stratification. Nevertheless, even in properly insulated vessels, stratification declines with time because of parasitic transient internal heat conduction. Apart from water, the liquid storage medium usually contributes a dominant share of the cost of the whole storage system. For expensive storage media, cheap solid materials can be used to reduce the proportion of costly liquid storage materials. Solid fillers are in direct contact with the liquid medium, and therefore, it is necessary to ensure the compatibility of the two media.

For a fully mixed liquid storage medium, the amount of stored thermal energy (Q_{st}) depends on the difference between the initial (T_i) and final (T_f) temperatures of the medium, and it can be computed from:

$$Q_{st} = m_{sm} \int_{T_i}^{T_f} C_{p,sm}(T)dT \tag{6.2}$$

TABLE 6.1 Major characteristics of sensible heat storage liquid materials.

Storage medium	T_{cold} (K)	T_{hot} (K)	ρ_m (kg m^{-3})	k_m (Wm^{-1} K^{-1})	$C_{p,m}$ (Jkg^{-1} K^{-1})	C_v (kWh m^{-3})	CPUM (US$ kg^{-1})	CPUE (US$ kWh^{-1})
Solar salt	493	–	1899	–	1495	–	1.3	–
Hitec	415	–	1860	–	1560	–	1.93	–
Hitec XL	393	406	1990	0.6	1495*	–	1.66	–
Mineral oil	473	573	770	0.12	2600	55	0.3	4.2
Synthetic oil	523	623	900	0.11	2300	57	3	42
Silicon oil	573	673	900	0.1	2100	52	5	80
Nitrite salts	523	723	1825	0.57	1500	152	1	12
Nitrate salts	538	838	1870	0.52	1600	250	0.5	3.7
Carbonate salts	723	1123	2100	2	1800	430	2.4	11
Liquid sodium	543	803	850	71	1300	80	2	21

$C_{p,m}$, average mass specific heat capacity; *CPUE*, cost per unit energy; *CPUM*, cost per unit mass; C_v, volume specific heat capacity; k_m, average thermal conductivity; *T*, temperature; ρ_m, average density.

Where $C_{p,sm}$ is the specific heat capacity of the storage material, and m_{sm} is the mass of storage medium. When $C_{p,sm}$ is independent of the temperature of the storage medium within the range T_i to T_f, then Eq. (6.2) reduces to:

$$Q_{st} = m_{sm} C_{sm} \left(T_f - T_i\right) \tag{6.3}$$

For an unstratified storage vessel, the energy balance of the system:

$$m_{sm} C_{p,sm} \frac{dT_{sm}}{dt} = Q_{ch} - Q_{hr} - UA(T_{sm} - T_a) \tag{6.4}$$

Where Q_{ch} is the rate of charging, Q_{hr} is the rate of heat removal, U is the coefficient of heat loss from the surface of the vessel, A is the area of the vessel, and T_a is the ambient temperature.

Example 6.1

A thermal power plant uses 50 kg of Therminol VP-1 as a storage medium in a fully mixed tank. Calculate the amount stored heat if the initial and final temperatures of the storage oil are 290 and 500 K, and $C_p = 2.82T + 716$.

Solution

From Eq. (6.2), the amount of stored heat is:

$$Q_{st} = m_{sm} \int_{T_i}^{T_f} C_{p,sm}(T) dT = 50 \int_{290}^{500} (2.82T + 716) dT = 50 \left[\frac{2.82T^2}{2} + 716T \right]_{290}^{500}$$

$$= 50\left(\frac{2.82 * 500^2}{2} + 716 * 500\right) - 50\left(\frac{2.82 * 290^2}{2} + 716 * 290\right) = 19.2 \text{ MJ}$$

6.2.1.1 *Low-temperature water systems*

Water is the most widely used liquid for low-temperature thermal storage. The advantages of this liquid are: (a) abundance and cost effectiveness, (b) relatively high thermal effusivity, (c) low thermal (temperature) diffusivity, which enhances thermal stratification within a hot-water storage tank, (d) possible utilization of water without heat exchangers, (e) flexibility of the water flow control systems that are usually state-of-the-art, (f) ease of storage in different kinds of containers, (g) ease of handling, and abundance of experience with water, (h) ease of mixing water with additives (such as antifreeze, anticorrosive), (i) environmental friendliness, and (j) nontoxic and nonflammable characteristics.

However, like other types of storage media, water has also some disadvantages that include: (a) limited range of operational temperature (273–373 K), (b) corrosiveness, and (c) high vapor pressures at $P > .5$ MPa and $T > 428$ K. This range of temperature is suitable for space heating and provision of domestic and industrial hot water. The CSP technology operates at higher temperatures, and so it requires storage of high-grade heat.

6.2.1.2 High-temperature water systems

For TES at high temperature, the steam accumulator technology (Ruths storage system) is commonly applied (Votava et al., 2019). A steam accumulator is a device that allows stocking of water steam to reduce the dependence of steam production on steam appliances under certain circumstances and vice versa. It uses a pressure vessel to store steam for instantaneous or later use. Pressure vessels are cylindrical or spherical vessels designed to store pressurized fluids with a pressure gradient between inside and outside. Nevertheless, spherical vessels are difficult to fabricate, and so the cylindrical design is commonest on the market. Cylindrical vessels may be vertical or horizontal. The pressure gradient varies from low to high, depending on application. High pressure applications require thicker shells of the vessels. Consequently, most of the costs of deploying steam accumulators are contributed by the pressure vessel to meet safety and other requirements.

The steam accumulator system has certain physical limitations that predominantly result from the properties of water and steam. This device stores energy and steam although its name is steam accumulator. The accumulator may be dry or wet. In the dry system, steam is introduced into a storage vessel without water. The advantages of the dry accumulator are more associated with boiler protection from low water shutdowns and carryover prevention, than with a substantial storage of steam for the load equipment. Saturated water and saturated steam with relatively low temperature and pressure are stored in a wet accumulator. Basically, steam from a high-temperature steam source (such as a solar receiver) is introduced into the saturated water (Fig. 6.3). The pressure of the high-temperature steam is higher than that inside the steam accumulator. Due to the pressure gradient, steam automatically flows into the steam accumulator when the steam inlet valve is opened. There is fast heat exchange that leads to the (a) cooling of the high-temperature steam and (b) heating of the saturated water and steam to a higher temperature to attain a new equilibrium state. The temperature of the water and steam inside the steam accumulator rises as the steam charging process continues, followed by an increase in pressure and water levels. Once the internal pressure attains the stipulated maximum value, the steam inlet valve closes, thereby ending the steam charging process. Steam discharging is the reverse process of steam charging. The valve at the steam outlet is opened during the discharge process. During this process, the saturated steam flows out of the accumulator into the pipe system because of the pressure gradient, thereby reducing the pressure inside the accumulator. Instantly, the water boils to evaporate as saturated steam when the internal pressure falls below the saturation pressure corresponding to the temperature of the stored water. As the water boils, its temperature and level in the accumulator drop down until they reach the specified minimum values at which the steam outlet valve closes, thereby ending the steam release process. Wet accumulators benefit from the high volumetric storage capacity of liquid water for sensible heat. It is possible to achieve high discharge rates because water can be utilized both as a working fluid and storage medium. The volume of the pressure vessel limits the storage

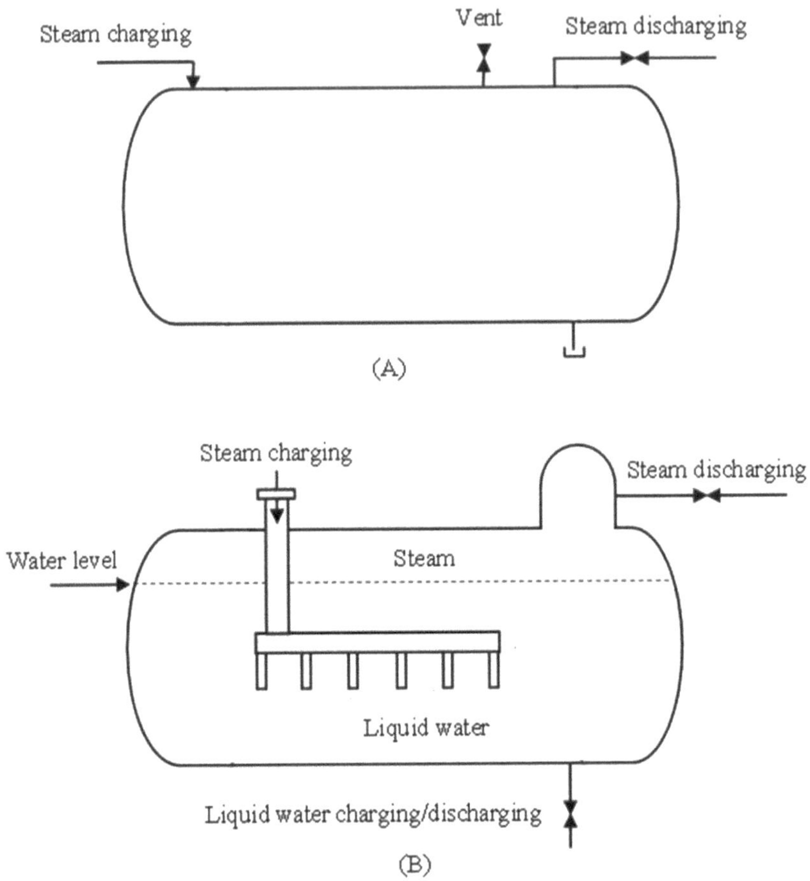

FIGURE 6.3 Schematic representation of: (A) dry steam accumulator and (B) wet steam accumulator.

capacity while the maximum temperature is restricted by the critical point of saturated water (T = 647.04 K and $P = 22.1$ MPa). Most of the costs are contributed by the pressure vessel, and the cost-effective capacity of steam accumulators is controlled by the size of the pressure vessel.

The capacity of storage depends on the change of sensible energy of the liquid volume during the discharging process. The water in the vessel is in the saturation state, and therefore variation in temperature corresponds to a change of pressure of the saturated steam that is provided during the discharge process. Although a larger reduction of pressure is advantageous pertaining to the volumetric storage capacity, the acceptable pressure variation may be restricted by considerations of the system efficiency. There is a logarithmic correlation between saturation pressure and saturation temperature. In this context, the saturation water vapor pressure can be estimated from Eq. (6.5) (Shallcross, 2004). Therefore, the absolute decrease of the pressure during the discharge process is increased at higher pressure levels to provide the same amount of energy. This storage system has short reaction time, and it can therefore be utilized as a buffer storage to compensate for cloud passing or to support other systems that have a larger storage capacity and require a longer procedure to start up.

$$\ln\left(\frac{P_v}{P_{cr}}\right)=\frac{T_{cr}}{T}\left(a_1\Theta+a_2\Theta^{3/2}+a_3\Theta^3+a_4\Theta^{7/2}+a_5\Theta^4+a_6\Theta^{15/2}\right) \quad (6.5)$$

Where P_v is the saturation vapor pressure in MPa, P_{cr} is the critical pressure in MPa, T is the temperature ($273.16\ K < T < 645\ K$), T_{cr} is the critical temperature, and the values of the coefficients are:

$a_1=-7.85951783$, $a_2=1.84408259$, $a_3=-11.7866497$, $a_4=22.6807411$, $a_5=-15.9618719$ and $a_6=1.80122502$.

$$\Theta=\left(1-\frac{T}{T_{cr}}\right)$$

The required mass of water in the accumulator is given by (Votava et al., 2019):

$$m_w=\frac{m_{sac}\left(i_{ste,in}-i_{ste,out}\right)}{\left(i_{w,in}-i_{w,out}\right)} \quad (6.6)$$

Where $i_{ste,in}$ is the enthalpy of the inlet saturated steam, $i_{ste,out}$ is the outlet enthalpy of saturated steam, $i_{w,in}$ is the inlet enthalpy of saturated water, $i_{w,out}$ is the outlet enthalpy of saturated water, m_{sac} is the mass of accumulated steam, and m_w is the required mass of water in the accumulator.

Example 6.2

A solar receiver directly generates saturated steam and 20 kgs of the steam is stored in a wet accumulator. The enthalpy of the inlet steam is 2500 kJ kg^{-1}, while the enthalpy of the outlet steam is 1500 kJ kg^{-1}. If the required mass of water is 25 kg, calculate the change in the enthalpy of the water.

Solution

From Eq. (6.6), the change in water enthalpy is:

$$i_{w,in}-i_{w,out}=\frac{m_{sac}\left(i_{ste,in}-i_{ste,out}\right)}{m_w}=\frac{20(2500-1500)}{25}=800\ kJ\ kg^{-1}$$

A few operational power plants use steam as a storage medium. In this case, the same fluid (water) can be utilized as a HTF, a storage medium, as well as a working fluid in steam Rankine-based power plants. The Planta Solar 10 (PS10) solar tower power plant (11 MW) utilizes a steam accumulator with a storage capacity of 1 h for compensation of cloud transients. This plant is located at Sevilla in Spain, and it represents the first European commercial solar thermal power plant that was commissioned in 2007. The Khi Solar One power plant (50 MW) is the largest operating commercial facility that utilizes a steam accumulator. It is located at Upington in South Africa and was commissioned in 2016, with a storage capacity of 2 h.

6.2.1.3 Thermal oil systems

Thermal oils include mineral and synthetic oils. Mineral oils are refined petroleum-based hydrocarbons produced through distillation. These oils comprise paraffins, napthenes, and aromatic oils (Rowe, 2009, pp. 113−144). A broad spectrum of mineral oil compositions is employed in cutting and grinding fluids. Additives are utilized to modify the lubrication properties for specific applications. Mineral oil is stable up to a maximum temperature of 613 K, but it is not currently attractive as a liquid storage medium for large-scale CSP applications. Some of its drawbacks are: (a) high capital costs for a large oil volume, (b) high environmental hazards of oil leakages that would require extra investments, and (c) limited maximum operational temperature for mineral oils restricts the thermal efficiency of the system. Synthetic oils can be obtained from hydrocarbons or from other chemical substances. The common commercial synthetic oils are Therminol VP-1 and Dowtherm A. These two oils are eutectic mixtures of diphenyl oxide and biphenyl. They are thermally stable up to 673 K, and therefore they are more attractive than mineral oils. This category of oils must be pressurized for use at higher pressure which is usually expensive.

Thermal oils are used as HTFs and storage media in CSP power plants. The hot oil is used to indirectly drive a thermodynamic cycle (Fig. 6.4). In this context, the first commercial Solar Electric Generating Station (SEGS-1) plant was built in California in 1984, and it used a mineral oil as a heat transfer medium in the solar receivers. The mineral oil was heated from 513 to 578 K and used to generate steam to operate a turbine at a maximum power of 13.8 MW. Two carbon steel tanks at ambient pressure each with a volume of 4700 m^3 were exploited to separate the cold oil from the hot oil. This CSP system had a thermal storage capacity of 120 MWh that allowed the operation of the turbine for about 3 h. The storage system of the SEGS-1 plant was destroyed by fire in 1999, and it was not renovated, and the plant is currently not operational.

6.2.1.4 Molten salt systems

A molten salt is a synthetic mixture of salts composed mainly of nitrates that can be used as a HTF and thermal storage medium. The major advantages of molten salt are high thermal stability, low vapor pressure, and low costs. Due to the low vapor pressure, it is possible to design storage systems without pressurized vessels. Moreover, a lot of experience has been gained in molten salts from various industrial applications pertaining to heat transfer, electrochemical reactions, and heat treatment. When contemplating to use molten salts, it is necessary to consider the melting point that defines the lower limit of the operational temperature. One of the main challenges with molten salts is

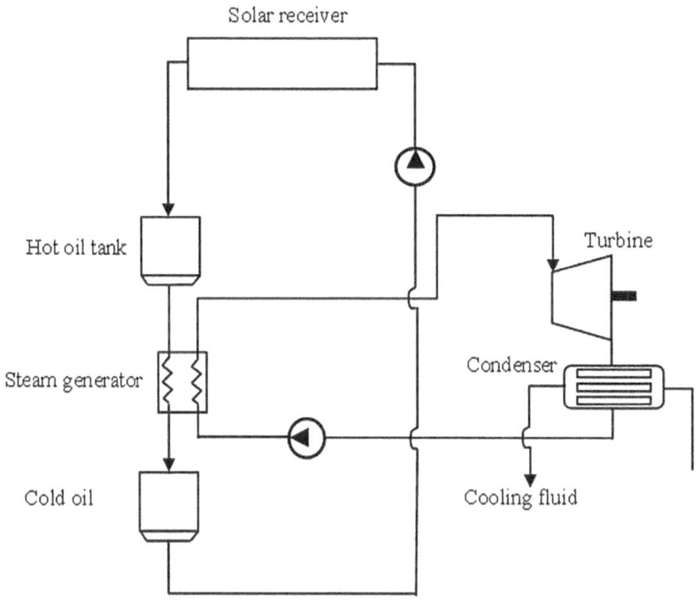

FIGURE 6.4 Simplified representation of a Rankine cycle-based concentrated solar power plant with direct storage by using thermal oil.

undesirable freezing during operation. So, it is critical to prevent it from occurring in the pipes, heat exchanger, and storage tanks. In view of this, auxiliary heating systems are usually required. Other drawbacks of molten salt-based storage systems arise from aspects such as high costs of the storage media, risk of corrosion, and challenge of handling hygroscopic salts. For an indirect two-tank storage system, the storage medium contributes the highest share of costs as shown in Fig. 6.5. The upper limit of operational temperature is defined by the thermal stability of the salts. Mixtures of salts have lower melting points than single salts. These mixtures can have similar limits of thermal stability as the individual salts of the mixtures. Consequently, salt mixtures can give a wider range of operational temperature than single salts. In CSP technology, alkali nitrate salt mixtures and alkali nitrite salts are the preferred candidate fluids for TES (see Section 5.2.3).

At present, molten salt is the commonest thermal storage liquid in operational CSP plants, and it can be exploited at higher temperatures (as high as 1123 K) than water or thermal oils. The salt is exploited in direct or indirect storage systems and indirectly heats the working fluid of a thermodynamic cycle such as the steam Rankine cycle (Fig. 6.6).

The first commercial CSP plant to use molten salt (solar salt) both as a HTF and a storage medium was the Gemasolar plant (19.9 MW). This power plant is located near Seville in Spain, and it started operating in 2011. It has a

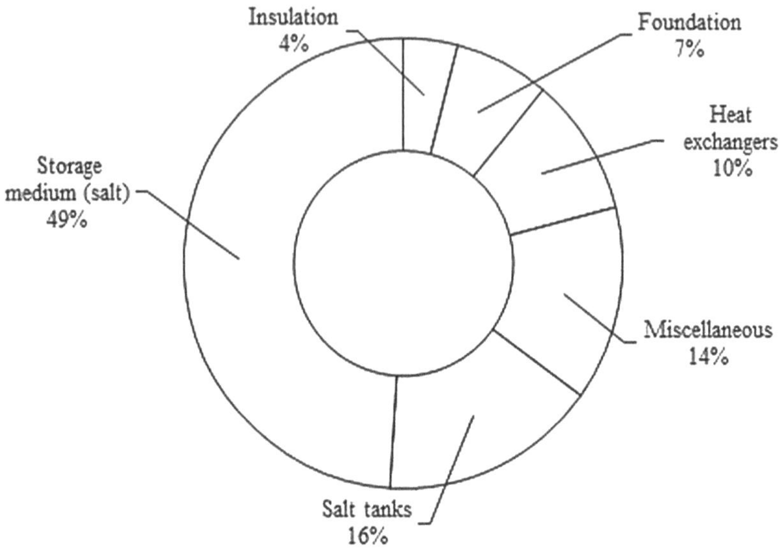

FIGURE 6.5 Distribution of investment costs for an indirect two-tank storage system. *Drawn by author using data from Kelly and Kearney (2006).*

twin-tank molten salt storage system, with a capacity of 15 h for full-load operation during hours with no sunshine. Due to the large storage capacity, a capacity factor of 75% is achieved, with daily operation for 24 h during the summertime. According to the data published by the National Renewable Energy Laboratories (NREL, 2021), about 82% of operational power plants with thermal storage exploited molten salt as the storage medium in 2021. Fortunately, this medium can also be used as a HTF in solar receivers. Materials that can be used as both heat transfer and thermal storage media are more cost effective. Heat is added to and extracted from this kind of storage system through the transportation of the storage medium. Consequently, this arrangement avoids the drop of temperature between the HTF and storage medium.

6.2.1.5 Liquid sodium

Liquid sodium (Na) has a very wide range of operational temperature (371−1155 K) and chemically stable up to near 1173 K with high thermal conductivity (Liu et al., 2019; Pacio et al., 2013). It is also possible to alloy this metal with potassium (K), thereby producing a eutectic mixture with melting and boiling points of 260 and 1058 K at normal pressure, respectively (Kotzé et al., 2014). This liquid metal can be used as a HTF and a TES medium, and it exhibits a significantly higher coefficient of heat transfer than molten salts. Consequently, it has great potential to replace molten salts in the next-generation CSP technology. Nevertheless, it is difficult to collect, transport, and store liquid metal at extreme temperatures exceeding 1300 K.

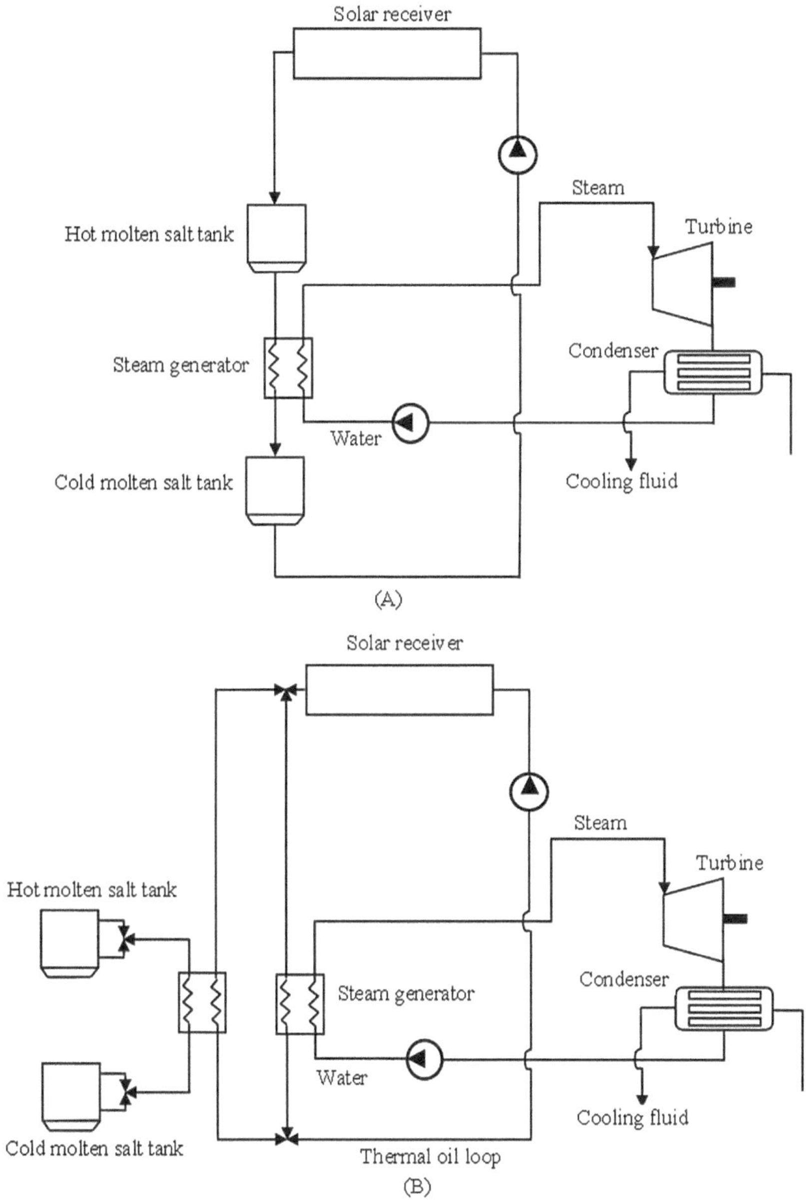

FIGURE 6.6 Simplified representation of a Rankine cycle-based concentrated solar power plant with: (A) direct and (B) indirect storage by using molten salt.

Due to some desirable characteristics, liquid sodium has found application in the CSP technology. The Jemalong Solar Thermal Station (1.1 MW) is currently the only operational power that was commissioned in 2017 as a pilot project at Jemalong in Australia. In this project, liquid sodium is used as a primary HTF in a steam Rankine cycle–based power plant driven by a solar tower (see Section 5.3.2.1). The same fluid is used as a storage medium in a two-tank direct type of thermal storage with a capacity of 3 h.

6.2.2 Sensible heat storage in solids

6.2.2.1 Materials

A solid is a phase of matter that has a defined shape and volume. Particles (molecules or atoms) in a solid material are closely packed. Consequently, solid materials have generally higher density than their corresponding liquid and gaseous forms. High density is one of the desirable characteristics of TES materials. In addition, solid materials have a wide range of operational temperature and very high upper limit of operational temperature, and they are usually chemically inert with low vapor pressure. For instance, magnesia bricks in Cowper regenerators can be heated up to 1273 K (Bauer et al., 2012). Moreover, it is usually possible to use a simpler containment of solid media than that of liquids. Solid storage materials fall into broad categories of metals and nonmetals. A selected list of them is presented in Table 6.2 (Tamme et al., 2009).

Solid materials may occur naturally or be produced. Rocks and pebbles are abundant and inexpensive natural materials. Rock and soil can be utilized as underground storage media for low-temperature applications. The thermo-mechanical stability of solid materials is a critical parameter for high-temperature applications. In this context, granite, basalt, quartzite, and pebbles possess suitable properties for thermal storage. Manufactured solid materials include various ceramics that are commonly used as heat storage media. Bricks function as a buffer for the climatization of buildings in the low-temperature range. At elevated temperatures, refractory bricks based on oxides (such as silica, alumina, magnesia, and iron oxide), carbonates (such as magnesite), and their mixtures are commercially available for use as TES media. Plentiful solid salts (such as sodium chloride and sodium sulfate) have also potential for utilization as storage media.

Graphite is another very promising solid material with great potential for use as a sensible heat storage medium. It is a mineral made up of carbon with a layered structure of rings of six carbon atoms arranged in widely spaced one-atom-thick sheets. This material has a high thermal diffusivity, and it can be heated up to 673 K (Bauer et al., 2012). Nevertheless, above 673 K, an inert atmosphere is required to prevent oxidation. Generally, metals have high thermal conductivity, but they are not economically competitive with

TABLE 6.2 List of selected solid thermal energy storage materials.

Material	T (K)	ρ (kg m^{-3})	C_p (kJ $kg^{-1}K^{-1}$)	k (W $m^{-1}K^{-1}$)	ξ ($10^{-6}m^2s^{-1}$)	e ($Jm^{-2}K^{-1}s^{-1/2}$)
Aluminum (99.99%)	293	2700	0.945	238.4	93.3	24.66
Copper (commercial)	293	8300	0.419	372	107	35.97
Iron	293	7850	0.465	59.3	16.3	14.7
Lead	293	11,340	0.131	35.25	23.6	7.24
Brick (dry)	293	1800	0.84	0.5	0.33	0.87
Concrete (aggregates)	293	2200	0.72	1.45	0.94	1.52
Granite	293	2750	0.89	2.9	1.18	2.67
Graphite	293	2200	0.61	155	120	14.41
Limestone	293	2500	0.74	2.2	1.19	2.02
Sandstone	293	2200	0.71	1.8	1.15	1.68
Slag	293	2700	0.84	0.57	0.25	1.13
Sodium chloride	293	2165	0.86	6.5	3.5	3.5
Soil (clay)	293	1450	0.88	1.28	1	1.28
Soil (gravelly)	293	2040	1.84	0.59	0.16	1.49

C_p, specific heat capacity; *e*, effusivity; *k*, thermal conductivity; *T*, temperature; ξ, thermal diffusivity; ρ, density.

nonmetals. The high thermal conductivity promotes quick charge and discharge processes of the storage system. Metals with high conductivity, such as aluminum and copper, can be exploited in electronic devices as a heat sink and for thermal management. An alternative metallic material is cast iron, which has high thermal conductivity and volumetric heat capacity. It is utilized in high-temperature applications of thermal storage together with oil as a HTF.

6.2.2.2 Heat transfer concepts

Often, a fluid is added to a solid material during heat charging and discharging processes. The heat carrier fluid may be operated either in direct or indirect contact with the solid storage medium. There is no partitioning wall between the HTF and the storage medium in the direct contact system of thermal storage. The direct contact design is achieved by using a packed bed of solid particles (such as a pebble bed or a rock pile) in a container or in a more defined layout (such as a regular array of checker bricks). Heat must be carried to or from such stores by a HTF. Beds and checkers offer a large surface area for heat transmittance. The effective cross-section of flow can be large, thereby decreasing pressure losses. Internal heat losses can be minimal due to, for instance, small contact surface of the particles with surrounding particles. Low insulation around the storage container is enough when air (low thermal conductivity) is used as a HTF and fills the gaps between the particles. Nevertheless, particle beds of this design produce large pressure drops in the fluid flow that must be compensated by pumps, fans, or compressors. Another disadvantage is the requirement for large storage volume. Since the solid storage materials are in direct contact with the HTF, it is necessary to take into consideration the material properties (including container and particle sizes, mechanical durability against abrasion, and thermal cycling and hygroscopy) that affect fluid flow and heat transfer. Thermal stratification is properly conserved in a heat store because free convection in the HTF is curbed, and internal heat conduction is diminished. For low-temperature applications, the mixture of air and hollow bricks is called a hypocaust-heating system while pebble beds flooded with water are referred to as artificial aquifers (Bauer et al., 2012).

6.2.2.3 Packed bed storage system

A packed-bed system of TES system (also known as a pebble bed or rock pile) utilizes the heat capacity of a bed of lightly packed solid particles to store heat. Heat is added to or removed from the bed by passing a fluid (such as air) through the bed. Essential components of a packed bed include a container, a screen to support the bed, support for the screen, and inlet and outlet ducts. During the heat charging process, hot air enters the packed bed unit through the hot air opening and exits through the cold air opening until the bed is fully charged (Fig. 6.7). The direction of flow is reversed during the heat discharging process when cold air is passed through the hot bed to extract the stored heat. Thus, the hot and cold fluids flow through the same passage and exchange heat alternately. A heat exchanger in which thermal energy is exchanged between hot and cold fluids through storage and release of the thermal energy is called a regenerator (Kakaç & Liu, 1998). Properly designed packed beds of rocks possess several desirable characteristics for solar energy applications. These good characteristics are (Duffie and Beckman, 2013): (a) high heat transfer coefficient between the air and rock particles, which boosts thermal stratification (the high degree of stratification constitutes is the main advantage of packed-bed storage systems), (b) low costs of the storage material and container, (c) low conductivity of the bed when there is no airflow, and (d) possibility of low pressure drop through the bed.

Due to the high coefficient of heat transfer, the product of the heat transfer coefficient and the area between the air and pebbles implies that heat is quickly transferred from the hot air entering the bed to the pebbles. Pebbles near the hot air opening are heated while those near the exit remain at the same temperature, with the exit air temperature remaining close to the initial temperature of the packed bed (Fig. 6.8). As the time of charging increases, a temperature front advances through the bed such that it reaches the end of the bed and the exit air temperature begins to rise. The temperature of the bed is uniform when it is fully charged. To discharge the bed, the flow direction is reversed such that cold air enters through the cold air opening and exits through the hot air opening (Fig. 6.9). This flow pattern results in the declination of pebble temperature at the entrance. The temperature front progresses toward the hot air opening until the bed is fully discharged.

Many researchers have investigated the heating and cooling processes of packed beds. Schumann (1929) conducted the first theoretical study on packed beds. In this study, it was assumed that: (a) plug flow was one-dimensional,

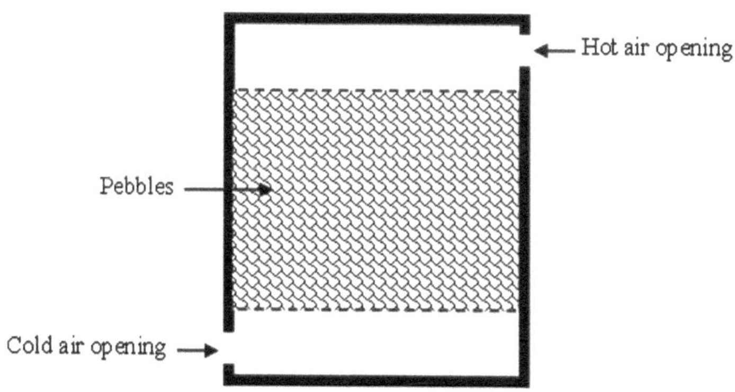

FIGURE 6.7 Schematic representation of a packed bed.

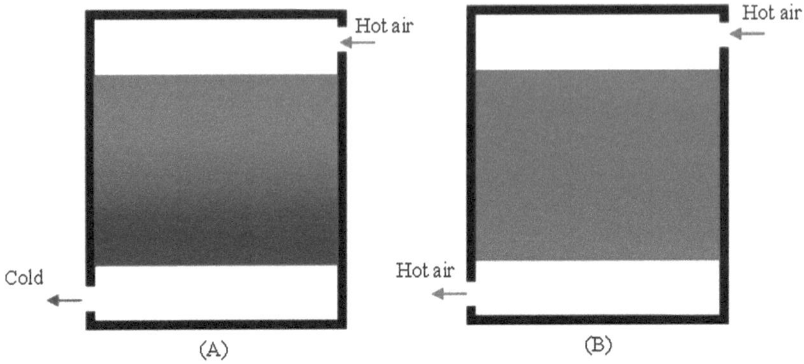

FIGURE 6.8 Schematic representation of a packed bed during charging: (A) partially charged and (B) fully charged.

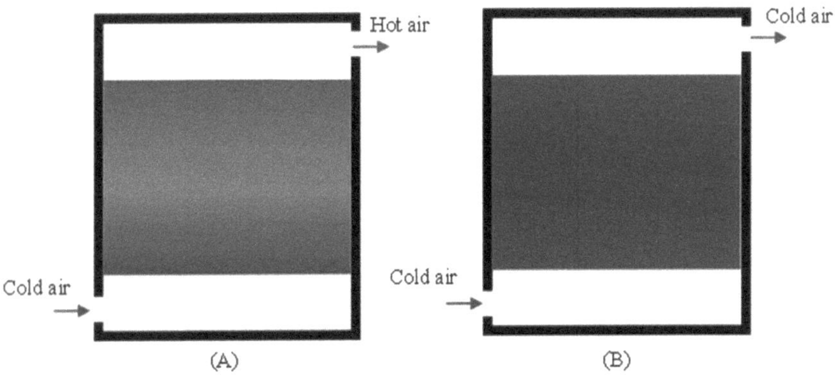

FIGURE 6.9 Schematic representation of a packed bed during the discharging process: (A) partially discharged and (B) fully discharged.

(b) there was no axial conduction or dispersion, (c) thermophysical properties remained constant, (d) there was no mass transfer, (e) there was no heat loss to the environment, and (f) temperature gradients were absent from the solid particles. With these assumptions, the Schumann model is expressed as follows:

$$\rho_f C_{p,f}\, \varepsilon \frac{\partial T_f}{\partial t} = -\frac{\dot{m}_f}{A_{be}} \frac{C_{p,f} \partial T_f}{\partial x} + h_v \left(T_{be} - T_f\right) \tag{6.7}$$

$$\rho_{be} C_{p,be} (1 - \varepsilon) \frac{\partial T_{be}}{\partial t} = h_v \left(T_f - T_{be}\right) \tag{6.8}$$

Where A_{be} is the bed cross-sectional area, $C_{p,f}$ is the specific heat capacity of the fluid, h_v is the volumetric heat transfer coefficient between the bed and the fluid (product of the area heat transfer coefficient and the bed particulate surface area per unit bed volume), $\dot{m}_f$ is the mass flow rate of the fluid, t is the time taken, T_{be} is the temperature of the bed, T_f is the temperature of the fluid, x is distance in the x-direction, ρ_f is the density of the fluid, ρ_{be} is the density of the pebbles in the bed, and $\in$ is the bed void fraction (porosity). When air is used as a HTF, the first term on the left-hand side of Eq. (6.7) can be ignored.

The volumetric heat transfer coefficient can be computed from (Löf & Hawley, 1948):

$$h_v = 650\left(\frac{J}{D}\right)^{0.7} \tag{6.9}$$

Where J is the mass velocity (product of fluid density and linear velocity), and D is the average diameter of the solid particles.

Air is used as a HTF in some solar tower receivers, and it offers great potential for cost-effective solutions with high conversion efficiencies. Moreover, it is a suitable HTF in packed-bed storage systems. These systems have a simple setup, are applicable to very high temperatures, and have the most attractive prospects for a deployment in large installations. Such characteristics show good opportunities for possible commercialization soon. Nevertheless, for this storage technology to be economically competitive on the market, it is necessary to develop efficient and up-scalable systems for the heat storage. Regenerative storage systems are well suited to the thermal storage needs, but they need further improvements pertaining to design and materials. Recently, Trevisan et al. (2022) experimentally investigated an innovative high-temperature packed bed TES system with radial flow. The system was of laboratory scale with an energy capacity of 49.7 kWh_{th} and working temperatures of 298-973 K. They found that reduced pressure drops were the key advantage of radial-flow packed-bed designs. Nevertheless, the degradation of thermocline degradation is the major drawback of this novel TES design.

Some initiatives have been made to integrate CSP technology with packed-bed storage systems. The German TSA consortium investigated initial designs and tests for air receivers. In this project, a test facility, including a thermal storage, was constructed in Spain in 1991. The TES system was based on a packed bed of aluminum oxide with spherical shape, and a capacity of about 1 MWh. Nevertheless, this concept is not amenable to a direct scale-up to commercial level due to: (a) high costs of the aluminum oxide of high purity and (b) increased thermal mechanical stresses during cyclic operation. A demonstration CSP plant (1.5 MW) with packed-bed TES was constructed at Jülich in Germany. This plant is driven by a solar tower and it has a storage capacity of 1.5 h. Air at atmospheric pressure is used as a heat transfer medium, and it is heated by a volumetric receiver to 953 K. The hot air indirectly

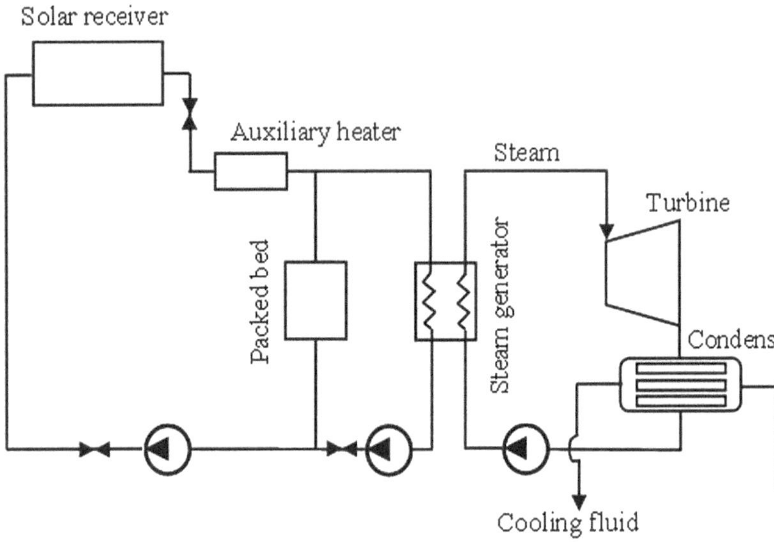

FIGURE 6.10 Schematic representation of a power plant layout with packed-bed heat storage unit and auxiliary heater.

heats a steam generator that produces steam at 10 MPa and 773 K and drives a turbine-generator set (Bauer et al., 2012). The heat storage unit is in parallel connection with the steam generator (Fig. 6.10). The plant was commissioned in 2008 and it is operational.

6.2.2.4 High temperature indirect contact concrete storage systems

It is possible to store sensible heat in solid media through indirect means of contact. In this case, a heat exchanger is embedded into the storage material. A hot fluid is passed through the heat exchanger to charge the storage medium with heat, while a cold fluid is used during the discharging process. The hot and cold fluids flow through the same passage alternately. Unfortunately, the inclusion of a heat exchanger tends to raise the costs of the system because tubes and manifolds contribute a significant share of the total costs (Bauer et al., 2012). Consequently, the geometry of the heat exchanger must be optimized to achieve a cost-effective TES system. The thermal conductivity and volumetric storage capacity of the storage medium are determinants of the surface area of the heat exchanger. This area decreases with increasing the thermal conductivity of the material, but it increases with the volumetric capacity of the storage unit. Other factors are the thermal expansion rate of the

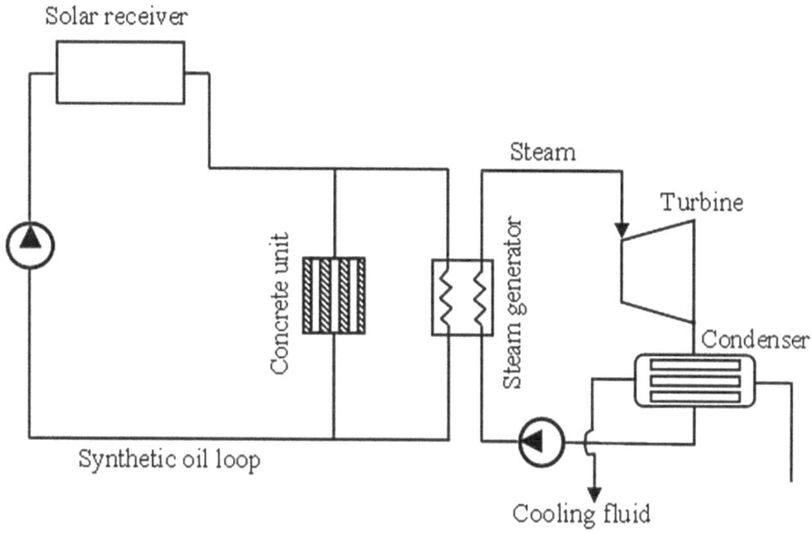

FIGURE 6.11 Schematic representation of a simplified power plant layout with a concrete heat storage unit.

storage material, which should be comparable with that of the heat exchanger material to minimize thermal stresses. In addition, the storage material should have a high degree of thermal cycling stability.

Concrete is an attractive material for sensible heat storage at high temperatures because of the following characteristics: (a) its low cost arising from cheap aggregates, (b) its availability worldwide, and (c) availability of a simple processing technique of casting. The density of concrete is weakly dependent on temperature due to the specific heat capacity, thermal conductivity, and coefficient of thermal expansion.

A concrete storage system has been developed for parabolic trough concentrator power plants with synthetic oil as a HTF (Fig. 6.11). The operational temperature range of the storage unit is 563–663 K, and the unit is charged and discharged daily (Bauer et al., 2012). The most economic design for this storage concept is a parallel steel tube heat exchanger embedded in a concrete structure made of quartz aggregates and a low level of binder made of blast furnace cement and flue ash.

6.3 Latent thermal energy storage

Latent heat is the energy that is added or removed from a substance to change its phase at almost constant temperature. Thus, in this case, the heat-charging and -discharging processes are almost isothermal. So, a phase change material (PCM) can store thermal energy through the latent heat of phase transitions. Generally, phase transitions can be classified into solid–solid, solid–liquid, solid–gas, and liquid–gas changes (Yang et al., 2022). The rearrangement of crystalline forms without a physical phase change may also be regarded as storage of latent heat. The heat of solid–solid phase changes (change of crystalline form) is lower than that of the heat of melting and solidification. The liquid–gas phase change is not used in latent heat storage because of the large volume of the gas phase, in spite of the high enthalpies associated with this change. Consequently, the solid–liquid phase change is often used in practical thermal storage systems. The amount of stored heat in this phase change is equal to the latent heat (Q_{latent}) of melting:

$$Q_{latent} = \Delta i_{melting} \tag{6.10}$$

Where $\Delta i_{melting}$ is the change in enthalpy of the medium at constant pressure.

If a PCM is heated from T_1 to T_2 and it undergoes a phase transition at $T = T^*$ (where $T_1 < T^* < T_2$), then the storage capacity of the system is the sum of the sensible heat of the solid (from $T = T_1$ to $T = T^*$), latent heat (at $T = T^*$), and sensible heat (from $T = T^*$ to $T = T_2$):

$$Q_{st} = m_{so} C_{p,so} (T^* - T_1) + \varsigma m_{so} + m_l C_{p,l} (T_2 - T^*) \tag{6.11}$$

Where $C_{p,l}$ is the specific heat capacity of the liquid phase at constant pressure, $C_{p,so}$ is the specific heat capacity of the solid phase at constant pressure, m_l is the mass of the liquid phase, m_{so} is the mass of the solid phase, and ς is the specific latent heat the solid phase.

The method of latent TES has a very high storage density, exceeding $100\ kJ\ kg^{-1}$ with small associated volumes (Elmaazouzi et al., 2020), and there is a low temperature change between storage and recovery (Abhat, 1983). In general, a phase change material is suitable for latent heat storage. Nevertheless, the major drawback of this method is the low thermal conductivity of PCMs, which results in low rates of heat transfer (Singh et al., 2019). In this connection, research is underway to augment the conductivity of these materials, through various techniques which include: (a) the addition of high-conductivity particles and powders, (b) using fins or extended surfaces, (c) employing metal foams, (d) using multiple PCMs and (e) embedding heat pipes in PCM structures.

6.3.1 Materials

It is necessary to characterize PCMs by using some general material characteristics reported in Section 6.1.2. In addition to these characteristics, PCMs specifically require: (a) appropriate phase change temperature, (b) high level of enthalpy of phase change, (c) appropriate thermal stability with a low vapor pressure at the maximum operational temperature, (d) small variation of volume during the melting process, (e) negligible or no subcooling during the freezing process, negligible, or no supersaturation during melting, nor segregation, and (f) high heat capacities of the solid and liquid phases if sensible heat is also exploited.

The selection of a suitable PCM is the starting point of developing a latent heat storage system. It is important for the temperature of the PCM to be compatible with the specified range of application. Organic PCMs can be used in the temperature range below 393 K. For application at higher temperatures, important characteristics of organic PCMs include the reactivity with oxygen, long-term thermal stability, and high vapor pressure. Some drawbacks can be eliminated by using hermetically sealed storage systems. Inorganic materials have inherently higher thermal stability. In this vein, inorganic anhydrous salts have great potential for application in temperatures from 393 to 1273 K in the solid–liquid phase change. Selected solid-liquid PCMs are presented in Table 6.3 (Janz et al., 1978). They include salt hydrates, metals, and anhydrous salts.

6.3.1.1 *Paraffins*

Paraffins are aliphatic saturated straight- or branched-chain hydrocarbons with the general chemical formula C_nH_{2n+2}. They fall in the series $C_{14}H_{30}$–$C_{40}H_{82}$ with melting points in the range 279–353 K. These melting points and heat of fusion increase with the molar mass of a paraffin. Paraffins exhibit low thermal conductivity, low electrical conductivity, and low chemical reactivity (Selvaganapathy, 2007). In addition, there is volumetric expansion during the solid–liquid phase change.

Advantages of using paraffins as storage materials (Bauer et al., 2012): (a) they are wide available at low cost, (b) they are noncorrosive and nontoxic, (c) they are chemically stable with a low density, (d) the phase transformation proceeds fast, and (e) there is insignificant supercooling. However, these materials have the following disadvantages: (a) low volumetric energy density, (b) volumetric change during melting (about 10%), (c) low thermal conductivity and diffusivity, (d) lack of distinct melting point because inexpensive technical paraffins exploited in thermal storage systems are mixtures of many chemical compounds, (e) lack of compatibility with polymers, and (f) a high degree of flammability.

6.3.1.2 *Salt hydrates*

A salt hydrate is a solid compound that comprises moles of a salt and water. The water molecules are attracted by the salt ions because of the polar nature

TABLE 6.3 Selected solid–liquid phase change materials.

Phase change material	T (K)	$\Delta i_{melting}$ (kJ kg^{-1})	k_{so} (W m^{-1}K^{-1})	k_l (W m^{-1}K^{-1})	$C_{p,so}$ (kJ kg^{-1})	$C_{p,l}$ (kJ kg^{-1})	ρ_l (kg m^{-3})
$CaCl_2 \cdot 6H_2O$	303	180	1.1	0.5	1.4	2.1–2.3	1560
$Na_2HPO_4 \cdot 12H_2O$	308	280	0.5	0.5	1.5–1.7	2.0–3.2	1440
$Na(CH_3COO) \cdot 3H_2O$	331	250	0.7	0.4	2.1	3	1280
Na	371	113	122	87	1.36	1.38	930
Zn	692	112	100	50	0.45	0.48	6760
Al	933	397	211	91	1.18	1.18	2380
KNO_3–$NaNO_2$–$NaNO_3$ (53-40-7)	415	80	0.5	0.5	1.3	1.6	1980
$LiNO_3$	527	360	1.4	0.6	1.8	1.6–2.0	1780
$NaNO_3$	579	175	0.6	0.51	1.66	1.66	1910
NaOH	595	210	0.9	0.8	2	2.1	2130

$C_{p,l}$, specific heat capacity of medium in liquid phase; $C_{p,so}$, specific heat capacity of medium in solid phase; k_l, thermal conductivity of medium in liquid phase; k_{so}, thermal conductivity of medium in solid phase; T, temperature; $\Delta i_{melting}$, change in enthalpy at constant pressure; ρ_l, density of medium in liquid phase.

of water. Consequently, the water molecules get enclosed within the salt crystal lattice. The phase change transformation of salt hydrates involves hydration or dehydration processes. A finite number of hydrates can be formed depending on the ionic structure of the salt, and usually only a few of them (one or two) are thermodynamically stable. For example, calcium chloride ($CaCl_2$) can form calcium chloride monohydrate ($CaCl_2 \cdot H_2O$), calcium chloride dihydrate ($CaCl_2 \cdot 2H_2O$), calcium chloride tetrahydrate ($CaCl_2 \cdot 4H_2O$) and calcium chloride hexahydrate ($CaCl_2 \cdot 6H_2O$) (Gaeini et al., 2018). Solid hydrate crystals liberate their water of crystallization at a certain temperature or in a certain temperature range, and the anhydrous salt dissolves in this water. The hydrate resolidifies when heat is withdrawn.

Many salt hydrates are cheap, and they are slightly corrosive and toxic. Moreover, they show minor changes in volume during phase change and possess a relatively high heat of fusion at low melting temperature. The general chemical formula of a hydrated salt is $X_iY_j \cdot nH_2O$ where X_iY_j represents the anhydrous salt and n represents the number of moles of water. When heated to its transition temperature, the salt hydrate undergoes the dehydration processes. Usually, salt hydrates are converted either to the anhydrous form (congruent dehydration) or to a salt hydrate with fewer moles of water (incongruent dehydration) (Onder and Sarier, 2015). In a congruent dehydration process, the salt hydrate melts to produce water and an anhydrous salt where the two products are completely separated. A salt hydrate melts to water and a hydrated salt with fewer moles of water in an incongruent dehydration process. Incongruent melting is an irreversible process. Nevertheless, reversibility is an important requirement for high efficiency of TES. Thus, incongruent dehydration reduces the efficiency of TES. The two types of reactions can be described by Eqs. (6.12) and (6.13). The amount of water inside a salt complex X_iY_j is called the loading of the salt. The formation reaction of X_iY_j is exothermic, i.e., it produces energy that can be used in heating processes.

$$X_iY_j \cdot nH_2O + \text{energy} \leftrightarrow X_iY_j + nH_2O, \text{ congruent dehydration} \quad (6.12)$$

$$X_iY_j \cdot nH_2O + \text{energy} \leftrightarrow X_iY_j \cdot mH_2O + (n - m)H_2O, \text{ incongruent dehydration} \quad (6.13)$$

Most of the salt hydrates have two main disadvantages for thermal storage: (a) incongruent melting and (b) supercooling. Melting is incongruent when insufficient water of crystallization is given out during the melting process, and so, the anhydrous salt is unable to completely dissolve because of the inadequacy of the solvent. Consequently, the remaining solid component settles at the bottom of the container due to its higher density. During recrystallization, a fraction of the precipitated solid cannot get into contact with the necessary amount of water for its recrystallization. In view of this, the useful heat of fusion declines after several charge–discharge cycles. Possible ways of preventing incongruent melting are: (a) addition of a thickening agent that holds the anhydrous solid component in suspension and averts its settling, (b) addition of surplus water, (c) stirring, vibrating, or rotating the material, and (d) encapsulation of the storage material (to prevent leakage of liquid and vapor from volatile polyol materials, and the corrosion of inorganic hydrated salt accumulator material, and to extend heat transfer area, to enhance heat transfer and the overall mechanical properties of storage media).

Supercooling takes place during the heat discharge process when there are insufficient nucleation sites. Crystallization does not take place at the melting point, the liquid is cooled below this point, and the stored heat is not discharged at the expected melting point but at another temperature. In certain cases, the stored heat cannot be extracted because of low temperature difference between the subcooled melt and the required fluid temperature. It is possible to avoid supercooling by the addition of small quantities of a nucleating agent that has similar crystal structure to that of the salt.

6.3.1.3 Anhydrous salts

An anhydrous salt is a salt that does not contain water, but it is made up of cations and anions. The cations are commonly alkali and alkaline earth metals, while the anions include bromides, carbonates, chlorides, fluorides, hydroxides, nitrates, nitrites, and sulfates.

Most of the anhydrous salts are miscible, which yields a wide variety of potential single salts and salt mixtures (Table 6.3). Alkali metal nitrates and nitrites, and their mixtures are particularly suitable PCMs for temperatures of up to 623 K, if requirements such as handling and steel compatibility are taken into consideration (Tamme et al., 2008).

6.3.2 Heat transfer concepts

Metals have some attractive thermophysical properties (distinctly high thermal conductivity) for heat storage. However, most of them (such as tin or lead) are characterized by high costs and low melting enthalpy. In view of this, metals are not often considered as PCMs. The other remaining candidate PCMs have low thermal conductivity. The thermal conductivity of widely exploited PCMs is below 1.0 $Wm^{-1}K^{-1}$ (Bauer et al., 2012). This parameter significantly affects the power density and the heat transfer design of the system.

The charging and discharging of a storage system corresponds to the melting and solidification time of the PCM. Modeling of the melting and freezing processes has been examined by researchers, and analytical solutions of the problems have been established only for a few cases with simple geometries (Alexiades & Solomon, 1999; Yao & Prusa, 1989). Numerical techniques are required to solve problems with more complex geometries.

6.4 High-temperature latent heat storage applications

Thermodynamic cycles exploited in the CSP technology require heat of high grade. Thus, it is desirable to store heat at high temperature. The direct generation of steam in the absorber tubes of a solar receiver has triggered interest in systems of latent heat storage. The thermal system acts as a steam storage system for the two-phase HTF (water/steam). Candidate PCMs have melting temperatures in the range 393−613 K (Bauer et al., 2012). In this case, application of storage concepts utilizing sensible heat is often cost-ineffective. For HTFs in two phases, sensible heat storage systems have a low energy efficiency because of large temperature differentials between the storage media and HTFs. In contrast, both the PCM and HTF go through a phase change at about the same temperature in latent heat storage systems. It is therefore possible to avert higher temperature gradients between the HTF and storage medium.

A fined parallel pipe heat exchanger is embedded into the PCM of a latent heat storage system. During the heat charging process, steam is passed through the pipes where it condenses, thereby causing the PCM to melt down. During the heat discharging process, the freezing PCM releases thermal energy that heats the water inside the pipes, thereby generating steam. Steam storage systems find application in areas that include solar thermal power, combined heat and power, and industrial process steam.

6.5 Thermochemical energy storage

Chemical energy storage is based on the principle of reversible chemical reactions. The enthalpy change of reversible chemical reactions is used in chemical energy storage systems. A chemical reaction is a process in which atoms of the same or different substances are converted to one or more new substances. Chemical reactions may be reversible or irreversible. Reversible reactions can proceed forwards and backwards, while irreversible reactions proceed only forwards. A chemical reaction may absorb or release heat. A reaction that absorbs heat from the surroundings is called an endothermic reaction, while a reaction that produces heat and loses it to its surroundings is an exothermic reaction (Fegley, 2013). Thus, endothermic reactions can be used to charge a thermochemical energy storage system with heat, while exothermic reactions can be applied in the process of discharging the system.

Sources of energy for chemical storage may be thermal (thermochemical reactions), electrical (electrochemical reactions such as those which proceed in batteries), or electromagnetic (photochemical reactions and photosynthesis). In thermochemical energy storage, the storage medium is charged with heat to drive a forward chemical reaction. The stored heat is recovered by reversing the reaction, which may require the addition of catalysts. Frequently, the forward reaction proceeds at a higher temperature than the reverse reaction. An ideal thermochemical reaction for TES is an endothermic reaction in which the reaction products can easily be separated and do not undergo further reactions. Thermal losses from the storage system are confined to sensible heat effects that are often significantly lower than heats of reaction. Nevertheless, candidates of this type are practically scarce for low-temperature solar applications. The challenge of using chemical reactions with water vapor as one of the products is that the heat of condensation is often lost. In some reactions, energy of formation is also lost. For example, reactions in which atomic chlorine (Cl) is produced lose energy upon the formation of the chlorine molecule Cl_2.

Mechanisms for thermochemical energy storage fall into two types (Bauer et al., 2012): (a) chemical reaction storage and (b) sorption storage. In comparison with sensible and latent heat systems, thermochemical storage can have different temperature levels for the charging and discharging processes. It is therefore feasible to upgrade and downgrade heat and supply heat at an appropriate temperature level. Systems without storage capacities are called heat transformers and chemical heat pumps (Garg et al., 1985). A heat transformer absorbs heat at a low temperature level and supplies heat at a high temperature level. On the other hand, a chemical heat pump absorbs heat at a higher temperature level and supplies heat at a lower temperature level. Even though these systems are not aimed at heat storage, there are some connections between developments in the areas of heat transformation/chemical heat pumps and thermochemical storage of energy.

6.5.1 Heat of chemical reactions

Chemical reaction energy storage depends on the reversibility of the reaction with endothermic decomposition and exothermic synthesis processes. The products of the reaction are sealed and stored separately, which would theoretically give an unlimited duration of storage for this type of thermochemical storage.

When heated, some substances undergo reversible dissociation reactions. For example, when heated, a solid can decompose into solid and gas products. The gas is released while the depleted solid remains in the reactor (endothermic reaction). The dissociation products must be stored separately to prevent the occurrence of spontaneous reverse reaction when the equilibrium is altered by a temperature drop or pressure rise. During the discharging process, in the exothermic reaction, the gas is mixed with the solid. There are different types of gas–solid reaction systems that can be exploited in thermochemical energy storage systems. These gas–solid reaction systems include (with corresponding temperature range of application) (Bauer et al., 2012): (a) dehydration of metal salt hydrates (313–533 K), (b) dehydration of metal hydroxides (523–873 K), (c) dehydrogenation of metal hydrides (application in the range of 353–673 K), (d) decarboxilation of metal carbonates (373–1223 K), and (e) thermal deoxygenation of metal oxides (873–1273 K). It is evident that some of these storage systems can store heat at very high temperatures suitable for driving advanced thermodynamic cycles. Some solid–gas reactions proceed as:

$$A_{(s)} + \text{heat} \leftrightarrow B_{(s)} + C_{(g)} \tag{6.14}$$

Where the subscript (s) defines a solid and (g) is a gas. For example, when heated (at a reaction temperature of 1770 K), solid barium carbonate ($BaCO_3$) decomposes into solid barium oxide (BaO) and gaseous carbon dioxide (CO_2):

$$BaCO_3 + \text{heat} \leftrightarrow BaO + CO_2 \tag{6.15}$$

Where $BaCO_3$ and BaO are solids and CO_2 is a gas.

Simmons (1976) investigated the thermal decomposition of metal oxides for energy storage. One main advantage of these reactions is that the oxygen-gas product can be exploited in other applications or discarded, and oxygen from the atmosphere utilized in the reverse reactions. The thermal decomposition of potassium oxide is one example of this type of chemical reactions. It proceeds over a wide range of temperature range (573–1073 K) with a heat of decomposition of 2.1 MJ kg^{-1} (Duffie & Beckman, 2013).

$$4KO_2 + \text{heat} \leftrightarrow 2K_2O + 3O_2 \tag{6.16}$$

Another example is the decomposition of lead oxide, which occurs within the temperature range of 573–623 K with a decomposition heat of 0.26 MJ kg^{-1}.

$$2PbO_2 + \text{heat} \leftrightarrow 2PbO + O_2 \tag{6.17}$$

It is evident the thermal decomposition of metal oxides has enormous potential for storage of high-grade heat. Nevertheless, there are many practical challenges to the implementation of these heat storage concepts.

6.5.2 Heat of sorption

In a sorption process, a gas or a vapor (called sorbate) is absorbed/captured by a substance in a condensed state (solid or liquid) called sorbent. This phenomenon occurs on a surface at the interface between two phases such as solid–liquid and liquid–gas interfaces. Cohesive forces act among the molecules of all the participating substances regardless of their state of accumulation. Molecular configurations of the substances are not transformed in the sorption process, which includes absorption and adsorption.

In thermal storage, adsorption refers to a sorption process in which the sorbent is a liquid. Molecules of a vapor (or gas) attach themselves to the surface of a solid without creating a new material in the adsorption process. For example, when hydrate magnesium sulfate ($MgSO_4 \cdot 7H_2O$) is heated, it decomposes into magnesium sulfate ($MgSO_4$) and water (H_2O):

$$MgSO_4 \cdot 7H_2O + \text{heat} \leftrightarrow MgSO_4 + 7H_2O \tag{6.18}$$

Advantages of thermochemical materials include: (a) high density of heat storage, (b) significant temperature increase, and (c) the possibility of storing the reactants (sorbent and sorbate) at ambient temperature and without self-discharge. However, thermochemical heat storage is perceived to be expensive, and it is at an early stage of development (Fernandes et al., 2012). Through research and development, it is hoped that the cost of thermochemical energy storage will go down in future.

6.6 Configurations of concentrating solar power plants with thermal storage

Solar receivers can be classified based on parameters of geometric design and adapted heat transfer media (see Chapter 3). Different geometries have implications on the configuration a CSP plant with thermal storage. In a solar receiver, the HTF can be directly or indirectly irradiated with concentrated solar radiation. Direct absorption solar receivers utilize a quartz window that separates the internal section of the receiver from the external environment (Bashir et al., 2019). Some direct absorption receivers utilize transparent concentric tubes to allow concentrated solar radiation to reach the HTF inside the inner tube (see Section 5.3.3.2). Indirect absorption receivers use device surfaces to absorb heat from the concentrated solar radiation, and this heat is then transferred from the surfaces to HTF. Opaque tubes are exploited in indirect absorption solar receivers and their application is restricted by the maximum service temperature of tube material. An opaque tube is sometimes surrounded by a transparent tube to reduce heat loss. Transparent surfaces must be clear of any obstruction from other components such as storage medium (unit).

Receivers and storage components can be designed as separate or integrated units. In a CSP system with separate receiver and storage units, concentrated solar radiation is collected and converted to heat by the receiver, and this solar heat is transmitted to the power block and storage unit by using a HTF (Fig. 6.12). In some designs, the working fluid in the power block is heated via a heat exchanger. Design flexibility is the main advantage of separating receivers from storage units. For example, the volume and weight of the storage unit can be as large as possible. It is therefore not surprising that all operating CSP plants with storage systems around the world currently use this type of configuration.

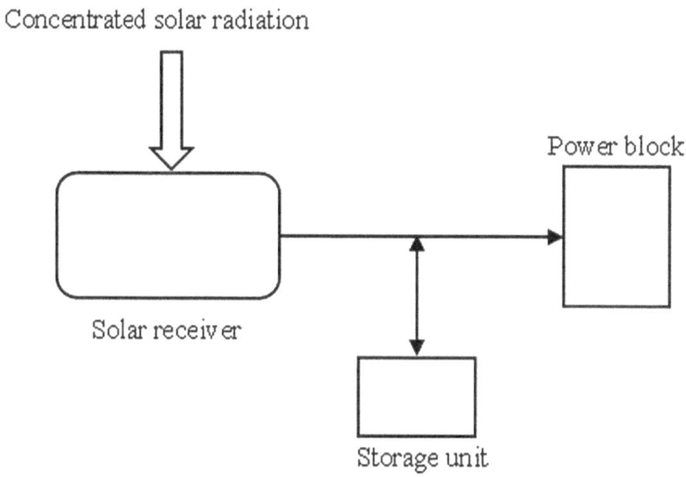

FIGURE 6.12 Basic configuration of a concentrating solar power plant with separate receiver and storage units.

In some CSP designs (such as the parabolic trough concentrator and parabolic dish concentrator), the receiver and concentrator are integrated and mounted on a sun-tracking system. Sun tracking systems can support a limited mass. Therefore, the weight of the storage unit is a critical mechanical consideration in receiver designs with integrated storage units. A large volume/mass of storage unit may be required depending on the characteristics of the storage medium and the required storage capacity, thereby posing a huge challenge of integrating receivers with storage units.

The thermal mass of solar receivers is relatively low. Consequently, the intermittent availability of solar radiation causes significant transients in the outlet temperature of the HTF exiting the receiver. Short-term unavailability of solar radiation is caused by passing clouds, and integration of solar receivers with storage can assist in stabilizing the outlet temperature of the receiver (Fig. 6.13). In this connection, it has been found that PCMs have great potential for utilization as storage media in integrated receiver-storage systems. The PCM is incorporated in solar receivers to make the volumetric receiver behave like a heat battery (Anwar et al., 2016). At the melting point, the temperature of a PCM remains constant during heat charging/discharging. Thus, the receiver ensures a uniform heat output during short-term discontinuities of solar radiation because of the thermal buffering effect (Senthil & Cheralathan, 2019). Nevertheless, solar radiation is also not availability over longer periods such as on overcast days or at night. In such cases, a CSP plant can be integrated with a large separate storage.

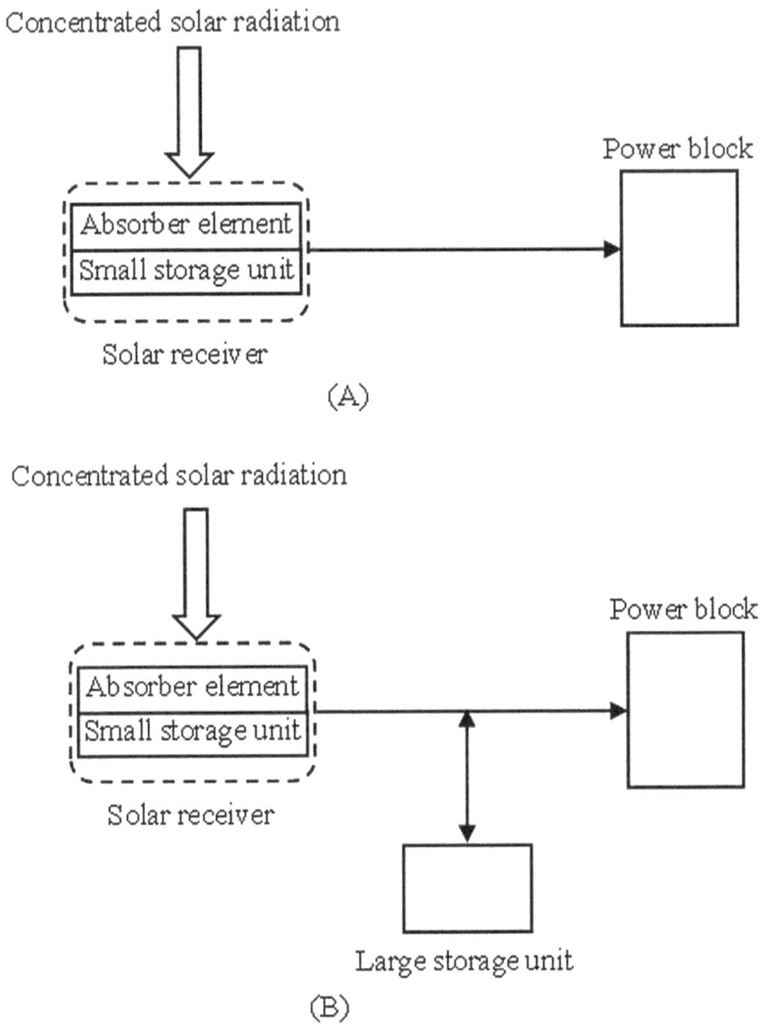

FIGURE 6.13 Configuration of a concentrating solar power plant with an integrated receiver storage system with: (A) short-term storage only and (B) short- and long-term storage.

Nomenclature

A area (m^2)
C_p specific heat capacity at constant pressure ($Jkg^{-1}K^{-1}$)
D diameter (m)
e effusivity ($Jm^{-2}K^{-1}s^{-1/2}$)
h_v coefficient of volumetric heat transfer ($Wm^{-3}K^{-1}$)
i enthalpy ($J\ kg^{-1}$)
k thermal conductivity ($Wm^{-1}K^{-1}$)

m mass (kg)
P pressure (Pa)
Q heat rate (W)
T temperature (K)
U coefficient of heat loss ($Wm^{-2}K^{-1}$)

Greek symbols

ρ density (kg m^{-3})
ς specific latent heat of the solid phase (J kg^{-1})

References

Abhat, A. (1983). Low temperature latent heat thermal energy storage: Heat storage materials. *Solar Energy, 30*(4), 313–332. https://doi.org/10.1016/0038-092X(83)90186-X

Alexiades, V., & Solomon, A. D. (1999). *Mathematical modelling of melting and freezing processes*. Hemisphere Publishing Corporation.

Anwar, M. K., Yilbas, B. S., & Shuja, S. Z. (2016). A thermal battery mimicking a concentrated volumetric solar receiver. *Applied Energy, 175*, 16–30. https://doi.org/10.1016/j.apenergy.2016.04.110

ASHRAE. (2017). *ASHRAE handbook – fundamentals*. American Society of Heating, Refrigerating and Air-Conditioning Engineers (ASHRAE).

Bashir, M. A., Giovannelli, A., & Ali, H. M. (2019). Design of high-temperature solar receiver integrated with short-term thermal storage for Dish-Micro Gas Turbine systems. *Solar Energy, 190*, 156–166. https://doi.org/10.1016/j.solener.2019.07.077

Bauer, T., Steinmann, W.-D., Laing, D., & Tamme, R. (2012). Thermal energy storage materials and systems. *Annual Review of Heat Transfer*, 131–177. https://doi.org/10.1615/annualrevheattransfer.2012004651

Bergman, T. L., Lavine, A. S., Incropera, F. P., & DeWitt, D. P. (2011). *Fundamentals of heat and mass transfer* (7th ed.).

Duffie, J. A., & Beckman, W. A. (2013). Solar engineering of thermal processes: fourth edition. In *Solar engineering of thermal processes* (4th ed.). John Wiley and Sons. https://doi.org/10.1002/9781118671603

Dutta, P. (2017). High temperature solar receiver and thermal storage systems. *Applied Thermal Engineering, 124*, 624–632. https://doi.org/10.1016/j.applthermaleng.2017.06.028

Elmaazouzi, Z., El Alami, M., Agalit, H., & Bennouna, E. G. (2020). Performance evaluation of latent heat TES system-case study: dimensions improvements of annular fins exchanger. In *Energy reports* (Vol. 6, pp. 294–301). Elsevier Ltd. https://doi.org/10.1016/j.egyr.2019.08.059

Fegley, B. (2013). Thermochemistry. In *Practical chemical thermodynamics for geoscientists* (pp. 127–172). Elsevier Inc.

Fernandes, D., Pitié, F., Cáceres, G., & Baeyens, J. (2012). Thermal energy storage: how previous findings determine current research priorities. *Energy, 39*(1), 246–257. https://doi.org/10.1016/j.energy.2012.01.024

Gaeini, A., Rouws, A. L., Salari, J. W. O., Zondag, H. A., & Rindt, C. C. M. (2018). Characterization of microencapsulated and impregnated porous host materials based on calcium chloride for thermochemical energy storage. *Applied Energy, 212*, 1165–1177. https://doi.org/10.1016/j.apenergy.2017.12.131

Garg, H. P., Mullick, S. C., & Bhargava, A. K. (1985). Solar thermal energy storage. In *Sol therm energy storage*. D. Reidel Publ Co.

Hidayat, M., Prahastuti, S., Anggie, S., Rakasiwi, A. S., Prisilia, S., & Hasan, K. (2022). Sub-chronic toxicity study of green peas protein hydrolysate in rats. *Toxicology Reports, 9*, 735–742. https://doi.org/10.1016/j.toxrep.2022.03.020

Janz, G. J., Allen, C. B., Downey, J. R., & Tomkins, R. P. T. (1978). *Physical properties data compilations relevant to energy storage. I. Molten salts: eutectic data.* Report National Standard Reference Data Series.

Kakaç, S., & Liu, H. (1998). *Heat exchangers: Selection, rating and thermal design.* Boca Raton: CRC Press.

Kelly, B., & Kearney, D. (2006). *Thermal storage commercial plant design study for a 2-tank indirect molten salt system.*

Kotzé, J. P., von Backström, T. W., & Erens, P. J. (2014). *NaK as a primary heat transfer fluid in thermal solar power installations.* Durban: SolarPACES 2012 International Conference.

Kuravi, S., Trahan, J., Goswami, Y., Rahman, M. M., & Stefanakos, E. K. (2013). Thermal energy storage technologies and systems for concentrating solar power plants. *Progress in Energy and Combustion Science, 39*, 285–319. https://doi.org/10.1016/j.pecs.2013.02.001

Liu, J., He, Y., & Lei, X. (2019). Heat-transfer characteristics of liquid sodium in a solar receiver tube with a nonuniform heat flux. *Energies, 12*(8), 1432. https://doi.org/10.3390/en12081432

Löf, G. O. G., & Hawley, R. W. (1948). Unsteady-state heat transfer between air and loose solids. *Industrial and Engineering Chemistry, 40*(6), 1061–1070. https://doi.org/10.1021/ie50462a017

NREL. (2021). *Concentrating solar power projects.* National Renewable Energy Laboratory (NREL). https://solarpaces.nrel.gov.

Onder, E., & Sarier, N. (2015). Thermal regulation finishes for textiles. In *Functional finishes for textiles: Improving comfort, performance and protection* (pp. 17–98). Elsevier Inc. https://doi.org/10.1533/9780857098450.1.17

Pacio, J., Singer, C., Wetzel, T., & Uhlig, R. (2013). Thermodynamic evaluation of liquid metals as heat transfer fluids in concentrated solar power plants. *Applied Thermal Engineering, 60*(1–2), 295–302. https://doi.org/10.1016/j.applthermaleng.2013.07.010

Rowe, W. B. (2009). *Principles of modern grinding technology.* Elsevier BV. https://doi.org/10.1016/b978-0-8155-2018-4.50014-8

Schumann, T. E. W. (1929). Heat transfer: a liquid flowing through a porous prism. *Journal of the Franklin Institute, 208*(3), 405–416. https://doi.org/10.1016/S0016-0032(29)91186-8

Selvaganapathy, P. R. (2007). Polymers. In *Comprehensive microsystems* (Vol. 1, pp. 75–105). Elsevier. https://doi.org/10.1016/B978-044452190-3.00004-5

Senthil, R., & Cheralathan, M. (2019). Enhancement of the thermal energy storage capacity of a parabolic dish concentrated solar receiver using phase change materials. *Journal of Energy Storage, 25*, 100841. https://doi.org/10.1016/j.est.2019.100841

Shallcross, D. (2004). *Physical property data book for engineers and scientists.* Institution of Chemical Engineers (IChemE).

Simmons, J. A. (1976). Reversible oxidation of metal oxides for thermal energy storage. In *Proceedings of the International Solar Energy Society (ISES) meeting* (Vol. 8). International Solar Energy Society.

Singh, R., Sadeghi, S., & Shabani, B. (2019). Thermal conductivity enhancement of phase change materials for low-temperature thermal energy storage applications. *Energies, 75*. https://doi.org/10.3390/en12010075

Tamme, R., Bauer, T., Buschle, J., Laing, D., Müller-Steinhagen, H., & Steinmann, W. D. (2008). Latent heat storage above 120°C for applications in the industrial process heat sector and solar power generation. *International Journal of Energy Research, 32*(3), 264–271. https://doi.org/10.1002/er.1346

Tamme, R., Bauer, T., & Hahne, E. (2009). *Heat storage media.* Wiley. https://doi.org/10.1002/14356007.a12_b30.pub2

Trevisan, S., Wang, W., Guedez, R., & Laumert, B. (2022). Experimental evaluation of an innovative radial-flow high-temperature packed bed thermal energy storage. *Applied Energy, 311*, 118672. https://doi.org/10.1016/j.apenergy.2022.118672

UN. (2011). *Globally Harmonized System of Classification and Labelling of Chemicals (GHS)* (4th). New York and Geneva: United Nations (UN).

Votava, J., Kyncl, J., Muller, Z., Vetoshkin, L., & Rimbala, J. (2019). Optimized use of the steam accumulator in the combined heat and power production. In *Proceedings of the 2019 20th international scientific conference on electric power engineering, EPE 2019.* Institute of Electrical and Electronics Engineers Inc. https://doi.org/10.1109/EPE.2019.8778141

Yang, L., Villalobos, U., Akhmetov, B., Onn, K. J., Gil, A., Tan, W. L., & Romagnoli, A. (2022). Active TES With PCM for Refrigeration Applications. In *Encyclopedia of Energy Storage* (1, pp. 479–497). Elsevier Inc.

Yao, L., Chen, L., Chen, B., Tang, Y., Zhao, Y., Liu, S., & Xu, H. (2021). Toxic effects of TiO2 NPs in the blood-milk barrier of the maternal dams and growth of offspring. *Ecotoxicology and Environmental Safety, 208*, 111762. https://doi.org/10.1016/j.ecoenv.2020.111762

Yao, L. S., & Prusa, J. (1989). Melting and freezing. *Advances in Heat Transfer, 19*(C), 1–95. https://doi.org/10.1016/S0065-2717(08)70211-9

Chapter 7

Thermodynamics of solar receivers

7.1 Introduction

Knowledge about thermodynamics is vital in the development concentrating solar power (CSP) technology. Consequently, the first major design step is the analysis of the thermodynamic performance of the CSP system or its components. A solar receiver is a critical component of the CSP system, and its thermodynamic parameters significantly affect the overall performance the system. Influential thermodynamic factors include fluid properties, flow characteristics, energy conversion, energy flows, and efficiency of the receiver.

Solar receivers convert solar radiation to heat that drives the thermodynamic cycles of CSP systems. Thermodynamics is the study of energy, its conversion from one form to another, and its relation to the states of matter (ASHRAE, 2017). A region in space or a quantity of matter bordered by a closed surface is called a thermodynamic system (Fig. 7.1). The surroundings of the system comprise everything that is outside the system, and the system is separated from its surroundings by movable, fixed, real, or imaginary

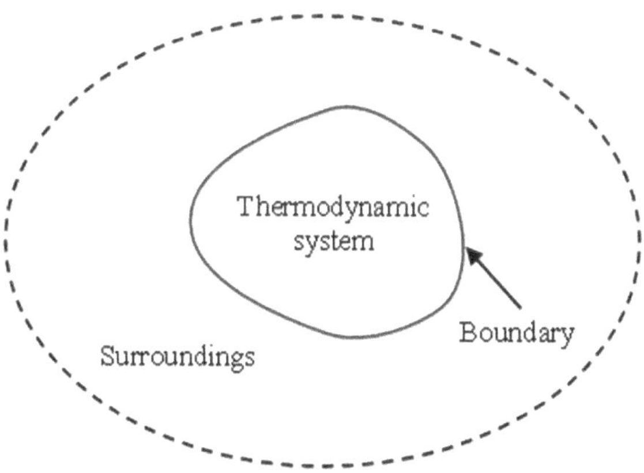

FIGURE 7.1 Schematic representation of a thermodynamic system and its surroundings.

Solar Receivers for Thermal Power Generation. https://doi.org/10.1016/B978-0-323-85271-5.00004-5

boundary. Heat can be transferred from the system to its surroundings or vice versa. In this vein, heat is lost from the receiver to the surroundings through conduction, convection and radiation. Heat losses reduce the efficiency of the receiver, and so, they need to be minimized.

The performance of a solar receiver is affected by thermodynamic properties that include temperature, pressure, and specific volume (see Section 1.5). Receivers operate in a wide range of temperature (low to high). A heat transfer fluid is used to evacuate heat from a receiver. In this context, the fluid must meet the required thermophysical properties and other characteristics for use as a heat transfer fluid (see Chapter 5). Thermophysical properties of fluids fall into two groups: transport and thermodynamic properties (see Section 6.1.2.1). The fluid flow is caused by a pressure gradient between the inlet and outlet points of the receiver. Pumps or compressors are utilized to drive heat transfer fluids through receivers. Materials with low specific volume (high density) are more preferred in both heat transfer and storage processes.

7.2 Laws of thermodynamics

Laws of thermodynamics are useful in the advancement of thermal technology. There are four of them (Szőllősi & Márkus, 2015; Tumen Ozdil et al., 2015): Zeroth, first, second, and third laws of thermodynamics.

7.2.1 Zeroth law

The Zeroth law explains the transitivity of an equilibrium state of systems. It is possible for a dynamic equilibrium to exist between two systems A and B that are thermally coupled by the same field. It states that if two bodies A and B are separately in thermal equilibrium with some third body C, then bodies A and B are also in equilibrium with each other.

The Zeroth law finds application in the measurement of the temperature of a system (such as a solar receiver) by using temperature sensors. If a thermometer (body C) is used to accurately measure the temperature (T) of bodies A and B, then the thermometer must separately be in thermal equilibrim with each body before readings are taken. A thermal equilibrium is attained where the temperatures of the sensor and system are equal. Thus, if the temperature readings of bodies A and B are respectively T_A and T_B, then the two bodies are in thermal equilibrim if T_A is equal to T_B. Otherwise, if T_A is not equal to T_B then the two bodies are not in thermal equilibrium with each other. Temperature sensors require calibration for meaningful comparison of measuerements taken by different sceintists and engineerins around the world. This calibration is done by using some standard. For example, the triple point of water can be used as a standard for calibrating thermometers (such gas-based thermometers). At the triple point, the solid, liquid and vapor states coexist only under a set of precisely controlled and invariant conditions that are determined by the physical charateristics of water. This thermodynamic point is assigned a temperature of $T_t = 273.16$ K (Honig, 2014). So, any other absolute temperature can be determined through the proportionality $T = 273.16\ (P/P_t)$, where P is the pressure of the gas at temperature T, and P_t is the pressure measured for the gas (such as helium He) in equilibrium with water at its triple point.

7.2.2 First law of thermodynamics

Knowledge about the energy balance of systems is vital in the design and testing of solar receivers. In this context, the first law of thermodynamics describes the energy balance of a given system. Fundamentally, it states that the total energy of a system is conserved, and consequently, the total energy (E) in a system can change if and only if energy crosses its boundaries. This law also addresses the ways in which energy can cross the boundaries of the system. There are only two ways in which energy can flow into or out of a closed system: heat (Q) and work (W). Heat is a mechanism that transfers energy across the borders of a system by virtue of temperature gradients, always from high to low temperature points. Heat is positive when energy flows into the system. For an adiabatic process, there is no flow of heat into or out of the system. Work is a mechanism that transmits energy across the boundaries of a system with a pressure gradient, from high to low pressure points. It is considered as positive when it is done by the system, whereby energy is transferred from the system. In differential form, the first law for a closed system can be expressed as follows:

$$dE = dQ - dW \tag{7.1}$$

Where dE is the change in total energy of the system.

The first law of thermodynamics is applicable in the design and performance evaluation of solar receivers and other thermal devices. Energy efficiency, which is computed from energy balances, is a critical indicator of the performance of receivers. Nevertheless, one of the shortcomings of the first law of thermodynamics is its inability to explain the reasons for which some thermodynamic processes proceed in a specified direction and cannot proceed in the reverse direction. This drawback is addressed by the second law of thermodynamics.

7.2.3 Second law of thermodynamics

The second law of thermodynamics dictates the direction of processes. This principle can be expressed by different statements (entropy balance, Clausius statement, Kelvin-Planck statement and Carnot statement) that are equivalent to one another (Tosun, 2013). The entropy balance expression states that any spontaneous process in any isolated system always results in a rise in the entropy of that system. More details about the concept of entropy are reported in Section 7.4. According to the Causius statement, it is impossible to construct a device that, operating in a cycle, will produce no effect other than the transfer of heat from a cold reservoir to a hot one. This statement shows that heat is often automatically transferred from a body at high temperature to a body at low temperature, and the reverse process cannot be performed spontaneously. The Clausius statement is about refrigeration and heat-pump devices (Sayyaadi, 2021). Work must be done on the device to transfer heat from a cold reservoir to hot one. The Kelvin-Planck statement postulates that it is impossible to construct a thermal engine that works in a thermodynamic cycle and generates power while it exchanges heat only with one thermal reservoir. This implies that heat engines must exchange heat with two thermal reservoirs that comprise a heat source (hot reservoir) and a heat sink (cold reservoir). The Kelvin-Planck statement is a statement about heat engines. A solar receiver generates heat and so it functions as a heat source for the CSP power block. The quality of energy, and the maximum possible useful work in a process (called exergy or irreversibility) are described by the second law of thermodynamics (Szőllősi & Márkus, 2015). In this context, Sadi Carnot (1796–1832) defined two other statements: The first statement is that the thermal efficiency of all reversible heat engines that operate between two thermal reservoirs are equivalent, regardless of the type of the cycle that each engine goes through. This statement implies that the thermal efficiency of the reversible heat engine only depends on the temperatures of the heat source and heat sink (see Section 1.5.1). The second Carnot statement is that the thermal efficiency of a reversible heat engine that operates between a hot reservoir and a cold reservoir has a higher thermal efficiency than an irreversible heat engine that operates between the same resrvoirs.

Exergy is a measure of the usefulness as well as the potential to cause change (Rosen, 2021). The exergy of a system can be analysed to establish how nearly the actual performance approaches the ideal one. Exergy analysis is a technique of thermodynamis analysis based on the second law of thermodynamics, and it identifies the causes and locations of thermodynamic losses more clearly than energy analysis (Dincer & Rosen, 2021). This kind of analysis helps in (a) revealing the type, location, and magnitude of energy losses and (b) discovering ways of reducing losses in order to maximize the thermal performance of the system. The exergy analysis of solar receivers is presented in Section 10.3.3.

7.2.4 Third law of thermodynamics

The third law of thermodynamics states that when the temperature of a system draws near to absolute zero the change in entropy of the system approaches zero. This implies that the entropy of a pure material is zero at absolute zero. No real system can attain absolute zero, and this is one reason for using the Kelvin temperature scale. Nevertheless, researchers at the Massachusetts Institute of Technology managed to achieve a temperature of 450 pK in 2003 (Zohuri, 2018), which is very close to absolute zero. Thus, it is possible that the absolute zero or lower emperature will be practically attained in future. The internal energy of a system is proportional to this temperature, and the use of this temperature scale eliminates the concern about dividing by zero in an equation. It should also be noted that there is currently no formula based on the third law of thermodynamics.

One important consequence of the third law of thermodynamics is that the absolute entropy of chemical materials (both elements and compounds) can be defined (Ekeren, 1998). Now, the absolute entropy of a substance can be determined calorimetrically by measuring the heat capacity and the latent heats of phase transitions of the substance over the whole temperature range from the absolute zero up to the desired temperature.

7.3 Energy analysis

Mass and energy flows can be steady or unsteady. Thus, flow systems can be classified into steady and unsteady categories.

7.3.1 Steady flow systems

In a steady flow process, a fluid flows through a control volume uniformly. Fluid properties can vary from one point to another within a control volume, but they remain constant at any point during the whole process (Çengel & Boles, 2006). Intensive or extensive properties within the control volume do not change with time in a steady flow process. Consequently, the volume, mass, and total energy content of the control volume remain constant. The implication of this is that the boundary work is zero for steady flow systems, and the total mass entering the control volume and the total mass leaving it must be equal. Similarly, the total energy entering the control volume and the total energy leaving it must be equal. These characteristics of steady flow systems significantly simplify the analysis.

7.3.1.1 Mass conservation

For a steady flow system with multiple inlets and outlets (multiple streams), the total inlet mass flow rate is equal to the total outlet mass. In this case, the conservation of mass flow rate can be expressed as follows:

$$\sum \dot{m}_{in} = \sum \dot{m}_{out} \tag{7.2}$$

Where $\dot{m}_{in}$ is the inlet mass flow rate and $\dot{m}_{out}$ is the outlet mass flow rate.

Example 7.1

Fig. 7.2 shows a steady flow system with one inlet and two outlets. If $\dot{m}_2 = \dot{m}_3 = 2.5$ kg s^{-1}, calculate the value of $\dot{m}_1$.

Solution

$$\dot{m}_1 = \dot{m}_2 + \dot{m}_3$$

But $\dot{m}_2 = \dot{m}_3 = 2.5$

Therefore,
$\dot{m}_1 = 2.5 + 2.5 = 5$ kgs^{-1}

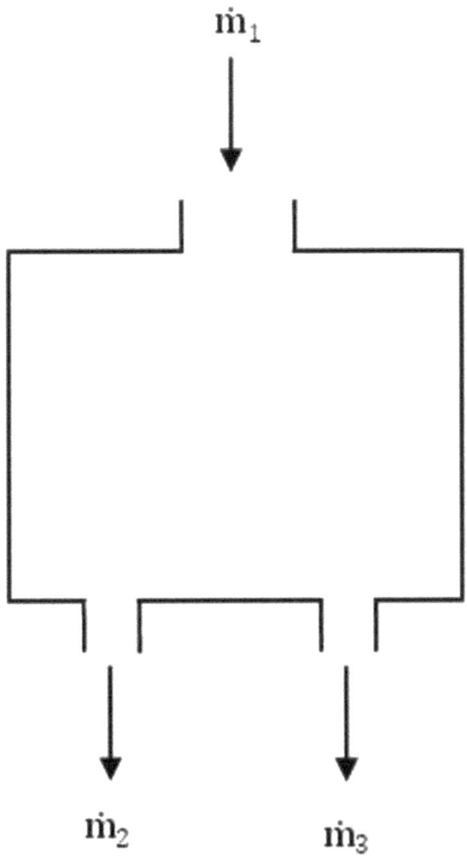

FIGURE 7.2 Steady flow system with one inlet and two outlets.

Many devices in engineering have a single stream. These devices include solar receivers, turbines, compressors, and pumps. For single-stream steady flow system, the mass balance is given by the following:

$$\dot{m}_1 = \dot{m}_2 \tag{7.3}$$

7.3.1.2 Flow work and energy of a moving fluid

There is mass flow across the boundaries of a control volume and some work is required to move the mass into or out of the control volume. This kind of work is called flow work, or flow energy, and it is necessary for sustaining a continuous flow through a control volume. The total energy of a simple compressible system comprises three parts: internal (e), kinetic ($u^2/2$), and potential (gz) energies. In addition to this, a flowing fluid possesses flow energy (Pv). Thus, the energy change of the system is the following:

$$E_2 - E_1 = \Delta E = Pv + e + \frac{u^2}{2} + gz \tag{7.4}$$

Where E_1 is the initial energy per unit mass of the system, E_2 is the final energy per unit mass of the system, g is the acceleration due to gravity, P is the pressure of the fluid, u is the velocity of the fluid, v is the specific volume of the fluid, and z is the height of the fluid above a reference point. The sum of flow and internal energies is defined as the enthalpy (i). Therefore, Eq. (7.4) can be expressed as follows:

$$\Delta E = i + \frac{u^2}{2} + gz \tag{7.5}$$

Consequently, when a fluid is flowing at a rate of $\dot{m}$, then the rate of energy flow $\left(\dot{m}\,\Delta E\right)$ transported by the fluid stream is as follows:

$$\dot{m}\,\Delta E = \dot{m}\left(i + \frac{u^2}{2} + gz\right) \tag{7.6}$$

When the kinetic and potential energies are negligible, then Eq. (7.6) can be simplified to:

$$\dot{m}\,\Delta E = \dot{m}i \tag{7.7}$$

Generally, a solar receiver has one inlet and one outlet. A heat transfer fluid enters the receiver at a low temperature and evacuates solar heat from the receiver. During normal operation, the rate of fluid flow through the receiver is steady. So, steady energy flow conditions can be applied in the analysis of receiver energy. For a receiver shown in Fig. 7.3, the rate of net heat from the receiver $\left(\dot{Q}_{rec}\right)$ can be expressed as:

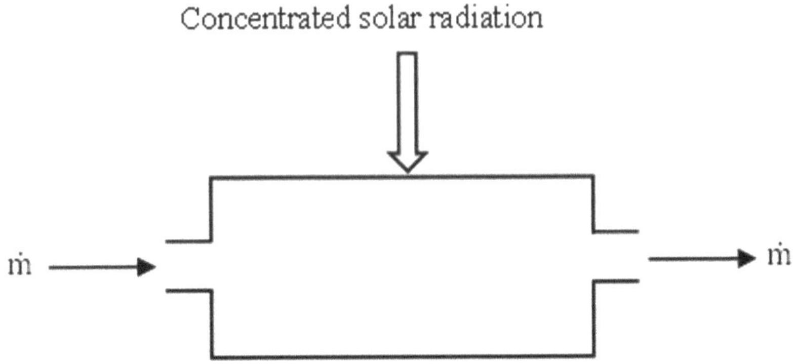

FIGURE 7.3 Schematic representation of steady flow in a solar receiver with one inlet and one outlet.

$$\dot{Q}_{rec} = \dot{m}_{htf}\left(i_{htf,out} - i_{htf,in}\right) \tag{7.8}$$

Where $\dot{m}_{hft}$ is the mass flow rate of the heat transfer fluid, and $i_{htf,in}$ and $i_{htf,out}$ are respectively the inlet and outlet enthalpies of the heat transfer fluid.

The net energy can also be expressed in terms of the difference between inlet (T_{in}) and outlet (T_{out}) temperatures of the receiver (see Section 1.5). Energy efficiency is an important indicator of the thermodynamic performance of a receiver. It is computed from the net energy and the amount of concentrated solar radiation intercepted by the receiver as shown Section 1.5. The energy efficiency of a receiver increases with the net heat generated by the receiver.

7.3.2 Transient flow systems

Many real processes entail changes of variables (such as mass and energy) with time within a control volume. Processes of this nature are said to be transient- or unsteady. When analyzing transient flow processes, it is essential to keep track of the mass and energy contents and the energy interactions across the boundary of the control volume. Transient flow processes commence and end over a period of finite time instead of prolonging indefinitely. In some respects, a transient flow system is similar to a closed system, except that the mass within the system boundaries changes during a process. Transient flow systems are not fixed in space, size, and shape. They are often stationary but may involve shifting of boundaries and therefore boundary work.

7.3.2.1 Mass conservation

For a transient-flow system with multiple inlets and outlets (multiple streams), the difference between the total inlet and outlet mass flow rates is equal to the rate of change of the mass of the system. In this case, the conservation of mass flow rate can be expressed by Eq. (7.9).

$$\sum \dot{m}_{in} - \sum \dot{m}_{out} = \frac{dm_{sys}}{dt} = \dot{m}_{sys} \tag{7.9}$$

Where $\dot{m}_{in}$ is the inlet mass flow rate, $\dot{m}_{out}$ is the outlet mass flow rate and $\dot{m}_{sys}$ is the rate of change of the mass of the system.

When $\frac{dm_{sys}}{dt} = 0$, mass flow is in a steady-state condition.

7.3.2.2 Energy balance

For a transient-flow system, the difference between the total inlet and outlet energy flow rates is equal to the rate of change of the energy of the system. In this case, the energy balance of the system can generally be given by:

$$\dot{E}_{in} - \dot{E}_{out} = \frac{dE_{sys}}{dt} = \dot{E}_{sys} \tag{7.10}$$

Where $\dot{E}_{in}$ is the flow rate of energy into the system, $\dot{E}_{out}$ is the flow rate of energy out of the system and $\dot{E}_{sys}$ is the rate of change of the energy of the system. When dE_{sys}/dt, energy flow is in a steady-state condition.

7.4 Entropy of a system

Entropy is a thermodynamic property that measures the molecular disorder of a system, and it increases with the extent of mixing. Generally, the term entropy refers to both total entropy and entropy per unit mass, with the context clarifying the meaning. The change of entropy between initial and final states of a system is given by Eq. (7.11).

$$\Delta s = s_2 - s_1 = \int_1^2 \left(\frac{\delta Q}{T}\right)_{ir} \tag{7.11}$$

Where δQ is the heat transfer to or from the system, Δs is the entropy change, s_1 is the initial entropy, s_2 is the final entropy, and T is the temperature of the system. The subscript ir stands for internally reversible process.

It should be noted that the change in entropy is defined instead of entropy itself, similar to the definition of the change in energy instead of the energy itself. Absolute values of entropy are found based on the third law of thermodynamics. In many engineering applications, changes in entropy are often of interest. So, the entropy of a system can be assigned a value of zero at an arbitrarily selected reference state, and the entropy values at other states can be computed by using Eq. (7.11). The system may transition from state 1 to state 2 via reversible or irreversible processes. However, the change in entropy is the same in both cases (Fig. 7.4).

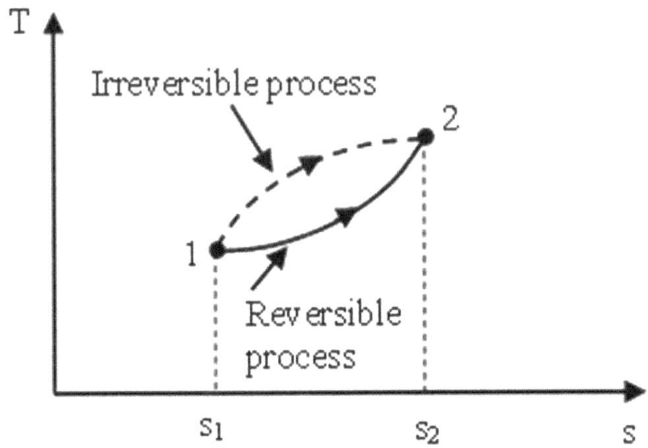

FIGURE 7.4 Reversible and irreversible processes between the same initial and final states have the same changes in entropy.

For an isothermal process, the temperature of the system is constant, and an isothermal heat transfer process is internally reversible. These processes can occur in thermal energy reservoirs that are able to indefinitely exchange heat at a constant temperature. Consequently, the entropy change of such a system is:

$$\Delta s = s_2 - s_1 = \frac{1}{T_{cons}} \int_1^2 (\delta Q)_{ir} = \frac{Q}{T_{cons}} \tag{7.12}$$

Where Q is the heat transfer for the internally reversible process and T_{cons} is the constant temperature of the system.

7.4.1 Clausius inequality

Clausius inequality was first defined by R. J. E. Clausius (1822–88). This is one of the important inequalities, and it is expressed as:

$$\oint \frac{\delta Q}{T} \leq 0 \tag{7.13}$$

The cyclic integral of dQ/T is always less than or equal to zero, and the inequality is valid for reversible as well as irreversible thermodynamic cycles. If irreversibilities are absent within the system with a reversible cyclic device, then the cycle undergone by the joined system is internally reversible. In view of this, the cycle can be reversed. In the reversed cycle, all the variables maintain their magnitudes but have an opposite sign. Consequently, the work (W_c), which could not be positive in the usual case, cannot be negative in the reversed scenario. The work cannot be positive or negative, and therefore, $W_{c,ir} = 0$. It follows that:

$$\oint \left(\frac{\delta Q}{T}\right)_{ir} = 0 \tag{7.14}$$

This shows that the equality in the Clausius inequality is true for completely or just internally reversible cycles and the inequality for the irreversible ones.

7.4.2 Entropy generation and increase

An arbitrary cycle shown in Fig. 7.5 is considered. It comprises two processes: Process 1−2 (reversible or irreversible) and process 2−1 (internally reversible). Using the Clausius inequality, we obtain the following:

$$\int_1^2 \frac{\delta Q}{T} + \int_2^1 \left(\frac{\delta Q}{T}\right)_{ir} \leq 0 \tag{7.15}$$

But $\int_2^1 \left(\frac{\delta Q}{T}\right)_{ir} = s_1 - s_2$, therefore

$$\int_1^2 \frac{\delta Q}{T} + s_1 - s_2 \leq 0 \tag{7.16}$$

which yields the following expression:

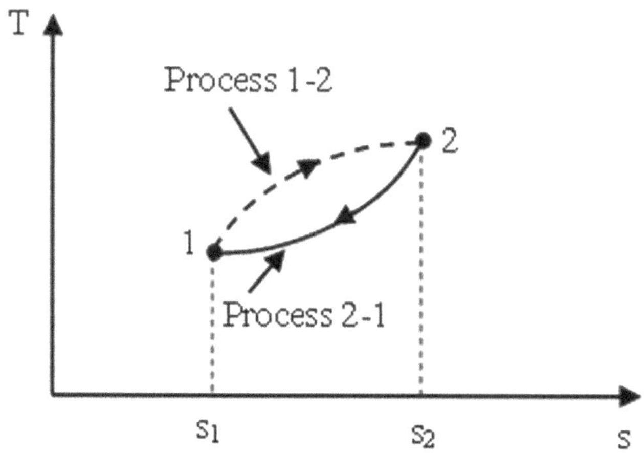

FIGURE 7.5 Schematic representation of an arbitrary cycle.

$$s_1 - s_2 \geq \int_1^2 \frac{\delta Q}{T} \tag{7.17}$$

In differential form, Eq. (7.17) can be expressed as follows:

$$ds \geq \frac{\delta Q}{T} \tag{7.18}$$

In Eq. (7.18), the equality holds when the process is internally reversible, while the inequality holds when the process is irreversible. Therefore, it is concluded that the entropy change of a closed system during an irreversible process is higher than the integral of dQ/T evaluated for that process. The limiting case is a reversible process, where these two quantities become equal. It should be noted that T is the thermodynamic temperature at the boundary where the differential heat dQ is transmitted between the system and its surroundings. The inequality sign in the preceding mathematical relations is a reminder that the entropy change of a closed system during an irreversible process is always higher than the entropy transfer. This shows that some entropy is generated during an irreversible process, which is entirely ascribed to the presence of irreversibilities. The entropy generated during a process is called entropy generation (s_{gen}), and it is always positive or zero. Now, the difference between the entropy change of a closed system and the entropy transfer is equal to entropy generation. Consequently, Eq. (7.17) can be expressed as:

$$\Delta s_{sys} = s_1 - s_s = \int_1^2 \frac{\delta Q}{T} + s_{gen} \tag{7.19}$$

Where Δs_{sys} is the entropy change of the system.

Eq. (7.17) has significant implications in the subject of thermodynamics. For an isolated system (adiabatic system), there is no exchange of heat between the system and its surroundings, and therefore, dQ = 0, and Eq. (7.17) reduces to Eq. (7.20)

$$\Delta s_{isolated} \geq 0 \tag{7.20}$$

Eq. (7.20) indicates that the entropy of an isolated system always rises during a process or it remains constant in case of a reversible process. This phenomenon is called the increase of entropy principle. In an adiabatic process, entropy change occurs because of irreversibilities only, and they always tend to raise entropy.

7.4.3 Entropy of pure substances

Solar receivers utilize various materials such as heat transfer fluids and storage media. These materials are used at specified temperature (T) and pressure (P).

Thus, it is often desirable to compute their entropies. For example, steam at $T \approx 523-573$ K and $P = 4.5$ MPa can be directly generated by a receiver to drive a Rankine cycle (Ho, 2017).

The value of entropy of a system is determined by fixing the state of the system. The state of a simple compressible system is fixed by specifying two intensive independent properties and the values of other properties at that state. It is possible to define entropy change of a substance in terms of other properties. Nevertheless, these mathematical relations are too complex and not amenable to hand calculations. Consequently, the entropies of substances are evaluated from measurable property data and the results are usually presented in a tabular form like the other properties such as internal energy and enthalpy.

Tabulated entropy values are presented relative to an arbitrary reference state. For instance, the entropy of saturated liquid (s_f) at 273.16 K is assigned the value of zero in steam tables (Çengel and Boles, 2006). For refrigerant-134a, the zero value is given to the saturated liquid at 233.15 K. It should be noted that the entropy of a substance assumes negative values at temperatures below the reference point. At a given state, the value of entropy is found like any other thermodynamic property. The values of entropy can be taken directly from the tables at the specified state, in compressed liquid and superheated vapor regions (Fig. 7.6). In the saturated mixture region, it is computed from Eq. (7.21).

$$s = s_f + x s_{fg} \tag{7.21}$$

Where x is quality, s_f and s_{fg} are values of entropy listed in tables.

Values of x can be computed from the following:

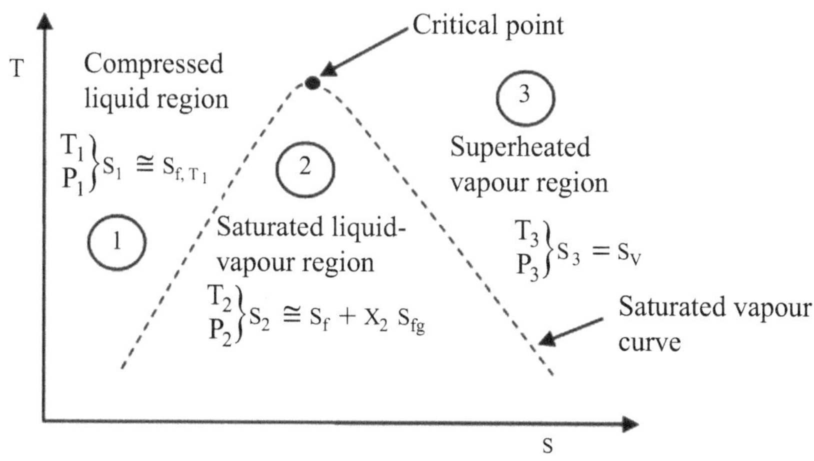

FIGURE 7.6 Entropy of a pure substance determined from the tables. *Adapted from Çengel and Boles (2006).*

$$x = \frac{V_m - V_f}{V_{fg}} = \frac{m_{vapour}}{m_{total}} \tag{7.22}$$

Where V_m is the average volume, V_f is the volume of liquid, $V_{fg} = V_v - V_f$, and V_v is the volume of vapor.

Example 7.2

A rigid tank contains 12 kg of water at 363 K. If 9 kg of the water is in liquid phase, and the liquid and vapor volumes are respectively 0.009 and 7.078 m^3, compute the average volume.

Solution

From Eq. (7.22),

$$x = \frac{V_m - V_f}{V_{fg}} = \frac{V_m - V_f}{V_v - V_f}$$

So,

$$V_m = x(V_v - V_f) + V_f$$

$$\text{But } x = m_{vapour}/m_{total} = 3/12 = 0.25.$$

Therefore,

$$V_m = 0.25 * (7.078 - 0.009) + 0.009 = 1.776 \text{ m}^3$$

The entropy of a compressed liquid can be approximated by the entropy of the saturated liquid at given temperature and pressure, in the absence of data on compressed liquid. The entropy change (ΔS) of a closed system with mass (m) during a process is given by Eq. (7.23).

$$\Delta S = m\Delta s = m(s_2 - s_1) \tag{7.23}$$

Where s_1 and s_2 are respectively the initial and final specific entropies of the system.

7.4.4 Isentropic processes

The entropy of a given mass can be varied by heat transfer and irreversibilities. Nevertheless, some processes are adiabatic and internally reversible. Thus, the entropy of a fixed mass does not change in such processes. An isentropic process is a process in which the entropy remains constant. In this process, the initial (s_1) and final (s_2) entropies of the system are equal:

$$s_1 = s_2 \tag{7.24}$$

Engineering systems or devices such as pumps, turbines, nozzles, and diffusers essentially operate without exchanging heat with the surroundings, and they perform best when the irreversibilities, such as the friction, are reduced. Consequently, an isentropic process can be used as a suitable model for real processes. Isentropic processes can also be applied in the definitions of ideal efficiencies for processes. For instance, the Carnot cycle is made up of two reversible isothermal and two reversible isentropic processes (Figure 1.11). The efficiency of a Carnot cycle is defined based on these ideal processes. Similarly, the Rankine steam cycle comprises two isentropic and two isobaric processes (see Sections 1.5.1 and 1.5.2). Efficiencies of these cycles are defined based on their corresponding ideal processes, and they can be compared with actual system performance.

7.5 Exergy of solar receivers

Exergy is the maximum useful work that can be obtained when the system is taken from its given state to the thermodynamic equilibrium with the environment (Querol et al., 2013). The rest of the energy is finally rejected to the environment as waste energy.

7.5.1 A system and its surroundings

In a dead state, a system is in thermodynamic equilibrium with its immediate surroundings. In this case, the system (a) is at the temperature and pressure of its surroundings, (b) has no kinetic or potential energy relative to the surroundings, and (c) does not chemically react with its surroundings. In addition, all magnetic, electrical, and surface tension effects between the system and its surroundings are balanced, where applicable to the system. Conventionally, the dead-state temperature and pressure are taken to be $T_0 = 298$ K and $P_0 = 101.325$ kPa (Chen, 2015). At the dead state, the exergy of a system is zero.

There is a distinction between the surroundings, immediate surroundings, and the environment. Surroundings refer to everything outside the system boundaries. Immediate surroundings are part of the surroundings that are affected by the process, and environment is the region outside the immediate surroundings (Fig. 7.7). Consequently, any irreversibilities during a process take place within the system and its immediate surroundings, while the environment is not affected by such irreversibilities. The concept of exergy is used to analyze the maximum possible work done by or on different systems.

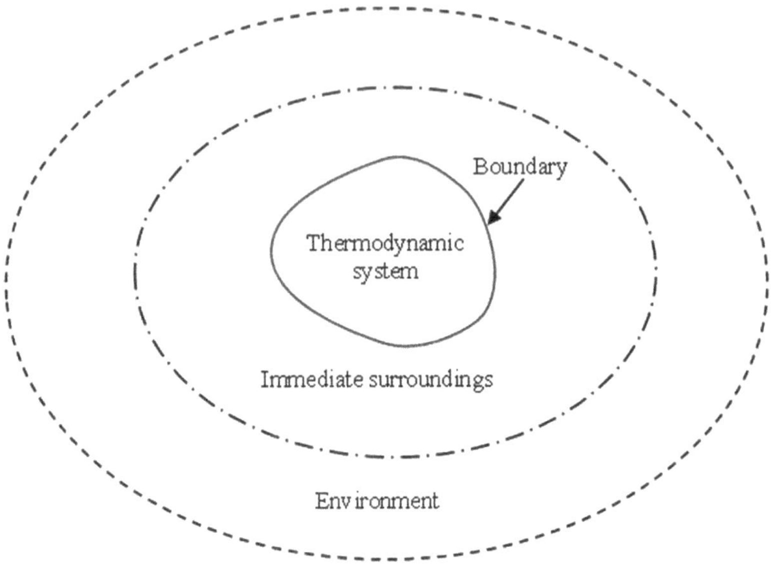

FIGURE 7.7 Schematic representation of a thermodynamic system and its immediate surroundings and environment.

7.5.2 Exergy analysis

Exergy may involve matter and nonmatter streams, and direct and indirect methods are available for its calculation. Matter-based exergy includes physical and chemical exergy, while the exergy of nonmatter streams involves work, electricity, heat flow, and solar radiation. A solar receiver converts solar radiation to heat that is evacuated by a heat transfer fluid. Thus, both matter and nonmatter-based exergies are encountered in the thermodynamic analysis of receivers. Only the exergy of solar radiation and heat flow are presented and discussed in this book. Details of other types of exergy are reported elsewhere (Madhlopa, 2018; Querol et al., 2013). Exergy analysis deals with the computation of the amount of irreversibility that takes place in each component of a thermodynamic cycle and aims at establishing effective parameters (Behbahaninia et al., 2017). For instance, exergy destruction varies from one component of a CSP plant to another.

7.5.2.1 Exergy of solar radiation

Solar radiation comes from the sun. Unlike thermal reservoirs, the sun is a radiation reservoir. Consequently, different approaches are employed in the

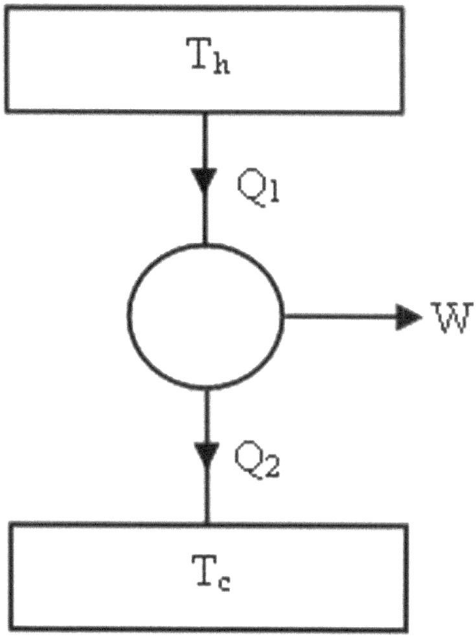

FIGURE 7.8 Schematic representation of a Carnot cycle.

computation of exergy for solar energy $\left(\dot{Q}_s\right)$. Often, the model proposed by Petela (2014) is applied in the calculation of exergy for solar energy:

$$\hat{E}_s = \dot{Q}_s\left[1 - \frac{4}{3}\left(\frac{T_a}{T_s}\right) + \frac{1}{3}\left(\frac{T_a}{T_s}\right)^4\right] \tag{7.25}$$

Where T_a is the ambient temperature and T_s is the temperature of the sun.

7.5.2.2 Exergy of heat flows

Fig. 7.8 is a schematic representation of the Carnot engine. The maximum work (W_{max}) that can be obtained from this engine is the available energy ($\hat{E}_{av}$), while the maximum energy that must be rejected is the unavailable energy ($\hat{E}_{ua}$). Using energy balances, the heat supply $\dot{Q}_1$ from the hot reservoir can be given by Eq. (7.26).

$$\dot{Q}_1 = \hat{E}_{av} + \hat{E}_{ua} \tag{7.26}$$

$$W_{max} = \hat{E}_{av} = \dot{Q}_1 - \dot{Q}_2 = \dot{Q}_1 - \hat{E}_{ua} \tag{7.27}$$

Where $\dot{Q}_1$ is the flow rate of energy from the hot reservoir, $\dot{Q}_2$ is the flow rate of energy into the cold reservoir.

The maximum energy efficiency (η_{max}) for this engine can be expressed as follows:

$$\eta_{max} = 1 - \frac{T_c}{T_h} \tag{7.28}$$

Where T_c is the temperature of the cold reservoir and T_h is the temperature of the hot reservoir.

Therefore, the maximum work output can be computed from:

$$W_{max} = \eta_{max}\dot{Q}_1 = \left(1 - \frac{T_c}{T_h}\right)\dot{Q}_1 \tag{7.29}$$

Eq. (7.29) shows that the quality of energy increases with the temperature of the hot reservoir for a given temperature of the cold reservoir. The minimum temperature at which an engine rejects heat is the ambient temperature (T_a). So, the exergy ($\hat{E}$) of heat $\dot{Q}$ at a temperature T is defined as the maximum possible output work at an ambient temperature and this variable can be computed from (Bellos et al., 2017):

$$\hat{E} = \left(1 - \frac{T_a}{T}\right)\dot{Q} \tag{7.30}$$

7.5.2.3 Exergy balance and efficiency

Exergy destruction in a component of the CSP plant can be defined as the difference between its inflow exergy and outflow exergy (Petrakopoulou et al., 2017). The total exergy is found by adding the exergies of the individual components. So, in general, the exergy balance of a system is described by Eq. (7.31) (Sansaniwal et al., 2018):

$$\sum \hat{E}_d = \sum \hat{E}_{in} - \sum \hat{E}_{out} \tag{7.31}$$

Where $\hat{E}_d$ is the exergy destruction, $\hat{E}_{in}$ is the inflow exergy and $\hat{E}_{out}$ is the outflow exergy.

In addition, the exergy destruction or irreversibility rate (I) can be computed from (Szargut, 2005):

$$I = \hat{E}_d = T_0 s_{gen} \tag{7.32}$$

Where T_0 is the reference temperature (dead state).

In general, there is exergy flow into and out of a solar receiver (Selwynraj et al., 2015). Some physical exergy flows into the receiver as the heat transfer

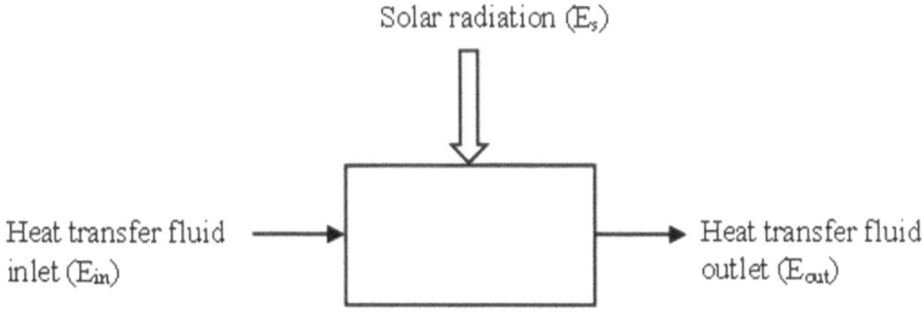

FIGURE 7.9 Exergy flow in a solar receiver.

fluid enters the receiver ($\hat{E}_{in}$). In addition, the receiver absorbs a fraction of the solar exergy ($\hat{E}_s$), thereby augmenting the exergy of the heat transfer fluid. The receiver delivers useful exergy ($\hat{E}_{out}$) from the outlet. Fig. 7.9 is a simplified representation of exergy flow in a solar receiver. Due to irreversibilities, some of the exergy is destroyed within the receiver. So, exergy destruction in the receiver can be expressed by Eq. (7.33).

Due to irreversibilities, some of the exergy is destroyed within the receiver. So, exergy destruction ($\hat{E}_d$) in the receiver can be expressed as follows:

$$\hat{E}_d = \hat{E}_{in} + \hat{E}_s - \hat{E}_{out} \tag{7.33}$$

$\hat{E}_{in}$ and $\hat{E}_{out}$ are physical exergies that can be computed from Eq. (7.34) (Querol et al., 2013).

$$\hat{E}_{ph} = \dot{m}[i - i - T_0(s - s_0)] = \dot{m}\left[C_p(T - T_0) - C_pT_0 \ln\left(\frac{T}{T_0}\right) + RT_0 \ln\left(\frac{P}{P_0}\right)\right] \tag{7.34}$$

Where i is the specific enthalpy, P is the pressure, R is the universal gas constant (8.314 J $mol^{-1}K^{-1}$), s is the specific entropy, and the subscript 0 indicates values at reference point.

The exergy delivered by a solar receiver at the outlet temperature (T_{out}) is given by (Tyagi et al., 2007):

$$\hat{E}_s = \left(1 - \frac{T_0}{T_{out}}\right)\dot{Q}_s \tag{7.35}$$

$$\dot{Q}_s = A_{ap}G_b \tag{7.36}$$

Where A_{ap} is the area of the aperture and G_b is the intensity of beam radiation intercepted by the concentrator.

From a thermodynamic perspective, a true measure of the system performance is given by the exergetic efficiency (also known as the second law efficiency). Exergetic efficiency (η_{ex}) can be used to evaluate the thermodynamic performance of a component as well as the entire power plant. It is given by the following:

$$\eta_{ex} = \frac{\hat{E}_{out}}{\hat{E}_{in} + \hat{E}_s} \tag{7.37}$$

Example 7.3

A solar receiver uses pressurized air as a heat transfer fluid. At a certain time of the day, the inlet and outlet temperatures of the air are, respectively, 550 and 900 K at a pressure P = 2 MPa and a flow rate of 5 kgs^{-1}. Calculate the exergy destruction and efficiency of the receiver, assuming the (a) solar heat flow is 6.5 MW, reference temperature is 273.15 K, reference pressure is 0.101325 MPa, and (b) specific heat capacity of air at constant pressure depends on temperature as follows:

C_p = 0.1805 T + 950.1, calculated at the mean of inlet and outlet temperatures.

Solution

From Eq. (7.33), exergy destruction, $\hat{E}_d = \hat{E}_{in} + \hat{E}_s - \hat{E}_{out}$

Mean temperature, T_m=(550 + 900)/2 = 725 K.

C_p = 0.1805 T + 950.1 = 0.1805 *725 + 950.1 = 1080.96.

$\hat{E}_{in}$ and $\hat{E}_{out}$ are physical exergies and they can be calculated from Eq.(7.34. Therefore,

$$\hat{E}_{in} = \dot{m}\left[C_p(T_{in} - T_0) - C_pT_0 \ln\left(\frac{T_{in}}{T_0}\right) + RT_0 \ln\left(\frac{P_{in}}{P_0}\right)\right]$$

$$= 5 * \left[1080.96 * (550 - 273.15) - 1080.96 * 273.15 * \ln\left(\frac{550}{273.15}\right)\right.$$

$$\left. + 8.314 * 273.15 * \ln\left(\frac{2}{0.101325}\right)\right] = 496,912$$

$$\hat{E}_{out} = \dot{m}\left[C_p(T_{out} - T_0) - C_pT_0 \ln\left(\frac{T_{out}}{T_0}\right) + RT_0 \ln\left(\frac{P_{out}}{P_0}\right)\right]$$

$$= 5 * \left[1080.96 * (900 - 273.15) - 1080.96 * 273.15 * \ln\left(\frac{900}{273.15}\right)\right.$$

$$\left. + 8.314 * 273.15 * \ln\left(\frac{2}{0.101325}\right)\right] = 1,661,539$$

$$\hat{E}_s = \left(1 - \frac{273.15}{900}\right)6,500,000 = 4,527,250$$

Therefore, $\hat{E}_d$ = 496,912 + 4,527,250 − 1,661,539 = 3.363 MW.

Exergy efficiency (η_{ex}) is:

$$\eta_{ex} = \frac{\hat{E}_{out}}{\hat{E}_{in} + \hat{E}_s} = \frac{1661539}{496912 + 4527250} = 0.33$$

Many studies have examined the exergy of CSP systems and components. Most of the exergy destruction occurs in the solar section, especially in the receiver because of the low conversion of solar radiation to thermal energy that ultimately gets transferred to the working fluid (Giostri, 2017). AlZahrani and Dincer (2018) also found similar results. It is therefore necessary to find ways of reducing exergy destruction in solar receivers.

Nomenclature

A area (m^2)
e internal energy ($J\ kg^{-1}$)
E energy per unit mass ($J\ kg^{-1}$)
g acceleration due to gravity ($m\ s^{-2}$)
i specific enthalpy ($J\ kg^{-1}$)
m mass (kg)
P pressure (Pa)
Q heat per unit mass (J)
s specific entropy ($J\ kg^{-1}K^{-1}$)
T temperature (K)
u velocity in the direction of x-coordinate (ms^{-1})
V volume (m^3)
W work per unit mass ($J\ kg^{-1}$)
$\hat{\mathbf{E}}$ exergy flow (W)
$\dot{\mathbf{m}}$ mass flow rate ($kg\ s^{-1}$)
$\dot{\mathbf{Q}}$ heat flow rate (W)
z distance along z-coordinate (m)

References

AlZahrani, A. A., & Dincer, I. (2018). Energy and exergy analyses of a parabolic trough solar power plant using carbon dioxide power cycle. *Energy Conversion and Management, 158*, 476–488. https://doi.org/10.1016/j.enconman.2017.12.071

ASHRAE. (2017). *ASHRAE handbook - fundamentals*. American Society of Heating, Refrigerating and Air-Conditioning Engineers (ASHRAE).

Behbahaninia, A., Ramezani, S., & Lotfi Hejrandoost, M. (2017). A loss method for exergy auditing of steam boilers. *Energy, 140*, 253–260. https://doi.org/10.1016/j.energy.2017.08.090

Bellos, E., Tzivanidis, C., & Antonopoulos, K. A. (2017). Parametric analysis and optimization of a solar assisted gas turbine. *Energy Conversion and Management, 139*, 151–165. https://doi.org/10.1016/j.enconman.2017.02.042

Çengel, Y. A., & Boles, M. A. (2006). *Thermodynamics: An engineering approach* (5th ed.). McGraw Hill.

Chen, Z. (2015). On the accuracy of laminar flame speeds measured from outwardly propagating spherical flames: Methane/air at normal temperature and pressure. *Combustion and Flame, 162*(6), 2442–2453. https://doi.org/10.1016/j.combustflame.2015.02.012

Dincer, I., & Rosen, M. A. (2021). Exergy and energy analyses. *In Exergy:Energy, Environment and Sustainable Development* (3rd ed., pp. 23–35). Elsevier Inc.

Ekeren, P. J., van (1998). Principles and Practice. *Handbook of Thermal Analysis and Calorimetry* (1). Elsevier Inc.

Giostri, A. (2017). Preliminary analysis of solarized micro gas turbine application to CSP parabolic dish plants. *Energy Procedia, 142*, 768–773. https://doi.org/10.1016/j.egypro.2017.12.124

Ho, C. K. (2017). Advances in central receivers for concentrating solar applications. *Solar Energy, 152*, 38–56. https://doi.org/10.1016/j.solener.2017.03.048

Honig, J. M. (2014). Fundamentals. *In Thermodynamics: Principles Characterizing Physical and Chemical Processes* (4th ed, pp. 1–115). Esevier Inc.

Madhlopa, A. (2018). *Principles of solar gas turbines for electricity generation.* Gewerbestras. https://doi.org/10.1007/978-3-319-68388-1

Petela, R. (2014). Exergy analysis of solar radiation. In N. Enteria, & A. Akbarzadeh (Eds.), *Solar thermal sciences and engineering applications* (pp. 7–96). CRC Press, Taylor & Francis.

Petrakopoulou, F., Sánchez-Delgado, S., Marugán-Cruz, C., & Santana, D. (2017). Improving the efficiency of gas turbine systems with volumetric solar receivers. *Energy Conversion and Management, 149*, 579–592. https://doi.org/10.1016/j.enconman.2017.07.058

Querol, E., Gonzalez-Regueral, B., & Perez-Benedito, L. (2013). *A practical approach to exergy and thermoeconomic analyses of industrial processes.* Springer.

Rosen, M. A. (2021). Exergy analysis. In M. E. H. Assad, & M. A. Rosen (Eds.), *Design and Performance Optimization of Renewable Energy Systems* (pp. 43–60). Elsevier Inc.

Sansaniwal, S. K., Sharma, V., & Mathur, J. (2018). Energy and exergy analyses of various typical solar energy applications: A comprehensive review. *Renewable and Sustainable Energy Reviews, 82*, 1576–1601. https://doi.org/10.1016/j.rser.2017.07.003

Sayyaadi, H. (2021). Thermal modeling and analysis. *In Modeling, Assessment, and Optimization of Energy Systems* (pp. 21–100). Elsevier Inc.

Selwynraj, A. I., Iniyan, S., Polonsky, G., Suganthi, L., & Kribus, A. (2015). Exergy analysis and annual exergetic performance evaluation of solar hybrid STIG (steam injected gas turbine) cycle for Indian conditions. *Energy, 80*, 414–427. https://doi.org/10.1016/j.energy.2014.12.001

Szargut, J. (2005). *Exergy method: Technical and ecological applications.* WIT Press.

Szőllősi, T., & Márkus, F. (2015). Searching the laws of thermodynamics in the Lorentz-invariant thermal energy propagation equation. *Physics Letters A, 379*(36), 1960–1965. https://doi.org/10.1016/j.physleta.2015.06.029

Tosun, I. (2013). Review of the First and Second Laws of Thermodynamics. *The Thermodynamics of Phase and Reaction Equilibria* (pp. 1–12). Elsevier Inc.

Tumen Ozdil, N. F., Segmen, M. R., & Tantekin, A. (2015). In Thermodynamic analysis of an Organic Rankine Cycle (ORC) based on industrial data. *Applied Thermal Engineering, 91*, 43–52. https://doi.org/10.1016/j.applthermaleng.2015.07.079

Tyagi, S. K., Shengwei, W., Singhal, M. K., Kaushik, S. C., & Park, S. R. (2007). Exergy analysis and parametric study of concentrating type solar collectors. *International Journal of Thermal Sciences, 46*, 1304–1310. https://doi.org/10.1016/j.ijthermalsci.2006.11.010

Zohuri, B. (2018). The Beginning and Concept of Cryogenics, Basic Principles. *In Physics of Cryogenics: An Ultralow Temperature Phenomenon* (pp. 387–473). Elsevier Inc.

Chapter 8

Hydrodynamics of solar receivers

8.1 Introduction

In concentrating solar power (CSP) plants, solar energy is focused onto a receiver through which a heat transfer fluid is passed. Pumps and compressors are usually utilized to create pressure differentials, move the fluid through the receiver, and distribute it to other components of the CSP system. The main function of a heat transfer fluid is to directly or indirectly transmit energy from the receiver to other CSP components, such as a storage unit and an power block. In direct designs, the heat transfer fluid is also used as a working fluid that transmits kinetic energy to a steam or gas turbine in the power block. For indirect systems, the heat transfer fluid transmits heat to the working fluid via a heat exchanger. Fig. 8.1 shows simplified layouts of direct and indirect heat transmission.

Some of the heat transfer media are conventional (various liquids and gases), while other media are nonconventional (such as nanofluids and suspended solid particles). These media exhibit a wide spectrum of flow and heat-transfer characteristics. In solar receivers, heat transfer fluids flow through

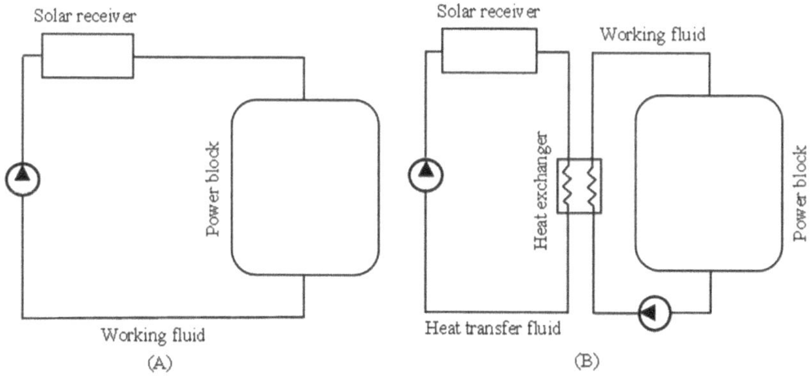

FIGURE 8.1 Schematic representation of simplified layouts of: (A) direct heating concentrating solar power plant and (B) indirect heating concentrating solar power plant.

Solar Receivers for Thermal Power Generation. https://doi.org/10.1016/B978-0-323-85271-5.00008-2

conduits of different geometries, such as tubes and porous materials (see Chapter 3). Some geometries present higher resistance to fluid flow than others. Advanced heat transfer fluids can be exploited in a wide range of temperature (low to high). Nevertheless, there are some hydrodynamic challenges associated with exploitation of these heat transfer materials. For example, particle receivers have the problems of particle conveyance, attrition, and transport (Teng & Xuan, 2018).

Hydrodynamics is a branch of science that deals with fluids in motion. Flowing fluids can transfer mass, momentum, and energy. The motion of a fluid is caused by a pressure gradient, in which the fluid flows from high to low pressure. A fluid is a continuous medium that deforms itself if it is submitted to shear stress. In contrast, a solid material exhibits finite deformation under stress. So, the major difference between solids and fluids is the way they react to shear stress. Fluids are classified into liquids and gases that have different properties.

8.2 Fluid properties

Liquids have a nearly fixed volume, while gases occupy the entire volume of a space. The natures of their molecular interactions significantly differ in both extent of compressibility and formation of a free surface in liquid (ASHRAE, 2017). Liquids are generally considered to be incompressible fluids, while gases range from compressible to almost incompressible. Liquids possess unbalanced molecular cohesive forces at or close to the surface (interface), and therefore, the liquid surface tends to contract and exhibit properties that are analogous to a stretched elastic membrane. Consequently, a liquid surface is under tension called surface tension.

Fluids flow due to pressure differentials, and various simple models are available for describing fluid motion. The ideal-fluid model is the simplest one, in which, it is assumed that the fluid does not resist shearing. Flow analysis of ideal fluids is well developed (Schlichting, 1979), and it is valid in a wide spectrum of applications.

8.2.1 Density

The density (ρ) of a substance is its mass per unit volume. Fluid density is influenced by other state properties such as temperature and pressure via the equation of state. In this context, the density of real gases can be estimated by using the ideal gas equation (Benoit et al., 2016):

$$\rho = \frac{P}{R_{ga}T} \tag{8.1}$$

Where p is pressure, R_{ga} is gas constant for a specific gas, and T is temperature. For instance, $R_{air} = 287\ \mathrm{J\ kg^{-1}K^{-1}}$.

Example 8.1

During a certain time of the day, the ambient air temperature is 293 K when the atmospheric pressure is 101,325 Pa. Calculate the density of air under these weather conditions.

Solution

From Eq. (8.1), the density of air is given by:

$$\rho = \frac{P}{R_{ga} T}$$

Therefore,

$$\rho = \frac{101325}{287 * 293} = 1.20\ \text{kgm}^{-3}$$

8.2.2 Viscosity

Viscosity is a measure of fluid resistance to shear. It is the internal friction, which molecular and ionic particles apply upon one another. The adhesion forces of the fluid particles influence the viscosity, and there are two types of viscosity used in science and engineering: dynamic (μ) and kinematic (ν) viscosity. Dynamic viscosity is the resistance to the movement of one layer of a fluid over another. It is commonly used as a factor in fluid dynamics problems. Kinematic viscosity is the ratio of dynamic viscosity to the density of the fluid ($\nu = \mu/\rho$), and it is occasionally used in fluid dynamics.

Some flows are significantly affected by viscosity while others are marginally influenced by this fluid property. A quantity that indicates the effect of viscosity on flow is the Reynolds number (Re), given by (ASHRAE, 2017):

$$\text{Re} = \frac{\text{inertial forces}}{\text{viscous forces}} = \frac{V_r D_h}{\nu} \tag{8.2}$$

Where D_h is the hydraulic diameter, V_r is the reference velocity of the fluid, and ν is the kinematic viscosity. For circular tubes and ducts, the hydraulic diameter is equal to the internal diameter of the tube. For noncircular conduits, the hydraulic diameter is equal to four times the ratio of cross-sectional area of flow to the total wetted perimeter (4× cross-sectional area/wetted perimeter).

Based on the Re, a criterion for establishing the effect of viscosity on flow is as follows:

(a) When Re>>1, the flow is very inviscid. An inviscid flow is a flow in which the dissipative, transport phenomena of viscosity, mass diffusion and thermal conductivity are neglected. Usually, inviscid flows can easily become turbulent.

(b) When R<<1, the flow is very viscous. There is a tendency for viscous flows to be laminar.

8.2.3 Newtonian and non-Newtonian fluids

Flow characteristics of a heat transfer fluid influence the performance of a solar receiver as well as the entire CSP plant. For instance, high resistance to flow requires more pumping power to circulate the heat transfer fluid, thereby reducing the overall energy efficiency of the plant. Consequently, it is necessary to analyze the flow of heat transfer fluids in solar receivers. During flow analysis, the effects of viscosity are taken into consideration by classifying a fluid as either Newtonian or non-Newtonian (Bar-Meir, 2013; Niño et al., 2022).

8.2.3.1 Newtonian fluids

A Newtonian fluid is a fluid in which viscosity is independent of the shear rate (Palagi et al., 2017). In this type of fluids, the rate of deformation is directly proportional to the shear stress. Most of the fluids (such as water, air, oil, refrigerants) exploited in the heating, ventilation, and air-conditioning industry can be classified as Newtonian. Conventional fluids, such as water and air, are commonly used as heat transfer fluids in solar receivers (see Chapter 5). For instance, a majority of the operating CSP plants is currently based on the Rankine cycle in which water is used as a working fluid (Dunham & Iverson, 2014). Consider a fluid flow between two parallel plates or coaxial cylinders. Fig. 8.2 shows a fluid flow between two parallel plates. The two plates are separated by distance h, and each one has an area (A). The bottom plate is stationary, while the top plate is moving, thereby causing a shearing force in the fluid. Similarly, in case of flow between coaxial cylinders, one cylinder may be stationary while the other one is moving. The

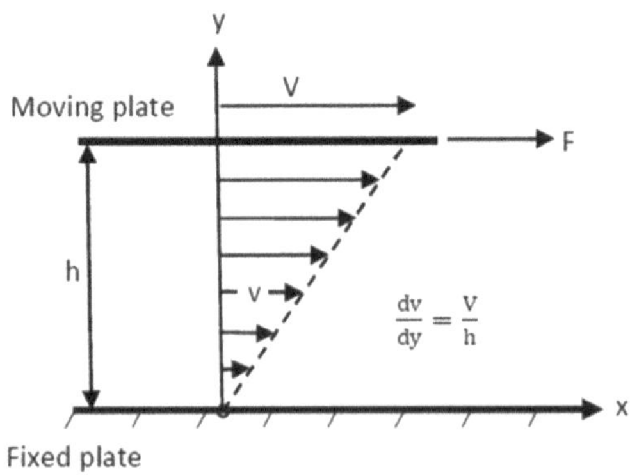

FIGURE 8.2 Velocity profile and gradient of a simple flow. *Adapted from ASHRAE (2017).*

flow of a viscous fluid in the space between two surfaces, one of which is moving tangentially relative to the other, is called a Couette flow (Kundu & Cohen, 2007). This type of flow was named after the French physicist Maurice Couette (1858–1943) and it may be extended to the case in which the flow is driven by a pressure difference (Katopodes, 2019). The theory of Couette flow is important in problems of heat transfer, lubrication, food processing and other applications. From a thermodynamics perspective, it provides the simplest model for analysing the transfer of heat in fluid flows between two coaxial cylinders or parallel plates (Demirel, 2002). In this vein, steady or transient conditions can be assumed when modelling a Couette flow.

For a Newtonian fluid, the tangential force (F) per unit area required to glide one plate with velocity V parallel to the other is proportional to the lateral velocity gradient dv/dy = V/h (ASHRAE, 2017). Therefore,

$$\frac{F}{A} = \mu\left(\frac{V}{h}\right) \tag{8.3}$$

Where the constant of proportionality factor (μ) is the absolute or dynamic viscosity of the fluid, and F/A is the shearing stress (σ).

In complex flows, the velocity and shear stress may change across the flow field, and the shear stress is given by:

$$\sigma_{xy} = \mu\frac{dv}{dy} \tag{8.4}$$

Where the subscript x represents the direction of the shear stress while the y subscript represents the direction of the area (at right angles to the area A).

Dynamic viscosity is predominantly dependent on temperature. Except at critical point, the viscosity of gases increases with the square root of the absolute temperature, as envisaged by the kinetic theory of gases. Some polynomial corrections of viscosity and temperature have also been developed for various fluids, including gases (Benoit et al., 2016). On the other hand, the viscosity of liquids decreases with rising temperature. Liquid metals can be considered as Newtonian fluids for many applications. In this context, liquid sodium is a promising heat transfer medium in solar receivers. In addition, many aluminum alloys behave like Newtonian liquids until the first solidification appears.

8.2.3.2 Non-Newtonian fluids

In non-Newtonian fluids, the rate of deformation and shear stress are related in a more complex way. In Eqs. (8.3) and (8.4), the relation between the stress and velocity gradient was assumed to be linear. Not all materials behave according to this model. There are many materials that exhibit nonlinear relationship between shear stress and velocity gradient. In general, the shear stress can be expressed by (Bar-Meir, 2013):

$$\sigma_{xy} = K\left(\frac{dV}{dy}\right)^{n-1}\frac{dV}{dy} \tag{8.5}$$

Where K and n are constants, and the term $K\left(\frac{dV}{dy}\right)^{n-1}$ represents viscosity.

When $n = 1$, Eq. (8.5) stands for a Newtonian fluid, and K becomes the usual coefficient of dynamic viscosity μ (which is always positive). When n exceeds one, the fluid is dilettante. When n is less than one, the fluid is pseudoplastic. The fluids that satisfy Eq. (8.5) are known as purely viscous fluids.

8.3 Hydrodynamic equations

Usually, the flow characteristics of a fluid are modeled by taking into consideration the governing equations of the flow: continuity, momentum, and energy equations (Anderson, 1991). These equations can be expressed in integral or differential forms. Nevertheless, the integral analysis has a limited accuracy. In view of this, the differential form is widely applied. It allows the investigation of the flow field in more detail. Differential analysis is performed on infinitesimal scale, and therefore, it yields higher accuracy. Partial differential equations of momentum are known as the Navier–Stokes equations, named after Claude-Louis Navier-Marie and George Gabriel Stokes who derived them. These equations were initially derived by Claude-Louis-Marie Navier in 1827. Using the same arguments as Navier, Simon-Denis Poisson independently derived these equations in 1831. These derivations were based on a molecular perspective of the way in which stresses are applied between fluid layers. Later, based on the relationship between stress and rate-of-strain, Barré de Saint Venant and George Gabriel Stokes derived these equations in 1843 and 1845, respectively. Navier–Stokes equations are nonlinear and possess more than one solution in many cases. One of the solution procedures is to reduce the equations by neglecting the effects of viscosity. The equations without viscosity effects are known as the ideal flow equations.

There are two types of forms of the governing equations: Conservation and nonconservation. In the conservation form, there is continuity after a shock wave, and the equations have a divergence term on the left-hand side. On the other hand, the nonconservation form does not see continuity after a shock wave. The conservation form of the governing equations is recommended for a shock-capturing method (Anderson, 1991). Application of the conservation form yields computed flow-field results that are generally smooth and stable. Nonetheless, the computed flow-field results often exhibit unsatisfactory spatial oscillations upstream and downstream of the shock wave, the shocks may be misallocated and the solution may become unstable when the nonconservation form is applied in the computation of a shock-capturing solution. On the other hand, satisfactory results are often found for both the conservation and nonconservation forms. In this book, both forms of the governing equations are presented. Details of the derivation of these equations fall outside the scope of this book but they can be found elsewhere in books on fundamentals of hydrodynamics or fluid dynamics.

8.3.1 Equations for viscous flow

8.3.1.1 Continuity equation

For a viscous flow, the law of conservation of mass states that mass cannot be created nor destroyed (Bergman et al., 2011). So, the mass flow rate into a section of the conduit should be equal to the mass flow rate out of that conduit section if no mass is stored or lost (say, through leakage). In this context, the flow of a fluid into and out of a three-dimensional infinitesimal control volume is considered. At a specific time, this control volume can be regarded as a system. The mass conservation for this infinitesimal tiny system is zero. In the Cartesian coordinates, the law of conservation of mass in a cubical control volume (dx $\times$ dy $\times$ dz) can be expressed as:

$$\frac{\partial \rho}{\partial t}+\frac{\partial(\rho u)}{\partial x}+\frac{\partial(\rho v)}{\partial y}+\frac{\partial(\rho w)}{\partial z}=0 \tag{8.6}$$

Where ρ is the density of the fluid, $\frac{\partial}{\partial t}$ is the local derivative at a fixed point, and u, v, and w are respectively velocities in the directions of x, y and z.

Geometry is an important factor of solar receiver design. In this context, some receivers have cylindrical geometry while others are hemispherical in shape (Bellos et al., 2019; Rodriguez-Sanchez et al., 2015). For instance, there are cylindrical and hemispherical cavity receivers, and some designs of external receivers are cylindrical. Consequently, it is convenient to analyze heat and mass transfer in such systems by using cylindrical or spherical coordinates. The continuity equation in cylindrical coordinates is given by:

$$\frac{\partial \rho}{\partial t}+\frac{1}{r}\frac{\partial(r\rho v_r)}{\partial r}+\frac{1}{r}\frac{\partial(\rho v_\theta)}{\partial \theta}+\frac{\partial(\rho v_z)}{\partial z}=0 \tag{8.7}$$

Where r is the radius of the cylinder, z is distance from the base of the cylinder along its axis, v_r, v_θ, and v_z are velocities in the r, θ, and z directions, respectively.

In spherical coordinates, the continuity equation is expressed as:

$$\frac{\partial \rho}{\partial t}+\frac{1}{r^2}\frac{\partial(r^2\rho v_r)}{\partial r}+\frac{1}{r\sin\theta}\frac{\partial(\rho v_\theta \sin\theta)}{\partial \theta}+\frac{1}{r\sin\theta}\frac{\partial(\rho v_\phi)}{\partial \phi}=0 \tag{8.8}$$

Where r is the distance from the origin to a point (r,θ,ϕ), θ is the angle between the x-axis and the radial line from the origin (similar to θ in polar coordinates), ϕ is the angle between the z-axis and the line from the origin to the point, and v_r, v_θ, and v_ϕ are velocities in the r, θ, and ϕ directions, respectively.

Compressible and incompressible fluids are utilized in solar receivers. For incompressible fluids, the continuity equation in Cartesian, cylindrical, and spherical coordinates can respectively be given by Eqs. (8.9)–(8.11):

$$\frac{\partial u}{\partial x}+\frac{\partial v}{\partial y}+\frac{\partial w}{\partial z}=0 \tag{8.9}$$

$$\frac{1}{r}\frac{\partial (rv_r)}{\partial r}+\frac{1}{r}\frac{\partial v_\theta}{\partial \theta}+\frac{\partial v_z}{\partial z}=0 \tag{8.10}$$

$$\frac{1}{r^2}\frac{\partial (r^2 v_r)}{\partial r}+\frac{1}{r\sin\theta}\frac{\partial (v_\theta \sin\theta)}{\partial \theta}+\frac{1}{r\sin\theta}\frac{\partial v_\phi}{\partial \phi}=0 \tag{8.11}$$

Except for the first term, the sum of the other terms of Eqs. (8.6)–(8.8) is the dot product of vectors (divergency). Consequently, the continuity equation with no source or sink terms can be expressed in the general nonconservation form (Anderson, 2009; Fanchi, 2010):

$$\frac{D\rho}{Dt}+\rho\nabla\cdot\mathbf{V}=0 \tag{8.12}$$

Where $D\rho/Dt$ is the substantial rate of change of fluid density (time rate of change of density of a given fluid element as it moves through space), $\rho\nabla.\mathbf{V}$ is the convective term of density, $\mathbf{V}$ is the velocity vector, and ∇ is the vector differential operator (del operator).

Eq. (8.12) is the continuity equation in nonconservation form, and it can also be expressed in conservation form (Anderson, 2009; Sandá et al., 2019):

$$\frac{\partial \rho}{\partial t}+\nabla\cdot(\rho\mathbf{V})=0 \tag{8.13}$$

Where $\partial\rho/\partial t$ is local rate of change of the fluid density, $\nabla.(\rho\mathbf{V})$ is the convective term of density.

8.3.1.2 Momentum equation

Momentum is the product of mass and velocity. The rate of change of momentum is equal to force, and forces acting on an infinitesimal cube are surface, gravitational (body), and internal forces. Shear stresses acting on different surfaces of the cube are shown in Fig. 8.3.

The body force that acts on the cube in the x-direction is:

$$\hat{I}\cdot f_B=f_{Bx}dxdydz \tag{8.14}$$

Where $\hat{I}$ is a unit vector.

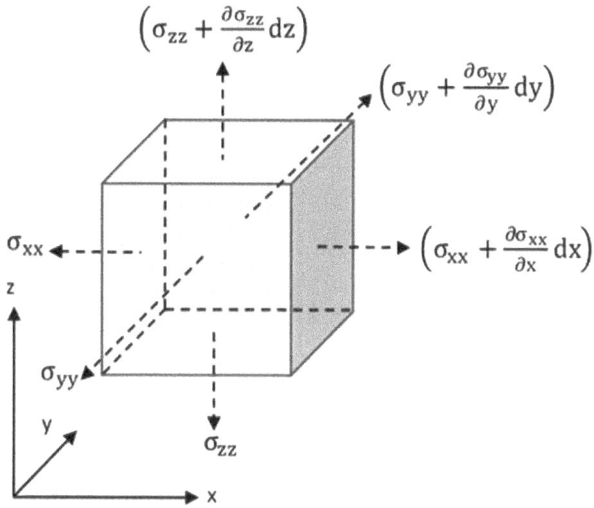

FIGURE 8.3 Schematic representation of shear stresses on different surfaces. *Adapted from Bar-Meir (2013).*

The total net surface force resulting from the shear stress in the x direction is:

$$f_x = \left(\frac{\partial \sigma_{xx}}{\partial x} + \frac{\partial \sigma_{yx}}{\partial y} + \frac{\partial \sigma_{zx}}{\partial z}\right) dxdydz \tag{8.15}$$

Internal forces in x direction amount to $\frac{Du}{Dt} \rho$ dx dy dz, and they are equal to the sum of the surface and body forces.

$$\frac{Du}{Dt} \rho dxdydz = \left(\frac{\partial \sigma_{xx}}{\partial x} + \frac{\partial \sigma_{yx}}{\partial y} + \frac{\partial \sigma_{zx}}{\partial z}\right) dxdydz + f_{Bx} dxdydz \tag{8.16}$$

Consequently,

$$\rho \frac{Du}{Dt} = \left(\frac{\partial \sigma_{xx}}{\partial x} + \frac{\partial \sigma_{yx}}{\partial y} + \frac{\partial \sigma_{zx}}{\partial z}\right) + f_{Bx} \tag{8.17}$$

Where $\frac{D}{Dt}$ is the substantial derivative.

$$\rho \frac{Dv}{Dt} = \left(\frac{\partial \sigma_{xy}}{\partial x} + \frac{\partial \sigma_{yy}}{\partial y} + \frac{\partial \sigma_{zy}}{\partial z}\right) + f_{By} \tag{8.18}$$

$$\rho \frac{Dw}{Dt} = \left(\frac{\partial \sigma_{xz}}{\partial x} + \frac{\partial \sigma_{yz}}{\partial y} + \frac{\partial \sigma_{zz}}{\partial z}\right) + f_{Bz} \tag{8.19}$$

Force causes relative motion of the layers of a fluid, and stress is the ratio of force to the area on which it is acting. In Cartesian coordinates, this division creates a stress tensor that can be expressed as:

$$\boldsymbol{\sigma} = \begin{pmatrix} \sigma_{xx} & \sigma_{xy} & \sigma_{xz} \\ \sigma_{yx} & \sigma_{yy} & \sigma_{yz} \\ \sigma_{zx} & \sigma_{zy} & \sigma_{zz} \end{pmatrix} \tag{8.20}$$

Where σ_{ij} is the stress on i^{th} surface in direction j.

The nondiagonal terms of the stress tensor are given by:

$$\sigma_{xy} = \mu\left(\frac{\partial u}{\partial y} + \frac{\partial v}{\partial x}\right) \tag{8.21}$$

The normal stress appears in the diagonal of the shear matrix, where $i = j$. The average of the normal stresses acting on an infinitesimal sphere is defined as the mechanical pressure (p_m):

$$P_m = \frac{\sigma_{xx} + \sigma_{yy} + \sigma_{zz}}{3} \tag{8.22}$$

Also, the normal stress on surface x can be expressed as:

$$\sigma_{xx} = -\left(P_m + \frac{2}{3}\mu\nabla\cdot\mathbf{V}\right) + \mu\left(\frac{\partial u}{\partial y} + \frac{\partial v}{\partial x}\right) \tag{8.23}$$

Eqs. (8.21) and (8.23) can be combined and expressed in index form to obtain:

$$\sigma_{ij} = -\left(P_m + \frac{2}{3}\mu\nabla\cdot\mathbf{V}\right)\delta_{ij} + \mu\left(\frac{\partial u_i}{\partial x_j} + \frac{\partial u_j}{\partial x_i}\right) \tag{8.24}$$

Where δ_{ij} is the Kronecker delta in which $\delta_{ij} = 1$ when $i = j$, and $\delta_{ij} = 0$ when $i \neq j$.

Eq. (8.24) indicates the need for a new definition of thermodynamical pressure (P):

$$P = P_m + \frac{2}{3}\mu\nabla\cdot\mathbf{V} \tag{8.25}$$

The coefficient $2\mu/3$ is empirical, and it relates to the viscosity of the fluid. Consequently, Eq. (8.25) can be expressed as:

$$p = P_m + \Lambda\nabla\cdot\mathbf{V} \tag{8.26}$$

where Λ is a coefficient obtained experimentally.

In literature, the coefficient Λ is called expansion viscosity, second coefficient of viscosity, or bulk viscosity (Bar-Meir, 2013). This parameter is positive, and the thermodynamic pressure continually tends to track the mechanical pressure during a change. The expansion rate of change and the fluid molecular structure influence the difference through Λ. Eq. (8.26) can be express in terms of the thermodynamic pressure as:

$$\sigma_{ij} = -\left[P + \left(\frac{2}{3}\mu - \Lambda\right)\nabla \cdot \mathbf{V}\right] + \mu\left(\frac{\partial v_j}{\partial x_i} + \frac{\partial v_i}{\partial x_j}\right) \tag{8.27}$$

The dot product ($\nabla.\mathbf{V}$) stands for the relative volume rate of change. For simple gases (dilute monoatomic gases), it can be shown that Λ disappears. In materials such as water, the bulk viscosity is very large (3 times μ), but the net effect is small because in these cases, $\nabla.\mathbf{V} \rightarrow 0$ (Bar-Meir, 2013). For complex liquids, the bulk viscosity can exceed 100 times μ. For incompressible flow, this coefficient or the whole effect vanishes. In most cases, the total effect of the dilation on the flow is very small, except for micro fluids, and small and molecular scales. Consequently, it is rather difficult to design experiments for measuring this effect. When this effect is neglected, the shear stress on the ith surface in the jth direction can be expressed as:

$$\sigma_{ij} = -P\delta_{ij} + \mu\left(\frac{\partial v_i}{\partial x_j} + \frac{\partial v_j}{\partial x_i}\right) \tag{8.28}$$

For the total effect in x direction, Eq. (8.27) is substituted into Eq. (8.16), which yields:

$$\rho\frac{Du}{Dt} = -\frac{\partial\left[P + \left(\frac{2}{3}\mu - \Lambda\right)\nabla \cdot \mathbf{V}\right]}{\partial x} + \mu\left(\frac{\partial^2 u}{\partial x^2} + \frac{\partial^2 u}{\partial y^2} + \frac{\partial^2 u}{\partial z^2}\right) + f_{Bx} \tag{8.29}$$

Similarly, it can be shown that:

$$\rho\frac{Dv}{Dt} = -\frac{\partial\left[P + \left(\frac{2\mu}{3} - \Lambda\right)\nabla \cdot \mathbf{V}\right]}{\partial y} + \mu\left(\frac{\partial^2 v}{\partial x^2} + \frac{\partial^2 v}{\partial y^2} + \frac{\partial^2 v}{\partial z^2}\right) + f_{By} \tag{8.30}$$

$$\rho\frac{Dw}{Dt} = -\frac{\partial\left[P + \left(\frac{2\mu}{3} - \Lambda\right)\nabla \cdot \mathbf{V}\right]}{\partial z} + \mu\left(\frac{\partial^2 w}{\partial x^2} + \frac{\partial^2 w}{\partial y^2} + \frac{\partial^2 w}{\partial z^2}\right) + f_{Bz} \tag{8.31}$$

Eqs. (8.29)–(8.31) are the Navier–Stokes (N–S) equations for a Newtonian fluid in stationary coordinates, and they can be expressed in vector form as:

$$\rho\frac{D\mathbf{V}}{Dt} = -\nabla P + \left(\frac{\mu}{3} + \Lambda\right)\nabla(\nabla \cdot \mathbf{V}) + \mu\nabla^2\mathbf{V} + \mathbf{f}_B \tag{8.32}$$

Where $\mathbf{f}_B$ is body force, which can be taken as ($\rho\mathbf{g}$), and $\mathbf{g}$ is the acceleration due to gravity.

In nonconservation form, the equation of momentum equation can be expressed as (Anderson, 2009):

$$\rho \frac{D\mathbf{V}}{Dt} = -\nabla P + \nabla \boldsymbol{\sigma} + \rho \mathbf{g} \tag{8.33}$$

Where ρDV/Dt is the substantial rate of change of momentum, $-\nabla P$ is the contribution of the pressure to surface forces per unit volume, $\nabla \boldsymbol{\sigma}$ is the shear stress contribution to surface forces per unit volume, and $\rho \mathbf{g}$ is the gravitational body forces per unit volume. The conservation form of the momentum equation for a viscous flow can be expressed as (Sandá et al., 2019):

$$\frac{\partial(\rho \mathbf{V})}{\partial t} + \nabla \cdot \left(\rho \mathbf{V} \mathbf{V}^T\right) = -\nabla P + \nabla \boldsymbol{\sigma} + \rho \mathbf{g} \tag{8.34}$$

Where $\partial(\phi \mathbf{V})/\partial t$ is the local rate of change of momentum, $\nabla.(\phi \mathbf{V} \mathbf{V}^T)$ is the convective term of momentum, $\mathbf{V}^T$ is the transpose of the velocity vector and $\boldsymbol{\sigma}$ is the the stress tensor.

It is possible to derive other forms of the Navier–Stokes equations based on appropriate assumptions. For instance, these equations can be simplified when the fluid is assumed to be inviscid (see Section 8.3.2).

8.3.1.3 Energy equation

To derive the energy equation, the first law of thermodynamics is applied to a moving fluid element. The rate of change of energy inside a fluid element is equal to the sum of the net flux of heat into the element and the rate of work done on the element due to body and surface forces. The net flux of heat into the element comprises of two parts: a) volumetric heating (such as through absorption, emission of radiation, or through chemical reactions) and b) heat transfer across the surface due to temperature gradients (thermal conduction). The heat transfer by thermal conduction is proportional to the local temperature gradient. In this regard, the nonconservation form of the energy equation for a viscous flow is (Anderson, 2009):

$$\rho \frac{D\left(e + \frac{V^2}{2}\right)}{Dt} = -\nabla \cdot (P\mathbf{V}) + \nabla \cdot (\boldsymbol{\sigma} \mathbf{V}) - \nabla \cdot \mathbf{Q} + \rho \mathbf{g} \cdot \mathbf{V} + q \tag{8.35}$$

Where e is the internal energy per unit mass, $\mathbf{Q}$ is the heat transfer by thermal conduction ($\mathbf{Q} = -k\,\nabla T$), $\rho\, D(e + V^2/2)/Dt$ is the substantial rate of change of energy per unit volume, $-\nabla.(P\mathbf{V})$ is the contribution to work of the pressure to forces per unit volume, $\nabla.(\boldsymbol{\sigma}\mathbf{V})$ is work derived by shear stress, which is often compacted as viscous dissipation function per unit volume, $-\nabla.\mathbf{Q}$ is the heat flow rate contribution by conduction per unit volume $[-\nabla.\mathbf{Q} = \boldsymbol{\nabla.(k\ \nabla T)}]$, $\rho\mathbf{g.V}$ is the work done by gravitational body forces per unit volume, k is the thermal conductivity of the fluid, T is the local temperature of the element and q is the volumetric heat source per unit volume (volumetric heating of the fluid element). Concentrated solar radiation is the source of heat for a solar receiver.

In conservation form, the energy equation for a viscous flow can be expressed as (Sandá et al., 2019):

$$\frac{\left[\left(e+\frac{V^2}{2}\right)\right]}{t}+\nabla\cdot\left[\left(e+\frac{V^2}{2}\right)\mathbf{V}\right]=-\nabla\cdot(P\mathbf{V})+\nabla\cdot(\mathbf{V})-\nabla\cdot\mathbf{Q}+\mathbf{g}\cdot\mathbf{V}+q \tag{8.36}$$

Where $\partial[\rho(e+V^2/2)]/\partial t$ is the local rate of change of energy per unit volume, and $\nabla.[\rho(e+V^2/2)\mathbf{V}]$ is the convective term of energy per unit volume.

8.3.1.4 Boussinesq approximation

Equations of mass, momentum, and energy conservation for a viscous fluid have been presented in Sections 8.3.1.1–8.3.1.3. These equations involve various properties of the flowing fluid, including density. In 1903, Boussinesq proposed that the density changes in the fluid can be ignored except in the buoyance term where the density (ρ) is multiplied by acceleration due gravity (g), under certain conditions. The most important condition is that flow speed should be smaller than the speed of sound in the medium. This condition is satisfied by most flows of liquids, and in flows of gases in which the speeds are less than 100 ms^{-1} (Kundu & Cohen, 2007). This approximation also assumes that the other properties of the fluid (such as viscosity, thermal conductivity, and specific heat capacity at constant pressure) remain constant. Based on these assumptions, the governing equations that correspond to the Boussinesq approximation are:

Continuity equation

$$\nabla\cdot\mathbf{U}=0 \tag{8.37}$$

Momentum equations

$$\frac{Du}{Dt}=-\frac{1}{\rho_0}\frac{\partial P}{\partial x}+v\nabla^2 u \tag{8.38}$$

Where ρ_0 is constant reference density at the reference temperature (T_0)

$$\frac{Dv}{Dt}=-\frac{1}{\rho_0}\frac{\partial P}{\partial y}+v\nabla^2 v \tag{8.39}$$

$$\frac{Dw}{Dt}=-\frac{1}{\rho_0}\frac{\partial P}{\partial z}+v\nabla^2 w-\frac{\rho g}{\rho_0} \tag{8.40}$$

$$\rho=\rho_0\left[1-\dot{\hat{\beta}}(T-T_0)\right] \tag{8.41}$$

$$\frac{DT}{Dt}=\xi\nabla^2 T \tag{8.42}$$

Where ξ is thermal diffusivity.

8.3.2 Equations for inviscid flow

An inviscid flow is a flow in which dissipative, transport phenomena of viscosity, mass diffusion, and thermal conductivity are ignored (Anderson, 2001). Inviscid flows do not really exist in nature but there are many practical aerodynamic flows in which the effect of transport phenomena is small, and so, they can be considered inviscid. Consequently, a high proportion of the aerodynamic theory deals with inviscid flows. An inviscid flow is theoretically attained as the Reynolds number approaches infinity because the Reynolds number increases with decreasing magnitude of viscous forces (Eq. 8.2). In this vein, for practical purposes, many flows with high but finite Re can be assumed to be inviscid. In such cases, the influence of friction, thermal conduction, and diffusion is confined to a very thin zone adjacent to the body surface (boundary layer), and flow outside the boundary layer is inviscid. The theory of boundary layer is presented and discussed in Section 10.5.2.2. Governing equations for transient, three-dimensional, and compressible inviscid flows are found by dropping the viscous terms from equations for viscous flows. The continuity equation for viscous flows does not contain viscous terms. So, there is no difference between the continuity equations for viscous and inviscid flows.

8.3.2.1 Continuity equation

In nonconservation and conservation forms, the continuity equations for inviscid flow can respectively be expressed by Eqs. (8.43) and (8.44).

$$\frac{D\rho}{Dt} + \rho\nabla\cdot\mathbf{V} = 0 \tag{8.43}$$

$$\frac{\partial\rho}{\partial t} + \nabla\cdot(\rho\mathbf{V}) = 0 \tag{8.44}$$

8.3.2.2 Momentum equations

In nonconservation form, the momentum can be represented by:

$$\rho\frac{D\mathbf{V}}{Dt} = -\nabla P + \rho\mathbf{g} \tag{8.45}$$

The conservation form of the momentum equation for an inviscid flow is:

$$\frac{\partial(\rho\mathbf{V})}{\partial t} + \nabla\cdot\left(\rho\mathbf{V}\mathbf{V}^{T}\right) = -\nabla P + \rho\mathbf{g} \tag{8.46}$$

8.3.2.3 Energy equation

In nonconservation form, the energy equation for an inviscid flow is given by:

$$\rho \frac{D\left(e+\frac{V^2}{2}\right)}{Dt} = -\nabla\cdot(P\mathbf{V}) + \rho\mathbf{g}\cdot\mathbf{V} + q \tag{8.47}$$

The conservation form, the energy equation for an inviscid flow is:

$$\frac{\partial\left[\rho\left(e+\frac{V^2}{2}\right)\right]}{\partial t} + \nabla\cdot\left[\rho\left(e+\frac{V^2}{2}\right)\mathbf{V}\right] = -\nabla\cdot(P\mathbf{V}) + \rho\mathbf{g}\cdot\mathbf{V} + q \tag{8.48}$$

8.3.3 Governing equations of two-phase flows

All the flows are a form of some multiphase flow. For instance, we sometimes consider air flow as a single-phase flow. Nevertheless, air is not a pure material but a mixture of gases and non-gaseous substances. Air is predominantly composed of nitrogen and oxygen gases, and so, many properties of this fluid are computed as if it is a good mixture of these two gases. Practically, for many applications, the homogeneous assumption is adequate and appropriate. This assumption is however not suitable when the air is stratified because of large body forces or a large acceleration (Bar-Meir, 2013). In such cases, the assumption may lead to significant errors. Consequently, under certain situations, air flow must be considered as a multiphase flow, and this effect should be taken into account in relevant computations.

Multiphase can be classified based on materials or phase flows. An example of materials flow is the flow of oil and water in one pipe. Nonetheless, it is common to classify the flow based on distinct phases that flow in the tube. Considering that there are three phases, phase flows can be solid–liquid, solid–gas, liquid–gas, and solid–liquid–gas flows. This classification approach ignores other flow categories that can and should fall in the category of multiphase flow. The category of liquid–gas should essentially be viewed as a special case of the liquid–liquid flow where the density ratio is significantly high. Multiphase flows occur in some designs of solar receivers, and therefore, research attention has also been given to this type of flow in the development of receivers.

Solar receivers use conventional and nonconventional materials as heat transfer media. Some of the heat transfer concepts involve two phase flows (see Chapter 5). For instance, water/steam is commonly used to transfer heat from the absorber element to other components of a CSP system. In volumetric solar receivers, air is often used to transfer heat from a porous medium. The air infiltrates through the porous element of the receiver, thereby yielding a system of solid and gas phases. Consequently, two-phase flow analysis is an essential aspect of the development of receivers.

The use of solid particles as a heat transfer medium is a promising pathway for the advancement of solar receivers. However, particle receivers have problems of particle conveyance, attrition, and transport (Teng and Xuan, 2018). Volumetric receivers are also gathering momentum in the development of CSP technology. These receivers commonly utilize air as the heat transfer fluid that flows through a porous medium. The flow of suspended solid particles through conduits, and the flow of a fluid through a porous material require special hydrodynamic equations.

8.3.3.1 Flow of suspended solids

In a particle receiver, solid particles are suspended in a gas (such as air). These particles absorb concentrated solar radiation and transfer the heat to the gas. So, this is a solid-gas flow. Similar equations presented in Section 8.3.2 can be applied to the gas phase flow. The particle flow is analyzed by using Newton's second law of motion. The net force exerted on a particle is directly proportional to the acceleration of the particle. So, the particle momentum equations (in Cartesian coordinates) can be given by (Evans et al., 1985):

$$\rho_{sp}\frac{\pi d_{sp}^3}{6}\frac{Du_{sp}}{Dt} = 3\pi d_{sp}\mu\frac{C_D}{C_{D,Stokes}}\left(u_f - u_{sp}\right) \tag{8.49}$$

$$\rho_{sp}\frac{\pi d_{sp}^3}{6}\frac{Dv_{sp}}{Dt} = 3\pi d_{sp}\mu\frac{C_D}{C_{D,Stokes}}\left(v_f - v_{sp}\right) \tag{8.50}$$

$$\rho_{sp}\frac{\pi d_{sp}^3}{6}\frac{Dw_{sp}}{Dt} = 3\pi d_{sp}\mu\frac{C_D}{C_{D,Stokes}}\left(w_f - w_{sp}\right) - \rho_{sp}\frac{\pi d_{sp}^3}{6}g \tag{8.51}$$

Where C_D is the coefficient of drag, d_{sp} is diameter of solid particle, u_f and v_f are respectively horizontal velocities of the fluid (gas) in the x- and y-directions, w_f is the vertical velocity in z-direction, u_{sp} and v_{sp} are respectively horizontal velocities of a solid particle in the x- and y-directions, w_{sp} is the vertical velocity of a solid particle in the z-direction, and ρ_{sp} is the density of solid particles.

The energy equation can be expressed as (Evans et al., 1985):

$$\rho_{sp}C_{sp}\frac{\pi d_{sp}^3}{6}\frac{DT_{sp}}{Dt} = Q_{rad} - Nuk\pi d_{sp}\left(T_{sp} - T_f\right) \tag{8.52}$$

Where Nu is the Nusselt number, and Q_{rad} is the radiation source term.

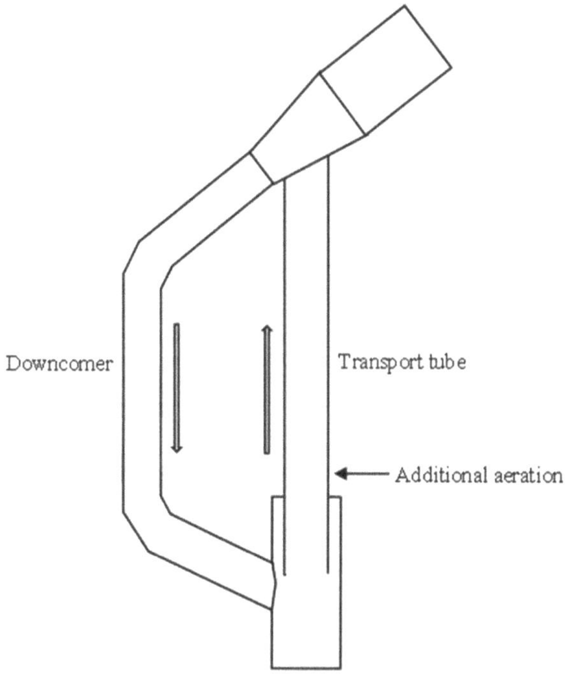

FIGURE 8.4 Schematic representation of suspended solid particle flow. Adapted from García-Triñanes et al. (2016).

García-Triñanes et al. (2016) experimentally investigated the hydrodynamic and particle motion in a real solar plant with tubular conduits at ambient temperature (Fig. 8.4). A dense particle suspension was passed upward through the transport tube, that has additional aeration, and ends up in a cyclone-like disengaging zone. From this zone, the disengaged solid particles fell under gravity into a downcomer. The governing parameters of the flow were identified as the solid feeding flow rate, the fluidization velocity, the solids holdup, the free board pressure, and the secondary air injection (aeration) velocity. To aerate the system, air was introduced into the uplift transport tube some way up its length. The motion of solid particles was measured using an on-invasive positron emission particle tracking (PEPT) technique to follow the movement of a 60 mm tracer particle, onto which was adsorbed the positron emitting 18F radioisotope. They found that the amount of the secondary air injection is the most important parameter for the stability and the uniform distribution of the solids flow in the tubes. In general, particle flux distribution of the upward flow was higher in the center but close to the walls in the downward flow. These observations suggest that this particular type of flow is suitable for transporting energy from the walls of the receiver tubes.

8.3.3.2 *Flow through porous media*

A volumetric receiver is made of a porous material (or a block of channels) that absorbs concentrated solar radiation. Air is utilized as a heat transfer fluid in this type of receivers. Governing equations for fluid flow in channels are similar to those of viscous and inviscid flows.

A porous medium is a combination of a stationary solid and a fluid. The fluid occupies the interstitial space between granules of the solid material. A porous medium is said to be saturated when the fluid is either a gas or a liquid (Bergman et al., 2011). On the other hand, all three phases coexist in an unsaturated porous medium. A saturated porous medium that consists of a stationary solid phase through which a fluid flows is called a packed bed, and it is discussed in Section 6.2.2.3. Flow through a porous receiver is significantly influenced by the internal structure of the absorber, and its study is difficult (Avila-Marin et al., 2019). Basic equations for viscous or inviscid flow can be applied to the gas phase. It is also possible to model the gas flow by taking into consideration the porosity of the porous medium. The solid phase is stationary, and therefore, only the energy equations are required for this phase. The energy conservation can be described by (Teng and Xuan, 2018):

$$\frac{\partial\left(\rho_{so}C_{p,so}T_{so}\right)}{\partial t} = \nabla(k_{so}\nabla T_{so}) + h_v\left(T_f - T_{so}\right) - \nabla\mathbf{Q}_{rad} \tag{8.53}$$

Where $C_{p,so}$, k_{so}, and ρ_{so} are respectively the specific heat capacity, effective thermal conductivity, and density of the solid medium, h_v is the volumetric coefficient of heat transfer, and $\nabla\mathbf{Q}_{rad}$ is the radiative heat transfer term.

8.3.3.3 *Liquid–gas flows*

Some heat transfer fluids are used in two phases in solar receivers. For example, in direct steam generation (DSG), water enters the receiver in liquid phase, and it is progressively heated until all of it becomes superheated steam (Fig. 8.5). So, there are three zones of flow field: (a) supercooled water, (b) mixture of water and steam, and (c) superheated steam (Zhang et al., 2017). The supercooled and superheated zones have one phase each, while the water-and-steam zone has two phases. A DSG power plant comprises one heater, or separate heaters (such as a pre-heater, an evaporator and a superheater) that are often connected in series. For solar-only systems, each heater is made of one or more receivers. The superheater may be a conventional backup heater (such as coal-fired boiler). It is also possible to use solar energy only for pre-heating the water, while conventional energy sources are utilized in the other heaters.

The total mass flow rate ($\dot{m}$) through the pipe is the sum of the mass flow rates of the liquid ($\dot{m}_l$) and gas ($\dot{m}_{ga}$) phases:

$$\dot{m} = \dot{m}_l + \dot{m}_{ga} \tag{8.54}$$

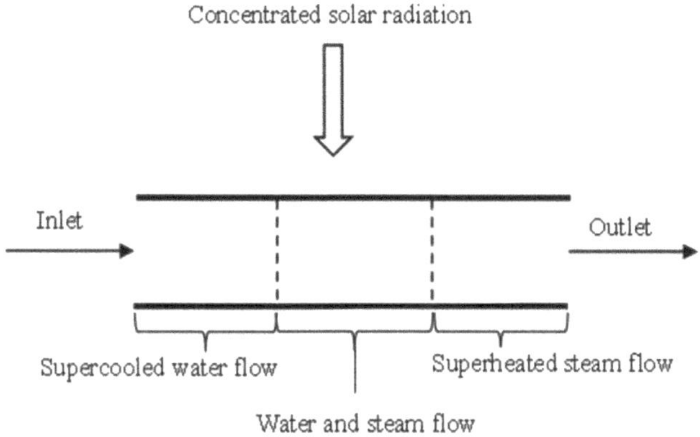

FIGURE 8.5 Schematic representation of two-phase flow in a heat absorbing pipe of a solar receiver. *Adapted from Zhang et al. (2017).*

Assuming the heat transfer fluid is an incompressible Newtonian fluid, and the fluid density change satisfies the Boussinesq approximation without heat source, then its two-dimensional flow momentum and energy equations for the receiver can be expressed as (Zang et al., 2017):

$$\rho\frac{\partial \mathbf{V}}{\partial t} + \rho(\mathbf{V}\cdot\nabla)\mathbf{V} = -\nabla P + \rho g\beta'(T - T_0) + \mu\nabla^2\mathbf{V} + \mathbf{f} \tag{8.55}$$

$$\frac{\partial(\rho C_p T)}{\partial t} + \nabla\cdot(\rho C_p T\ \mathbf{V}) = \nabla(k\nabla T) \tag{8.56}$$

Where C_p is the specific heat capacity, $\mathbf{f}$ is the body force, P is the pressure, T is the temperature, T_0 is the reference temperature, and $\mathbf{V}$ is the velocity of the fluid.

8.4 Characteristics of fluid flows

Viscous flows can generically be classified into two categories: laminar and turbulent (Kundu et al., 2012). Laminar flow is orderly, while turbulent flow is disorderly. Flow can be over a solid body or through a conduit bounded by solid surfaces. Fluid flow over a body is called external flow, and the flowing fluid completely or partially surrounds the solid body. For example, the flow of air over blades of a wind turbine is external. Internal flows are completely bounded by surfaces of a solid body, and they include flows through pipes, contractions and expansions, valves, and fittings. Flows of working and/or heat transfer fluids through receivers are generally internal. Characteristics of laminar and turbulent, and internal and multiphase flows are presented and discussed in this section.

8.4.1 Laminar flows

Under natural convection, flow is generally laminar, and it is characterized by low velocities and smooth paths of fluid particles in layers, and each layer moves smoothly past the adjacent layers with negligible or no mixing. Nevertheless, a boundary layer is created between the moving fluid and a solid surface (wall) that is in contact with the fluid under forced convention (Fig. 8.6). This layer is the region near the wall where wall friction affects the flow. Flow within the boundary layer close to the leading edge of the surface is laminar forced convection. The laminar flow continues until a transition zone is attained within which conversion from laminar to turbulent conditions occurs. At diminished velocities, flow remains laminar in long tubes or channels with a small hydraulic diameter, and it is called fully developed laminar flow (ASHRAE, 2017). For steady, fully developed laminar flow between two parallel plates, the shear stress varies with the distance y from the centerline (Fig. 8.7).

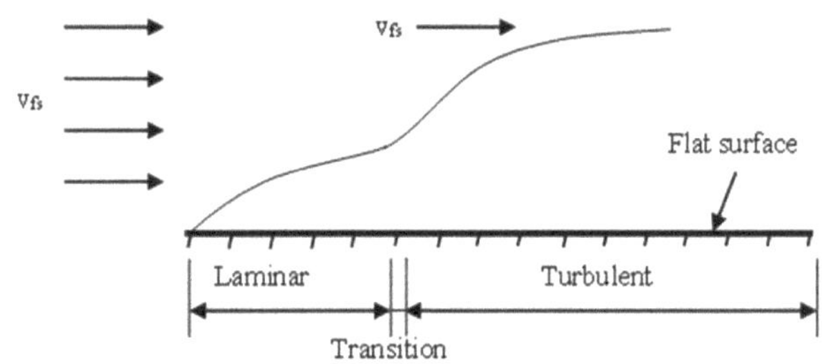

FIGURE 8.6 Formation of laminar and turbulent boundary layers when a fluid flows over a flat surface. The velocity of the fluid increases with the distance from the surface and the edge. v_{fs} is free stream flow velocity. *Adapted from Bergman et al. (2011).*

FIGURE 8.7 Fully developed laminar flow between two parallel plates separated by a distance of 2b. At the centerline, y = 0. *Adapted from ASHRAE (2017).*

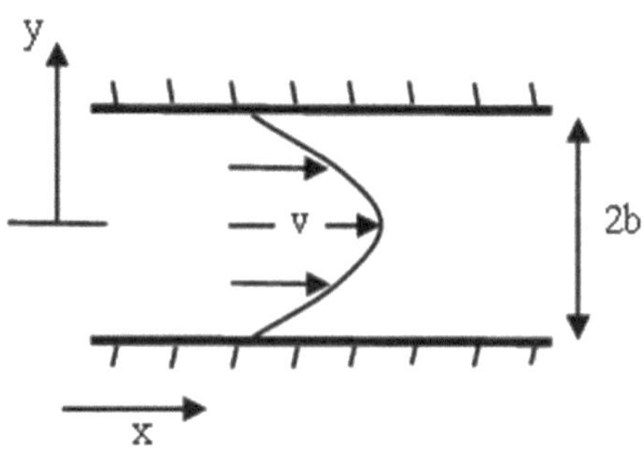

For a wide rectangular channel, the shear stress (σ) can be described by (ASHRAE, 2017):

$$\sigma = \left(\frac{y}{b}\right)\sigma_w = \mu\frac{dv}{dy} \tag{8.57}$$

Where σ_w is wall shear stress [b(dP/dx)], and x is the flow direction.

The velocity is zero at the wall (y = b), and so, Eq. (8.57) can be integrated to obtain:

$$v = \left(\frac{b^2 - y^2}{2\mu}\right)\frac{dP}{dx} \tag{8.58}$$

For a wide rectangular channel, the resulting parabolic velocity profile is usually known as Poiseuille flow. Maximum velocity is attained at the centerline (y = 0), and the average velocity v is 2/3 of the maximum velocity. Based on this, the longitudinal pressure drop in terms of v can be given by:

$$\frac{dP}{dx} = -\left(\frac{3\mu}{b^2}v\right) \tag{8.59}$$

For a pipe, v is 1/2 of the maximum velocity, and the pressure drop can be expressed as:

$$\frac{dP}{dx} = -\left(\frac{8\mu}{R^2}v\right) \tag{8.60}$$

Where R is the radius of the pipe.

8.4.2 Turbulent flows

As flow continues along the surface (Fig. 8.6), the thickness of the boundary layer increases up to a transitional point. Thereafter, turbulent flow commences, and the fluid flows irregularly with possible mixing of particles (Fig. 8.8). At a high fluid velocity or in a tube with a large diameter, transition to turbulent flow takes place, and flow is fully developed turbulent. In general, fluid flows are turbulent, involving random fluctuations of velocity and pressure, characterized by a broad spectrum of scales or frequencies (Robertson, 1963).

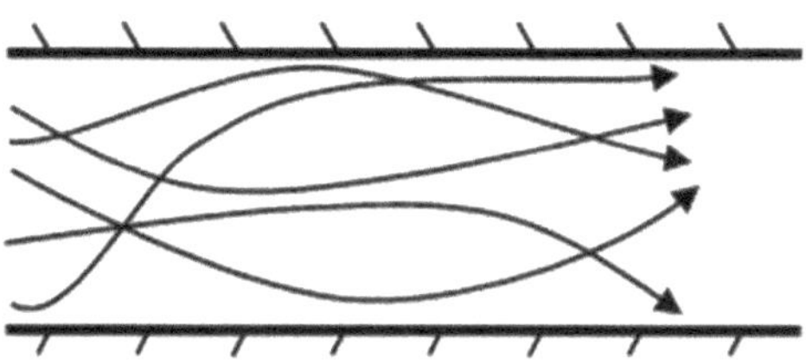

FIGURE 8.8 Onset of mixing in turbulent flow through a conduit.

It is possible to statistically quantify turbulence. Usually, the time-averaged velocity is used as the velocity of the fluid. Using the root mean square of the instantaneous variation in velocity, the strength of turbulence is characterized. Turbulence causes the fluid to quickly transmit momentum, heat, and mass across the flow. Laminar and turbulent flows can be distinguished by using the Re. When Re is less than 2300, then flow in pipes or ducts is laminar. Transitional flow exists in the range $2300 < Re < 10{,}000$, and predictions are unreliable in this flow regime while fully turbulent flow occurs when $Re > 10{,}000$.

8.4.3 Internal flows

Laminar boundary layers on each wall grow until they meet when the velocity of the fluid is adequately low for tubes, channels, or ducts of small diameter. This convergence occurs when the Re falls between 2000 and 2300. After this point, the velocity distribution remains constant, and there is no transition to turbulent flow. This type of fluid flow is referred to as fully developed laminar flow. The boundary layers become turbulent before they meet, and fully-developed turbulent flow sets in when the Re exceeds 10,000.

Three different flow regions exist in turbulent flow: Laminar, transition, and turbulent regions. The laminar region is located adjacent to the wall, where heat is transferred by thermal conduction. The transition region is known as the buffer layer, in which both eddy mixing and conduction effects are significant. The turbulent layer extends to the axis of the pipe, where the dominant mode of heat transfer is vortex mixing. In most flow devices, flow is turbulent. The inside diameter is the characteristic length for internal flow in pipes and tubes. For internal flow through noncircular tubes or ducts, the hydraulic diameter (D_h) is used to calculate the Re and Nu. The Nu is the ratio of convective to conductive heat transfer coefficients within a fluid, and it can be computed from:

$$Nu = \frac{hL_c}{k} = f(Re,\ Pr) \tag{8.61}$$

Where h is the coefficient of convective heat transfer, L_c is the characteristic length, and Pr is the Prandtl number.

Based on Eq. (8.61), various correlations of Nu have been established through experimental approaches. One of the useful forms for turbulent internal flow is called Colburn's analogy (Eq. 8.62). Selected specific correlations for internal flows in pipes or ducts are presented in Table 8.1.

$$j = \frac{Nu}{RePr^{1/3}} = \frac{f_F}{2} \tag{8.62}$$

Where f_F is the Fanning friction factor (1/4 of the Darcy-Weisbach friction

TABLE 8.1 Selected correlations for internal flows for pipes and ducts.

Flow regime	Correlation	Remarks
Laminar: Re <2300	$Nu = 1.86\left(\frac{RePr}{L/D_h}\right)^{1/3}\left(\frac{\mu}{\mu_s}\right)^{0.14}$ Where L is the length of the conduit, μ and μ_s are values of fluid viscosity computed at bulk and surface temperatures respectively.	$\frac{L}{D} < \frac{RePr}{8}\left(\frac{\mu}{\mu_s}\right)^{0.42}$
Laminar: developing	$Nu = 3.66 + \frac{0.065(D_h/L)RePr}{1+0.04[(D_h/L)RePr]^{2/3}}$	–
Laminar: fully developed, round	$Nu = 3.66$	Uniform surface temperature
Laminar: fully developed, round	$Nu = 4.36$	Uniform heat flux
Turbulent: fully developed	$Nu = 0.023\ Re^{0.8}Pr^{0.4}$	Heating fluid, $Re \geq 10{,}000$
Turbulent: fully developed	$Nu = 0.023\ Re^{0.8}Pr^{0.3}$	Cooling fluid, $Re \geq 10{,}000$

factor, f) and j is the Colburn's j-factor. For laminar flow, $f = 64/Re$. The value of f is almost constant for turbulent flow, falling in the range 0.01–0.05 (ASHRAE, 2017).

8.5 Flow stability

Different states of a fluid flow can be established by the way in which the fluid reacts to an infinitesimal disturbance in the initial state. Flow is stable if for all time, the perturbations to a basic flow do not grow. In contrast, small changes produce some noticeable effects on the state of the system in an unstable flow. These disturbances are related to the initial properties of the flow such as density, pressure, and velocity. A real flow may be stable under infinitesimal disturbances (linearly stable) but unstable under significantly large disturbances (nonlinearly unstable). Two fundamental reasons for studying the stability of fluid flow are to: a) understand the transition process from laminar to turbulent, and b) predict the onset of this transition (Kundu et al., 2012). The flow behaviour can be predicted by using mathematical models (see Chapter 10). In this context, models based on linear stability have yielded excellent agreement with empirical data. For instance, experiments have confirmed the accuracy of the Navier–Stokes equations.

When developing new flow devices, it is critical to account for instabilities (Roberts, 2020), and one way of accounting for them is through the minimization of turbulence onset or the understanding of different wavelengths that may be produced and possibly cause resonance (instability). Fluid instabilities occur in engineering and other applications, including solar receivers. Absorber elements of receivers are exposed to high levels of flux that thermally induce thermal stresses (see Chapter 9) and flow instabilities in these elements. Unfortunately, high performance absorbers are damaged by local overheating because of inhomogeneous flux distributions and flow instabilities (Roldán et al., 2014). Thus, it is important to consider flow instabilities when designing and testing receivers.

8.5.1 Method of normal modes for stability analysis

Fundamental analysis of linear stability involves the introduction of sinusoidal disturbances on a basic state (also known as the background, initial state or equilibrium state). The background flow is the flow whose stability is being investigated. For example, the velocity field of a basic state involving a flow parallel to the x-axis, and changing along the y-axis, is V = [V(y), 0, 0]. A sinusoidal disturbance of the following form is superposed on the background flow (Kundu & Cohen, 2007):

$$u(\mathbf{x}, t) = \hat{u}(y)e^{ikx+imz+\omega t} \tag{8.63}$$

Where i is a complex number ($i = \sqrt{-1}$), and $\hat{u}(y)$ is the amplitude of the complex function.

Solutions with exponential in x, z, and t are allowed in Eq. (8.57) because the coefficients of the differential equation governing the perturbation in this flow do not depend on (x, z, t). One assumption is that the flow field is unbounded in the x and z directions, and therefore the wavenumber components k and m can only be real so that the dependent variables remain bounded as x, $z \to \infty$; $\omega = \omega_r + i\omega_i$ is a complex number. During the analysis, the behavior of the system is investigated for all possible values of $\mathbf{K} = [k, 0, m]$. If ω_r is positive for any value of the wavenumber, the background flow is unstable to disturbances of this wavenumber. The background flow is stable if no such unstable state can be found. Based on this analysis, criterion for the stability of the system can be summarized as follows (Kundu & Cohen, 2007):

$\omega_r < 0$, stable
$\omega_r = 0$, neutrally stable
$\omega_r > 0$, unstable

The method of analysis that involves the examination of Fourier components, such as Eq. (8.63), is known as the normal mode method. An arbitrary perturbation can be decomposed into a complete set of normal modes. The stability of each mode is examined separately, as the linearity of the problem indicates that the different modes do not interact. This method leads to finding the solution to an eigenvalue problem.

The boundary between stability and instability is referred to as the marginal state, when $\omega_r = 0$. Depending on the values (zero or nonzero) of ω_i in the state, there are two kinds of marginal states. If $\omega_i = 0$ in the marginal state, then Eq. (8.63) indicates that the marginal state is characterized by a stationary pattern of motion such as cellular convection or secondary flow. In contrast, if $\omega_i \neq 0$ in the marginal state, then the instability starts as oscillations of increasing amplitude. Both the neutral and the marginal states have $\omega_r = 0$, but the marginal state has the extra constraint that it lies at the borderline between stable and unstable solutions. A minor change of parameters (such as the Reynolds number) from the marginal state can drive the system into an unstable regime where $\omega_r > 0$. In most cases, the stability criterion can be found by simply setting $\omega_r = 0$, without strictly showing that it is at the borderline of unstable and stable states.

8.5.2 Instability in parallel flows

Parallel flows are predominantly unidirectional and their velocity profiles are created by the no-slip condition at the boundary (Roberts, 2020). They include flows in which streamlines have some curvature, such as flows over a wing. In parallel flows, the distance between adjacent streamlines is uniform. These flows can occur in both viscous and inviscid fluids, and it is possible to obtain exact solutions for problems in this class of flows (Durbin & Medic, 2007). Consequently, where possible, the parallel approximation is applied to some flow problems. Investigation of the stability of inviscid fluids helps provide useful insights into the stability of viscous parallel flows. For both the derivation of inviscid and viscous flow stability problems, it is assumed that a flow profile varies only in the y-direction and is constant in both the x- and z-directions. The stability of selected parallel flows is presented and discussed in this section.

8.5.2.1 Stability of viscous parallel flows

The linear stability analysis method can be applied to find the criterion for the stability of a given system. Consider a primary flow directed along the x-direction and varying in the y-direction so that $V = [V(y), 0, 0]$. The total flow can be decomposed into basic flow and the disturbance: $\tilde{u} = [V + u, v, w]$ and $\tilde{p} = P + p$. Both the basic and perturbed flows satisfy the Navier–Stocks equations. The perturbed flow satisfies the x-momentum equation (Kundu & Cohen, 2007):

$$\frac{\partial u}{\partial t} + (V+u)\frac{\partial (V+u)}{\partial x} + (V+u)\frac{\partial (V+u)}{\partial y} = -\frac{\partial (P+p)}{\partial x} + \frac{1}{Re}\nabla^2 (V+u) \tag{8.64}$$

Where p is perturbed pressure, P is background pressure, Re is the Reynolds number ($Re = V_0 L/v$), u is the perturbed velocity of fluid, and V is background velocity. The variables are nondimensionalized by a characteristic length scale L (such as the width of the flow) and a characteristic maximum velocity U_0 of the basic flow; time is scaled by L/V_0 and the pressure is scaled by ρV_0^2.

The background flow can be expressed as:

$$-\frac{\partial P}{\partial x}+\frac{1}{Re}\nabla^2 V=0 \tag{8.65}$$

Subtracting Eq. (8.65) from Eq. (8.64) and ignoring terms that are nonlinear in the perturbations, yields the x-momentum equation for the perturbed flow:

$$\frac{\partial u}{\partial t}+V\frac{\partial u}{\partial x}+v\frac{\partial u}{\partial y}=-\frac{\partial p}{\partial x}+\frac{1}{Re}\nabla^2 u \tag{8.66}$$

The y- and z-momentum equations can be given by:

$$\frac{\partial v}{\partial t}+V\frac{\partial v}{\partial x}=-\frac{\partial p}{\partial y}+\frac{1}{Re}\nabla^2 v \tag{8.67}$$

$$\frac{\partial w}{\partial t}+V\frac{\partial w}{\partial x}=-\frac{\partial p}{\partial z}+\frac{1}{Re}\nabla^2 w \tag{8.68}$$

The continuity equation of the perturbed flow is expressed as:

$$\frac{\partial u}{\partial x}+\frac{\partial v}{\partial y}+\frac{\partial w}{\partial z}=0 \tag{8.69}$$

The coefficients in the perturbation Eqs. (8.66)–(8.69) depend only on y, and therefore, the equations admit solutions exponential in x, z, and t. We can assume normal modes of the form:

$$[v,p]=[\widehat{v}(y),\ \widehat{p}(y)]e^{i(kx+mz-kct)} \tag{8.70}$$

Considering that the flow is unbounded in x and z, the wavenumber components k and m should be real, while the wave speed $c = c_r + ic_i$ may be complex. Only positive values of k and m are considered without loss of generality. So, the direction of propagation is left open by not specifying the sign of c_r. The normal modes stand for waves that travel obliquely to the basic flow with a wavenumber magnitude of $\sqrt{k^2+m^2}$. These normal modes have an amplitude that varies with time. Consequently, the criterion for stability is (Kundu & Cohen, 2007):

$c_i < 0$: stable
$c_i > 0$: unstable

Substitution of the normal modes into Eqs. (8.66)–(8.69) yields:

$$ik(V-c)\widehat{u}+\widehat{v}V_y=-ik\widehat{p}+\frac{1}{Re}\left[\widehat{u}_{yy}-(k^2+m^2)\widehat{u}\right] \tag{8.71}$$

$$ik(V-c)\widehat{v}=-\widehat{p}_y+\frac{1}{Re}\left[\widehat{v}_{yy}-(k^2+m^2)\widehat{v}\right] \tag{8.72}$$

$$ik(V-c)\widehat{w} = -im\widehat{p} + \frac{1}{Re}\left[\widehat{w}_{yy} - \left(k^2+m^2\right)\widehat{w}\right] \tag{8.73}$$

$$ik\widehat{u} + \widehat{v}_y + im\widehat{w} = 0 \tag{8.74}$$

In Eqs. (8.71)–(8.74), the subscripts represent derivatives with respect to y. These equations are normal mode equations for three-dimensional perturbations. It can be shown that only two-dimensional perturbations need to be considered. In this vein, Herbert Brian Squire formulated a theorem stating that "for the study of the stability of flow between parallel walls it is sufficient to confine attention to disturbances of two-dimensional type" (Squire, 1933). This means that for each unstable three-dimensional disturbance, there corresponds a more unstable two-dimensional type. This theorem was proved by using the following transformations:

$$\overline{c} = c,\ \overline{k} = \left(k^2+m^2\right)^{1/2},\ \overline{k}\,\overline{Re} = kRe,\ \overline{k}\overline{u} = k\widehat{u} + m\widehat{w},\ \frac{\overline{p}}{k} = \frac{\widehat{p}}{k},\ \overline{v} = \widehat{v} \tag{8.75}$$

These transformations are substituted into Eqs. (8.71)–(8.74). Eqs. (8.81) and (8.73) are added, while the rest are just transformed. The result is a system of two-dimensional normal modes, given by:

$$i\overline{k}(V-c)\overline{u} + \overline{v}V_y = -i\overline{k}\overline{p} + \frac{1}{\overline{Re}}\left(\overline{u}_{yy} - \overline{k}^2\overline{u}\right) \tag{8.76}$$

$$i\overline{k}(V-c)\overline{v} = -\overline{p}_y + \frac{1}{\overline{Re}}\left(\overline{v}_{yy} - \overline{k}^2\overline{v}\right) \tag{8.77}$$

$$i\overline{k}\overline{u} + \overline{v}_y = 0 \tag{8.78}$$

Eqs. (8.76)–(8.78) are the same as Eqs. (8.71)–(8.74) but with $m = 0$ and $\widehat{w} = 0$. In addition, Squire's transformation indicates that the equivalent two-dimensional system of equations is associated with a lower Re because $\overline{k} > k$. This shows that the critical Re at which the instability sets in is lower for two-dimensional disturbances. Consequently, we just need to consider a two-dimensional perturbation to find the minimum Re for the occurrence of instability in a flow.

In Squire's theorem, the set of Eqs. (8.71)–(8.74) is considered with $m = \hat{w} = 0$. The two-dimensionality allows the definition of a stream function $\psi(x, y, t)$ for the perturbation field to be given by:

$$u = \frac{\partial\psi}{\partial y},\ v = -\frac{\partial\psi}{\partial x} \tag{8.79}$$

Let us assume normal modes of the form:

$$[u, v, \psi] = [\widehat{u}, \widehat{v}, \phi]e^{ik(x-ct)} \tag{8.80}$$

For consistency, the complex amplitude of ψ should be denoted by $\widehat{\psi}$. However, the symbol ϕ has been used for consistency with the standard notation of this variable in the literature. We then obtain:

$$\widehat{u} = \phi_y, \ \widehat{v} = -ik\phi \tag{8.81}$$

Elimination of the pressure from Eqs. (8.66)–(8.69) yields a single equation in terms of ϕ:

$$(V - c) + (\phi_{yy} - k^2\phi) - V_{yy}\phi = \frac{1}{ikRe}(\phi_{yyyy} - 2k^2\phi_{yy} + k^4\phi) \tag{8.82}$$

Where the subscripts denote derivatives with respect to y. Eq. (8.82) is a fourth-order differential equation with no-slip boundary conditions at the walls ($u = v = 0$) that require:

$$\phi = \phi_y = 0 \text{ at } y = y_1 \text{ and } y = y_2 \tag{8.83}$$

Eq. (8.82) is known as the Orr–Sommerfeld equation. It governs the stability of flows that are almost parallel and viscous, such as those in straight channels or boundary layers. Analytical solutions of the Orr-Sommerfeld equations are difficult to obtain, but it is possible to use numerical methods to solve the system of equations. However, solutions for simple flows can be found analytically. Table 8.2 shows results of linear stability of selected common viscous parallel flows.

8.5.2.2 *Stability of inviscid parallel flows*

Knowledge about the stability of inviscid flows is useful for obtaining insights into the viscous stability of parallel flows, assuming the perturbations obey the theory of inviscid dynamics. The stability governing equation of inviscid parallel flows can be found by letting $Re \rightarrow \infty$ in the Orr-Sommerfeld equation (Eq. 8.82), yielding (Kundu & Cohen, 2007):

$$(V - c) + (\phi_{yy} - k^2\phi) - V_{yy}\phi = 0 \tag{8.84}$$

TABLE 8.2 Linear stability of selected common viscous parallel flows.

Flow	$V(y)/V_0$	Re_{cr}	Remarks
Plane Poiseuille	$1-(y/L)^2$	5780	$L =$ half-width
Pipe flow	$1-(r/R)^2$	∞	Always stable
Plane Couette	y/L	∞	Always stable

Eq. (8.84) is known as the Rayleigh equation. If the inviscid flow is bounded by walls at $y = y_1$ and $y = y_2$ where $v = 0$, then the boundary conditions are expressed as:

$$\phi = 0 \text{ at } y = y_1 \text{ and } y = y_2 \tag{8.85}$$

The set of Eqs. (8.84) and (8.85) define an eigenvalue problem, with c(k) as the eigenvalue and ϕ as the eigenfunction. These equations do not involve i, and so taking the complex conjugate shows that if ϕ is an eigenfunction with eigenvalue c for some value of k, then ϕ^* is also an eigenfunction with an eigenvalue c^* for the same value of k. Consequently, there is an eigenvalue with a negative value of c_i corresponding to each eigenvalue with a positive value of c_i. This indicates that there is a decaying mode corresponding to each rising mode. Therefore, stable solutions of inviscid flows can have only real values of c.

Using Eq. (8.84), Rayleigh proved that the existence of an inflection point in the basic velocity profile V(y) was a necessary (but not sufficient) criterion for instability of an inviscid parallel flow around 1880. About 70 years later, the Swedish meteorologist Fjortoft derived a stronger necessary condition for the instability of inviscid parallel flows. He proved that a necessary condition for instability of inviscid parallel flows is that $V_{yy}\ (V - V_I) < 0$ somewhere within the flow, where V_I is the value of V at the inflection point. However, neither of the two conditions is sufficient for instability. There are some cases that have points of inflection but stable under certain conditions. For instance, consider the sinusoidal profile $V = \sin y$, with boundaries at $y = +b$ and $y = -b$. It has been proved that the flow is stable if the width is restricted to $2b < \pi$, even though there exists an inflection within the flow field at point $y = 0$. A useful criterion about the behavior of the complex phase speed c in inviscid parallel shear flows, valid both with and without stratification, was derived by Howard (1961). He showed that the complex wave velocity c of any unstable mode of a perturbation in parallel flows of an inviscid fluid must lie inside the semicircle in the upper half of the c-plane, which has the range of V as the diameter (Fig. 8.9). This concept is called the Howard semicircle theorem. The complex eigenvalue c must fall within the semicircle, in several inviscid parallel flows. This theorem states that the maximum rate of growth is limited by $kc_i < \frac{1}{2}(V_{max} - V_{min})$.

The Howard theorem is useful when looking for eigenvalues c(k) in numerical solutions of instability problems. Inviscid parallel flows satisfy this theorem. Growing and decaying modes are characterized by a nonzero c_i, while as neutral modes are confined to a real $c = c_r$. Consequently, neutral modes must have $V = c$ somewhere in the flow field. The vicinity of y around y_c, at which $V = c = c_r$, is known as a critical layer. The point y_c is a critical point of the inviscid governing Eq. (8.84), since the highest derivative vanishes out at this value of y. The solution of the eigenfunction is discontinuous across this layer. The full Orr–Sommerfeld Eq. (8.82) does not have such a critical

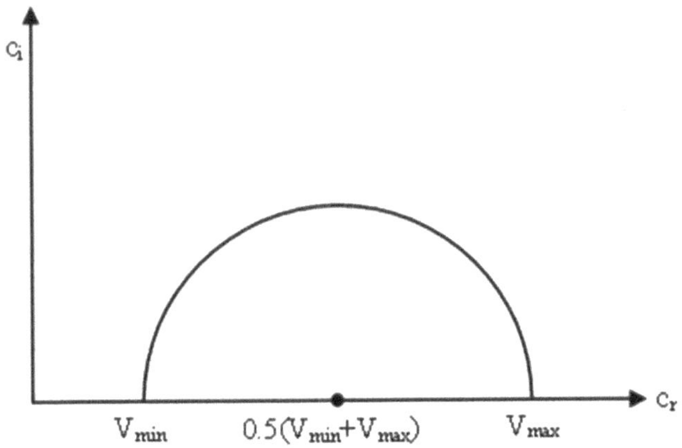

FIGURE 8.9 Schematic representation of the Howard semicircle theorem. V_{min} is the minimum velocity and V_{max} is the maximum velocity. *Adapted from (Kundu & Cohen, 2007).*

layer because the retention of highest order derivative when $V = c$. It is apparent that a viscous boundary layer must form at the location where $V = c$ in a real flow, and the thickness of the layer decreases as $Re \rightarrow \infty$.

8.5.3 Thermal instability

A layer of fluid heated from below has higher density on the top side, but it does not necessarily undergo a convective motion. This observation is ascribed to the viscosity and thermal diffusivity of the fluid that try to suppress the appearance of convection. However, the layer becomes unstable when temperature gradients are sufficiently large.

In 1900, Bénard conducted the first intensive experiments on the instability of a fluid layer heated from below (Kundu & Cohen, 2007). Only thin layers (1 mm or less) with a free surface were used in these experiments. Results showed hexagonal cells when the convection process developed. Later, it was established that the flow pattern observed by Bénard were caused by instabilities driven by the variation of surface tension with temperature and not the thermal instability (higher density at top surface). The influence of instabilities driven by surface tension decline with increasing the thickness of the layer. Further experimental work on thermal convection in thicker layers (with or without a free surface) revealed convective cells of many different shapes (not limited to hexagonal). Nonetheless, the phenomenon of thermal convection in a layer of fluid is still often known as the Bénard convection. Bénard's initial experiments formed the foundation for the work of Rayleigh in 1916 who derived the theoretical condition for the development of convective motion in a layer of fluid with two free surfaces. The works of Rayleigh and Benard led to the establishment of the phenomenon of Rayleigh-Benard convection that occurs at a specific temperature difference between two thermally conducting plates heated from below. In this phenomenon, there is coordinated movement of a fluid that is confined between the plates. Convection rolls are created in the Rayleigh-Benard convection and allow more efficient transportation of energy between the plates. Further increase in the temperature difference causes the fluid to rise very fast, and a transition from laminar to turbulent flow of the fluid may take place (Banzhaf, 2003). Rayleigh established that the instability would start when the temperature gradient was large enough to make the Rayleigh number (Ra) greater than a critical value (Ra_{cr}).

$$Ra = -\frac{dT}{dz}\left(\frac{g\widehat{\beta}d^4}{v\xi}\right) \tag{8.86}$$

Where g is the acceleration due to gravity, d is the depth of the layer, ξ is the thermal diffusivity, $\widehat{\beta}$ is the coefficient of thermal expansion, and v is the kinematic viscosity.

For a fluid layer between two rigid plates (upper and lower sides), it was found that instability occurred when Ra exceeded $Ra_{cr} = 1708$. For a layer with free surfaces and a rigid bottom wall, it was found that $Ra_{cr} = 1101$.

8.5.4 Centrifugal instability

In a Couette flow, the source of instability is the adverse gradient of angular momentum. The convection in a Couette flow is produced by the centrifugal forces that can overcome the viscous forces. Rayleigh discovered an inviscid stability criterion for this problem, and then Taylor found a solution of the viscous case in 1923. Experimental work shows that the instability initially appears in the form of axisymmetric disturbances (for which $\partial / \partial\theta = 0$).

8.5.4.1 Criterion for inviscid flow

In 1888, Rayleigh was the first investigator to consider the problem of instability in a Couette flow of an inviscid fluid. In his study, viscous effects were neglected, and he found the source of instability for this problem and established a necessary and sufficient condition for instability.

Suppose the velocity at any radial distance is V_θ (r). For inviscid flows, V_θ (r) is any function, but only certain distributions can be stable. Consider two fluid rings of equal masses at radial distances r_1 and r_2 (where $r_2 > r_1$) that are switched around. Rayleigh's criterion states that: An inviscid Couette flow is not stable if

$$\frac{d\Gamma^2}{dr} < 0 \tag{8.87}$$

Where Γ is the circulation ($\Gamma = 2\pi r V_\theta$).

Consequently, the "stratification" of angular momentum in a Couette flow is not stable if it diminishes with increasing radius. Consider a situation in which a viscous fluid is sandwiched between two concentric cylinders. The outer cylinder is fixed, while the inner cylinder is rotated. Then, Rayleigh's criterion implies that the flow is inviscidly unstable. Nevertheless, merely having a potentially unstable arrangement does not necessarily result in the instability of a moving viscous fluid. The Rayleigh criterion for inviscid Couette flow is modified by Taylor's solution, as shown in the next section.

8.5.4.2 Criterion for viscous flow

For a viscous fluid flowing between concentric cylinders, Taylor derived a criterion for stability of the flow. One cylinder may be stationary, or both may move. In addition, the cylinders can rotate in the same or different directions. A Taylor number (Ta) was defined by:

$$Ta = 4\left(\frac{\Omega_1 R_1^2 - \Omega_2 R_2^2}{R_1^2 - R_2^2}\right)\frac{\Omega_1 d^4}{v^2} \tag{8.88}$$

Where d is the lateral width of the flow cells ($d = R_2 - R_1$), v is the kinematic viscosity, Ω_1 and Ω_2 are respectively the angular speeds of the inner and outer cylinders, and R_1 and R_2 are their corresponding radii.

Taylor's solution for viscous Couette flow shows that the flow remains stable until a critical Taylor number (Ta_{cr}) is attained:

$$Ta_{cr} = \frac{1708}{(1/2)(1 + \Omega_2/\Omega_1)} \tag{8.89}$$

Example 8.2

A fluid flow between two concentric cylinders of radii R_1 and R_2. Determine the condition under which the Ta_{cr} approaches the critical Reynolds number (Re_{cr}).

Solution

When the critical Taylor and Reynolds numbers are equal, $Ta_{cr} = Re_{cr}$

From Eq. (8.89), $Ta_{cr} = \frac{1708}{(1/2)(1+\Omega_2/\Omega_1)}$

But $Re_{cr} = 1708$

Therefore, $\frac{1708}{(1/2)(1+\Omega_2/\Omega_1)} = 1708$

This implies that $(1/2)\ (1 + \Omega_2/\Omega_1) = 1$.

$1 + \Omega_2/\Omega_1 = 2$.

$\Omega_2/\Omega_1 = 1$.

Consequently, $Ta_{cr} \rightarrow Re_{cr}$ as $\Omega_2/\Omega_1 \rightarrow 1$.

8.6 Pressure loss

A pressure drop is the difference (ΔP) in the total pressure between two given points in a network that transports a fluid. A pressure loss (drop) occurs when a heat transfer fluid enters one end of a solar receiver and exits at the other. Pressure losses occur when the mechanical energy of the fluid undergoes irreversible transformation into heat (ASHRAE, 2017). These energy losses influence specifications for pumps, pipework, and other flow components. There are two types of pressure losses: (a) friction and (b) dynamic. These are briefly presented and discussed in this section.

8.6.1 Friction losses

Heat transfer fluids are commonly passed through ducts (such as tubes) to evacuate heat from receivers and convey it to other components of the CSP system (see Chapter 3). As the fluid flows through a duct, it loses energy via friction. Friction losses are caused by the viscosity of the fluid, and they are produced by the exchange of momentum between molecules in laminar flows or between individual particles of adjacent fluid layers moving at different velocities in turbulent flow. These losses occur along the whole length of the duct.

For fluid flow in conduits, the Darcy Equation can be used to compute friction losses (ASHRAE, 2017):

$$\Delta P_{fr} = \left(\frac{1000fL}{D_h}\right)\left(\frac{\rho V^2}{2}\right) \tag{8.90}$$

Where f is the friction factor, D_h is the hydraulic diameter of the duct, L is the duct length, V is the velocity of the fluid, ΔP_{fr} is the friction loss in terms of total pressure, and ρ is the density of the fluid.

As shown in Section 8.4.2, the friction factor is a function of the Re only in the region of laminar flow. For completely turbulent flow, this factor is a function of duct surface roughness and internal protuberances (such as joints).

A transitional zone exists between the bounding limits of hydraulically smooth behavior and fully rough behavior, where the friction factor is a function of both roughness and Re. In both the transitional and fully rough regions, the friction factor is computed by using Colebrook's equation (Eq. 8.91) (ASHRAE, 2017). Nevertheless, this equation cannot be solved explicitly for f, and so iterative techniques can be used to find solutions (Behls, 1971).

$$\frac{1}{\sqrt{f}} = -2\log\left(\frac{\varepsilon}{3.7D_h} + \frac{2.51}{Re\sqrt{f}}\right) \tag{8.91}$$

Where ε is the material absolute roughness factor.

bThe following equation can be used to compute the Re:

$$Re = \frac{D_h V}{1000v} \tag{8.92}$$

Where v is the kinematic viscosity.

8.6.2 Dynamic losses

Flow disturbances are caused by duct-mounted equipment and fittings that change flow direction (elbows), area variations (transitions), and converging/diverging junctions. Dynamic losses result from these disturbances, and a detailed discussion of hydraulic networks falls outside the scope of this book, but it is reported elsewhere (Idelchik et al., 1994).

8.6.2.1 Local loss coefficients

In dynamically similar flow streams, a dimensionless coefficient (C) has the same value and is therefore a useful parameter in the analysis of flow resistance. The fluid resistance coefficient is the ratio of total pressure loss to velocity pressure at the cross section of reference.

$$C = \frac{\Delta P_{to}}{\rho\left(V^2/2\right)} = \frac{\Delta P_{to}}{P_v} \tag{8.93}$$

Where ΔP_{to} is the total pressure loss and P_v is velocity pressure.

For all fittings, except joints, the total pressure loss is computed from:

$$\Delta P_{to} = C_o \Delta P_{v,0} \tag{8.94}$$

Where C_o is local coefficient of fitting, and $\Delta P_{v,0}$ is the velocity pressure of fitting.

8.6.2.2 Darcy-Weisbach equation

Total pressure loss in a duct section is computed by adding friction and dynamic losses. This is achieved by combining Eqs. (8.90) and (8.93) in terms of ΔP to obtain Eq. (8.89), known as the Darcy–Weisbach equation. The loss coefficient for each fitting must be referenced to the velocity pressure of that section.

$$\Delta P = \left(\frac{1000fL}{D_h} + \sum C\right)\left(\frac{\rho V^2}{2}\right) \tag{8.95}$$

A porous medium is used as a an absorber element in volumetric solar receivers. Nevertheless, this material creates some additional pressure drop when it is introduced into a conduit (Becker et al., 2006). The magnitude of this pressure drop depends on the properties (such as porosity, thermal conductivity, and specific surface) of the porous medium. Darcy's law in the bFbvorchheimer extension can be used to study instability in the porous medium (Ávila-Marín, 2011):

$$-\frac{dP}{dx} = \frac{\mu}{K_1}V + \frac{\rho_f}{K_2}V^2 \quad (8.96)$$

where K_1 is the permeability of porous medium, K_2 is the inertial coefficient, x is the coordinate in the direction of flow, V is the velocity of the fluid, ρ_f and μ are respectively the density and dynamic viscosity of the fluid.

Integration of Eq. (8.96) yields a quadratic form of pressure drop:

$$\frac{p_{in}^2 - p_{out}^2}{2} = \left[\dot{m}\frac{\mu(T_{in})}{K_1 T_{in}^{0.7}}T_{out}^{1.7} + \dot{m}^2\frac{T_{out}}{K_2}\right]RL \quad (8.97)$$

Where L is length of the solar absorber, R is gas constant, T_{in} and T_{out} are respective inlet and outlet temperatures, $\dot{m}$ is mass flow rate, and $\mu(T_{in})$ is the dynamic viscosity of the fluid as a function of the inlet temperature.

It has been found that the heat conductivity and permeability properties of the porous materials have significant influence on flow instabilities. Instability occurs only in highly porous honeycomb structures (Pitz-Paal et al., 1997). However, it is possible to prevent instabilities by using suitable receivers with a high thermal conductivity and a quadratic behavior of the pressure drop correlation and a high ratio of K_1 to K_2 (such as low-porosity ceramic foams).

Nomeclature

A Area (m^2)
C_D Coefficient of drag (dimensionless)
C_p Specific heat capacity at constant pressure (J kg^{-1} K^{-1})
D Diameter (m)
e Internal energy (J kg^{-1})
f Body force (N m^{-3})
F Force (N)
g Acceleration due to gravity (m s^{-2})
k Thermal conductivity ($Wm^{-1}K^{-1}$)
K_1 Permeability of porous medium (m^2)
K_2 Inertial coefficient (m)
L Length (m)
P Pressure (Pa)
Pr Prandtl number (dimensionless)
q Volumetric heat source per unit volume (W m^{-3})
Re Reynolds number (dimensionless)
t Time (s)
T Temperature (K)
u Velocity in x-direction (m s^{-1})
V Velocity (m s^{-1})
v Velocity in y-direction (m s^{-1})
w Velocity in z-direction (m s^{-1})

x Distance in x-direction (m)
y Distance in y-direction (m)
C′ Coefficient of fitting (dimensionless)
$\dot{\mathbf{m}}$ Mass flow rate (kg s^{-1})
z Distance in z-direction (m)
$\widehat{\beta}$ Coefficient of thermal expansion (K^{-1})
μ Dynamic viscosity of the fluid (Pa s)
ν Kinematic viscosity (m^2 s^{-1})
ρ Density (kg m^{-3})
σ Stress (N m^{-2})
ξ Thermal diffusivity (m^2 s^{-1})
Ω Angular speed (rad s^{-1})

References

Anderson, J. D., Jr. (1991). *Fundamentals of aerodynamics* (2nd ed.). McGraw-Hill Inc.

Anderson, J. D., Jr. (2001). Flight (Aerodynamics). In R. A. Meyers (Ed.), *Encyclopedia of Physical Science and Technology* (pp. 1–21). Elsevier Science Ltd.

Anderson, J. D., Jr. (2009). Governing equations of fluid dynamics. In J. F. Wendt (Ed.), *Computational fluid dynamics: An introduction*. Berlin: Springer-Verlag.

ASHRAE. (2017). *ASHRAE handbook - fundamentals*. American Society of Heating, Refrigerating and Air-Conditioning Engineers (ASHRAE).

Ávila-Marín, A. L. (2011). Volumetric receivers in solar thermal power plants with central receiver system technology: A review. *Solar Energy, 85*(5), 891–910. https://doi.org/10.1016/j.solener.2011.02.002

Avila-Marin, A. L., Fernandez-Reche, J., & Martinez-Tarifa, A. (2019). Modelling strategies for porous structures as solar receivers in central receiver systems: A review. *Renewable and Sustainable Energy Reviews, 111*, 15–33. https://doi.org/10.1016/j.rser.2019.03.059

Banzhaf, W. (2003). Self-organizing systems. In R. A. Meyers (Ed.), *Encyclopedia of Physical Science and Technology* (3rd, pp. 589–598). Elsevier Inc.

Bar-Meir, G. (2013). *Basics of fluid mechanics*. Potto. http://www.potto.org/downloads.php.

Becker, M., Fend, T., Hoffschmidt, B., Pitz-Paal, R., Reutter, O., Stamatov, V., Steven, M., & Trimis, D. (2006). Theoretical and numerical investigation of flow stability in porous materials applied as volumetric solar receivers. *Solar Energy, 80*(10), 1241–1248. https://doi.org/10.1016/j.solener.2005.11.006

Behls, H. F. (1971). Computerized calculation of duct friction. *Building Science Series, 39*.

Bellos, E., Bousi, E., Tzivanidis, C., & Pavlovic, S. (2019). Optical and thermal analysis of different cavity receiver designs for solar dish concentrators. *Energy Conversion and Management, X 2*, 100013. https://doi.org/10.1016/j.ecmx.2019.100013

Benoit, H., Spreafico, L., Gauthier, D., & Flamant, G. (2016). Review of heat transfer fluids in tube-receivers used in concentrating solar thermal systems: Properties and heat transfer coefficients. *Renewable and Sustainable Energy Reviews, 55*, 298–315. https://doi.org/10.1016/j.rser.2015.10.059

Bergman, T. L., Lavine, A. S., Incropera, F. P., & DeWitt, D. P. (2011). *Fundamentals of heat and mass transfer* (7th ed.). John Wiley & Sons.

Demirel, Y. (2002). Using the second law of thermodynamics. *Nonequilibrium Thermodynamics: Transport and Rate Processes in Physical and Biological Systems* (pp. 124–185). Elsevier BV.

Dunham, M. T., & Iverson, B. D. (2014). High-efficiency thermodynamic power cycles for concentrated solar power systems. *Renewable and Sustainable Energy Reviews, 30*, 758–770. https://doi.org/10.1016/j.rser.2013.11.010

Durbin, P. A., & Medic, G. (2007). *Fluid Dynamics with a Computational Perspective*. New York: Cambridge University Press.

Evans, G. H., Houf, W. G., Greif, R., & Crowe, C. (1985). *Numerical modeling of a solid particle solar central receiver.*

Fanchi, J. R. (2010). Reservoir Simulation. *Integrated Reservoir Asset Management* (pp. 223–241). Elsevier Inc.

García-Triñanes, P., Seville, J., Boissière, B., Ansart, R., Leadbeater, T., & Parker, D. (2016). Hydrodynamics and particle motion in upward flowing dense particle suspensions: Application in solar receivers. *Chemical Engineering Science, 146*, 346–356. https://doi.org/10.1016/j.ces.2016.03.006

Howard, I. N. (1961). Note on a paper of John W. Miles. *Journal of Fluid Mechanics, 13*, 158–160.

Idelchik, I. E., Steinberg, M. O., Malyavskaya, G. R., & Martynenko, O. G. (1994). *Handbook of hydraulic resistance.*

Katopodes, N. D. (2019). Viscous Fluid Flow. *Free-Surface Flow: Environmental Fluid Mechanics* (pp. 324–426). Elsevier Inc.

Kundu, P. K., & Cohen, I. M. (2007). *Fluid mechanics* (4th ed.). Elsevier Science & Technology.

Kundu, P. K., Cohen, I. M., & Dowling, D. R. (2012). Laminar Flow. In P. K. Kundu, I. M. Cohen, & D. R. Dowling (Eds.), *Fluid Mechanics* (5th, pp. 309–359). Elsevier Inc.

Niño, L., Gelves, R., Ali, H., Solsvik, J., & Jakobsen, H. (2022). Numerical determination of bubble size distribution in Newtonian and non-Newtonian fluid flows based on the complete turbulence spectrum. *Chemical Engineering Science, 253*, 117543. https://doi.org/10.1016/j.ces.2022.117543

Palagi, S., Walker, D., Qiu, T., & Fischer, P. (2017). Micro- and nanorobots in Newtonian and biological viscoelastic fluids. In M. Kim, A. A. Julius, & U. K. Cheang (Eds.), *Microbiorobotics: Biologically Inspired Microscale Robotic Systems* (2nd, pp. 133–162). Elsevier Inc.

Pitz-Paal, R., Hoffschmidt, B., Böhmer, M., & Becker, M. (1997). Experimental and numerical evaluation of the performance and flow stability of different types of open volumetric absorbers under non-homogeneous irradiation. *Solar Energy, 60*(3–4), 135–150. https://doi.org/10.1016/S0038-092X(97)00007-8

Roberts, M. S. (2020). *Fluid instabilities and transition to turbulence*. IntechOpen. https://doi.org/10.5772/intechopen.92219

Robertson, J. M. (1963). *A turbulence primer. University of Illinois; engineering experiment station circular 79.*

Rodriguez-Sanchez, M. R., Sanchez-Gonzalez, A., Marugan-Cruz, C., & Santana, D. (2015). Flow patterns of external solar receivers. *Solar Energy, 122*, 940–953. https://doi.org/10.1016/j.solener.2015.10.025

Roldán, M. I., Smirnova, O., Fend, T., Casas, J. L., & Zarza, E. (2014). Thermal analysis and design of a volumetric solar absorber depending on the porosity. *Renewable Energy, 62*, 116–128. https://doi.org/10.1016/j.renene.2013.06.043

Sandá, A., Moya, S. L., & Valenzuela, L. (2019). Modelling and simulation tools for direct steam generation in parabolic trough solar collectors: A review. *Renewable and Sustainable Energy Reviews, 113*, 109226. https://doi.org/10.1016/j.rser.2019.06.033

Schlichting, H. (1979). *Boundary layer theory* (7th ed.). McGraw-Hill.

Squire, H. B. (1933). On the Stability for Three-Dimensional Disturbances of Viscous Fluid Floio between Parallel Walls. *Proceedings of the Royal Society of London A, 142*, 621–628. https://doi.org/10.1098/rspa.1933.0193

Teng, L., & Xuan, Y. (2018). Thermal and hydrodynamic performance of a novel volumetric solar receiver. *Solar Energy, 163*, 177–188. https://doi.org/10.1016/j.solener.2018.01.087

Zhang, Y., Liu, P., Li, P., Chen, Y., & Pan, Y. (2017). Study on the instability of two-phase flow in the heat-absorbing tube of trough solar collector. *International Journal of Photoenergy, 2017*. https://doi.org/10.1155/2017/6547203

Chapter 9

Thermomechanical considerations in solar receivers

9.1 Introduction

High-grade heat is required to drive advanced thermodynamic cycles for higher efficiency of concentrating solar power (CSP) systems. A solar receiver is the source of heat for the heat engine of a CSP system. Consequently, it is necessary for the receivers to supply heat at high temperatures. This advancement requires structural materials that can resist extreme temperatures and their associated stresses.

When a structural member is heated (or cooled), it undergoes deformation (expansion or contraction). This process of thermal deformation can produce stress if the structural member is mechanically constrained by external forces or if the geometry of the structural member is incompatible with the tendency to expand or contract (Huston & Josephs, 2009). An example of the first condition is the thermal behavior of a straight bar fixed at both ends and subjected to a constant temperature gradient. The case corresponding to the second condition can be a cylinder having a temperature gradient across its wall. The thermal stresses are produced in the cylinder because the strains do not match with the free thermal deformation even though no external forces of constraint are applied to this case. These two cases show that a structural component can be constrained by external or internal forces.

Thermal stresses may lead to damage in system components and to the degradation of material properties, thereby adversely affecting the performance of the system. For instance, thermal stress can trigger cracking when the maximum tensile thermal stress in the material reaches the fracture stress level (Davidge, 1991). The amount of cracking predominantly depends on the mechanical properties of the material. Thus, it is generally necessary to use materials with suitable characteristics. When thermal stresses on a structure are compressive, the structure may undergo the instability, known as thermal buckling Fig. 9.1.

Solar Receivers for Thermal Power Generation. https://doi.org/10.1016/B978-0-323-85271-5.00009-4

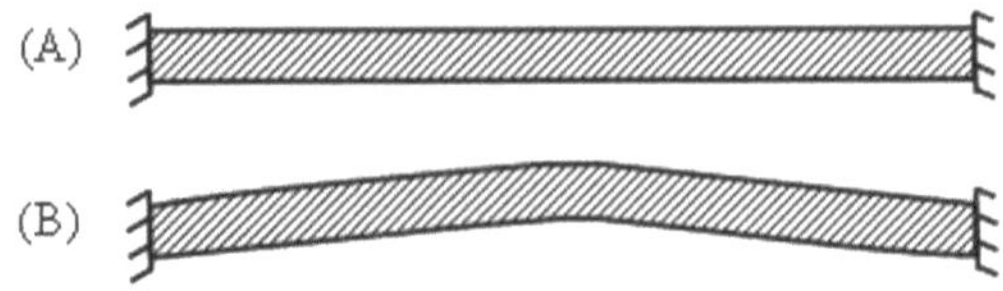

FIGURE 9.1 (A) normal shape and (B) buckled shape of a beam constrained on both ends.

Thermal stresses arise because of constraints in system components and these stresses may be effectively controlled by decreasing the extent of the restraint (Barron & Barron, 2012). For instance, flexible expansion bellows are utilized in cryogenic fluid transfer lines to decrease the constraint and forces between the cold inner line and the warm outer vacuum jacket to acceptable limits. The problem of thermal buckling of plates separated by gaps can be reduced by providing sufficiently large gaps to allow some unconstrained expansion of the plates. The aperture of some solar receivers is glazed, but glass is very brittle, and so it easily cracks when subjected to thermal stress. Consequently, it is critical to provide enough clearance between the edge of the glass plate and the frame that holds the plate. These and other thermomechanical problems are encountered in the process of developing receivers, and the use of structural materials with suitable characteristics is one strategy for overcoming these challenges.

9.2 Characteristics of structural materials

Uniform or nonuniform temperature changes cause thermal stresses in a structure that is somehow constrained against expansion or contraction (Sumelka & Łodygowski, 2013). Consequently, receiver components exposed to high temperatures require materials with high heat resistance to ensure the safety of the system. Other characteristics must also be considered when selecting materials for constructing solar receivers.

9.2.1 Operational temperature range and thermal stability

The efficiency of CSP power plants is predominantly restricted by the operating temperature of materials exposed to environments of high temperatures. This temperature limitation is applicable to working fluids and structural components. Required characteristics of working fluids have been discussed in Chapter 5. A structural component should therefore have a wide range of the operational temperature. This implies that the candidate solid material should retain a high percentage of its strength in a wide range of operational temperature. For instance, walls of conduits for hot working fluids must have high failure strength within the operational temperature range that permits the exploitation of advanced thermodynamic cycles.

Application of heat to a material can cause physical and chemical changes to the properties of the material. For example, a quartz glass cover may recrystallize under elevated temperatures, thereby compromising its optical properties. Therefore, the thermal stability of materials is a critical factor in the selection of structural materials for receiver components located in extreme temperature environments. Thermal stability is defined as the ability of a material to resist the action of heat and to maintain its properties at a given temperature. Solar receivers operate at elevated temperatures and therefore structural materials with high thermal stability are required for receiver parts that are in zones of extreme temperatures.

9.2.2 Thermophysical properties

Thermal conductivity influences temperature gradients within a structure or fluid. In this vein, high thermal conductivities of the absorber element and the heat transfer fluid are critical for the augmentation of the heat transfer coefficient. Solar radiation is intermittent, which can drastically alter the outlet temperature of the heat transfer fluid, thereby causing instability in the CSP system (see Chapter 6). One way of ensuring continuous supply of heat to the power block is through thermal energy storage. Nevertheless, most of the receivers have inherently low thermal inertia. In this connection, the thermal capacitance can be raised by integrating receivers with thermal storage units (Anwar et al., 2016). Thus, a high specific heat capacity of the structural members (such as the absorber element) can raise the thermal capacitance of a receiver. The thermal capacitance of a component also increases with the density of the construction material. Another thermal consideration is expansion/contraction of a structural component under temperature changes. In this regard, the coefficient of thermal expansion influences the extent of variation of the volume of the structure as temperature changes. When the structural component is constrained, thermal expansion/contraction creates stresses in the structure. Therefore, materials with low values of the coefficient of thermal expansion are desirable.

Costs associated with lifting, transportation, and mounting of a component increase with the density of materials for constructing a given volume of the component. Therefore, low-density materials would be preferred to minimize the mass of the system, which conflicts with the density requirement for thermal capacitance. An important aspect of the design process is the system optimization that aims at maximizing or minimizing a given function relative to a set of available choices under a specified situation. It happens that some of the optimization criteria may conflict with each other. For instance, increasing the efficiency of a receiver may raise the cost of the system. Consequently, multiobjective optimization can be used to obtain an optimal density of materials for construction of a component with structural and thermal functions (Cui et al., 2017). Another mechanical property of a structural component is the Young's modulus. It gives a measure of the characteristic strength of the material. Materials with high values of the Young's modulus can withstand higher levels of stresses and are therefore more preferred in structural components. Also related to the elasticity of a material is the Poisson's ratio. It also affects the magnitude of the stress exerted on a material.

9.2.3 Flow pressure

When a fluid stream passes through a conduit (such as pipe, duct, etc.), it exerts pressure on the walls of the conduit. For instance, the flow pressure of an ultrasupercritical steam cycle is 31.0 MPa at a temperature of 893 K (Boss et al., 2007).

In a solar receiver, heat transfer fluids are commonly pumped through tubes and other components of a CSP system. Consequently, it is necessary for the walls of the conduits to have a certain level of mechanical strength to withstand the flow pressure. Conduits with thin walls can be utilized when the flow pressure is low, which reduces the wall temperature gradient and consequently thermal stresses. Moreover, for a given material, the cost of conduits increases with the thickness of the conduit walls.

9.2.4 Operational aspects

Operational aspects, such as safety and corrosion behavior of materials, play an important role in the use of materials in different industries. Diminished hazard properties and sound compatibility of materials are highly advocated.

Another important operational consideration is the impact of a material on the environment. Some environmental concerns are the global warming potential (GWP) and the ozone depletion potential (ODP) of a material. Materials with low magnitudes of GWP and ODP are preferred (ASHRAE, 2017).

9.2.5 Availability and affordability of materials

A structural material should be readily available and affordable to assist in reducing the cost of the CSP system. At the time of writing this book, the cost of CSP-based electricity was relatively higher than that from conventional power plants such as coal. So, cheaper materials are desirable to achieve the goal of making the CSP technology economically competitive on the electricity market.

There is no single material that meets all the important requirements of a structural material. In view of this, research activities are continuing to develop new materials with many suitable characteristics. For instance, there has been significant advancement in the development of superalloys that have good thermal and mechanical properties. These alloys have already found application in solar receivers.

9.3 Major structural elements of solar receivers

Based on the number of dimensions in which solar radiation is absorbed, solar receivers can be classified into surface and volumetric receivers. Central to a solar receiver is the absorber element. A high proportion of the incident concentrated solar radiation is absorbed and converted to heat by this element. To reduce heat losses, some receivers are equipped with a transparent cover on the front and insulation elements on the back and side walls. Temperatures are highest in the absorber, which exposes the transparent and insulation elements to high temperatures. Consequently, it is important to consider the thermomechanical behavior of receivers.

9.3.1 Surface receivers

Surface receivers absorb solar radiation by using an opaque surface in two dimensions. The surface may be tubular or flat plate. One main advantage of the tubular geometry is its ability to resist pressure. Thus, tubular conduits have found application in both linear and point focus receivers (see Chapter 3). The parabolic trough concentrator (PTC) and linear Fresnel reflector utilize linear focus receivers, while the solar tower (ST) and parabolic dish concentrator (PDC) utilize point focus receivers. Most of the operating power plants in the world are based on the PTC concept with tubular receivers. On the other hand, cavity and external receivers are commonly exploited in ST and PDC systems. Circular conduits are also used in these types of receivers.

There are many designs of surface-based linear and focus receivers. Fig. 9.2 shows examples of tubular and spherical cavity receivers. A basic tubular receiver comprises concentric absorber and glass tubes. Tubular

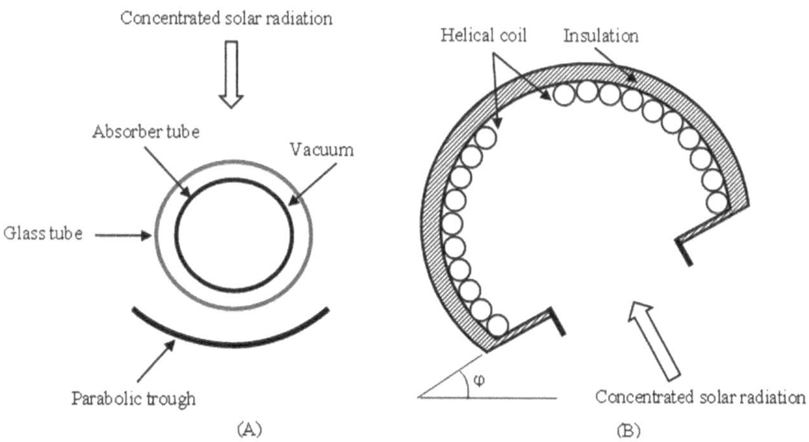

FIGURE 9.2 Schematic representation of: (A) evacuated tubular receiver and (B) spherical cavity receiver.

receivers can be classified into two categories: evacuated and nonevacuated. Evacuated receivers are suitable for application at temperatures above 573 K because they have a high vacuum (1.0 Pa) between the absorber and glass tubes, which reduces heat losses to the environment and augments the efficiency of the receiver at elevated temperatures (Moya, 2012). Nonevacuated receivers have neither a vacuum between the absorber and glass tubes nor glass-to-metal welds. This category of tubular receivers is suitable for applications with operational temperatures below 573 K, because thermal losses are not significant at these temperatures.

To increase absorptance, the outer surface of the absorber tube is coated with a thin black film that may be a selective or nonselective absorbing coating. This coating must withstand extreme temperatures that prevail in the absorber tube. Concentrated solar radiation is transmitted through the glass tube to the absorber coating where it gets absorbed and converted to heat. The generated heat is conducted from the outer surface to the inner surface of the tube wall. Thus, high thermal conductivity of materials for the tube wall is required. A heat transfer fluid is passed through the hot absorber tube to evacuate heat from the absorber to other components of the CSP system. Heat is transmitted from the inner surface of the absorber tube to the fluid through convection. For the receiver design shown in Fig. 9.2, solar radiation is concentrated on the bottom side of the tube, and so, this is the hottest zone of this receiver. The glass tube is also exposed to elevated temperatures and thermal stresses. Therefore, it is critical to consider the thermomechanical behavior of the glass tube, metallic tube, and selective coating elements.

In a spherical/hemispherical cavity receiver, the absorber is a helical metallic coil (tube) fitted inside a cavity, and a heat transfer fluid is passed through the coil. Like the tubular receiver, the coil wall absorbs concentrated solar radiation and it is the hottest part of the receiver, and therefore subjected to the highest levels of stress. To reduce heat loss, the cavity is insulated from the ambient environment. Thus, both the cavity and insulation elements are subjected to elevated temperatures and stresses. One main advantage of this design is the minimized external surface that leads to low thermal losses (Bellos et al., 2019). Heat is still predominantly lost from the cavity receiver via the aperture through convective and radiative heat transfer modes. Convective heat loss is influenced by the inclination angle (ϕ) of the receiver to the horizontal plane (Msaddak et al., 2018). It is therefore necessary to install the receiver at an optimal inclination angle to achieve the highest efficiency. The aperture of the receiver can be glazed to reduce heat losses from the front.

9.3.2 Volumetric receivers

Conventional surface receivers are commonly exploited in operating CSP plants worldwide. Nevertheless, they have limited thermal efficiency because of the following: (a) high heat loss during the process of heat transfer between the absorber surface and the heat transfer fluid and (b) their finite absorption area that is confined to two dimensions. In view of this, research attention has also been given to the development of volumetric receivers that directly or indirectly absorb concentrated solar radiation in three dimensions. These receivers are perceived to be the best alternative to tubular receivers (Ávila-Marín, 2011).

9.3.2.1 Conventional volumetric receivers

A conventional volumetric receiver utilizes a porous medium or a block of channels as an absorber (Fig. 9.3). The porous structure can be interlocking shapes, knit-wire packs, foam, or foil arrangements, made of metal, ceramic, or other adequate materials with a specific porosity. Such types of absorber elements raise the contact surface area between the absorber and the heat transfer fluid (Zaversky et al., 2020). The aperture of the receiver is covered with a glass window, while the sides are insulated to reduce heat loss from the system.

The basic operating principles of a conventional volumetric receiver are as follows: (a) a porous medium (or block of channels) is installed inside the volume of the receiver so that the concentrated solar radiation is absorbed in the depth of the structure, (b) the absorbed concentrated solar radiation is converted to heat and heats up the solid material in the volume and the heat transfer fluid is simultaneously passed through the receiver volume and heated up by forced convection, (c) heat is transmitted from the hot surface of the porous medium (or block of channels) to the heat transfer fluid, and (d) the

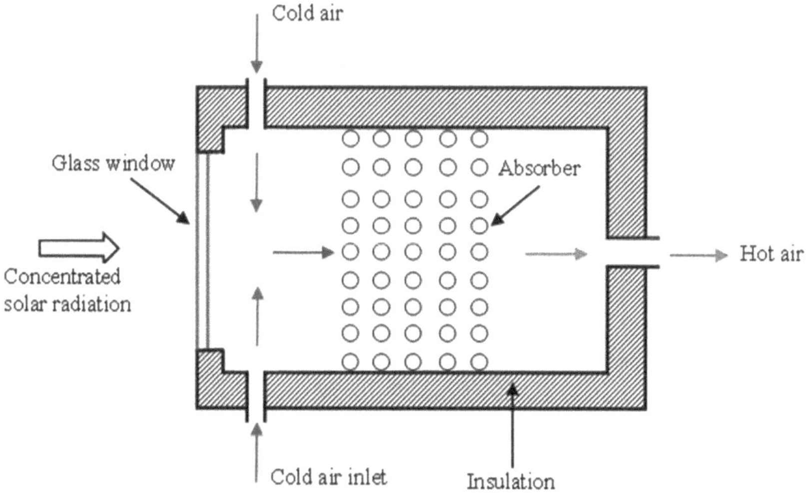

FIGURE 9.3 Volumetric solar receiver with porous absorber.

volumetric effect causes the temperature of the irradiated side of the absorber element to be lower than the outlet temperature.

The porous absorber is a solid structure and it must have good thermal and mechanical properties (in addition to optical properties) to achieve high efficiency. Important parameters of the structure are porosity, cell density, thermal conductivity, and resistance to temperature. Porosity is the proportion of the volume of voids over the total volume that characterizes the structure. A high value of the porosity is required to allow the concentrated solar radiation to penetrate the volume of the cellular material (Capuano et al., 2016). A high cell density is required to achieve large surface areas to transfer heat from the material to the fluid flowing through the channels. A high thermal conductivity of the structure is also necessary for high rates of heat transfer from the material to the fluid. The structure is exposed to high temperatures. Consequently, it must have high temperature resistance to maintain the desired structural strength at the operational temperature.

9.3.2.2 Advanced volumetric receivers

Although the conventional volumetric receiver increases the contact surface between the hot solid absorber and the heat transfer fluid, one of its major limiting factors is pressure loss across the receiver. More energy is required to pump the transfer fluid through a porous medium or block of narrow channels. This led to the development of a volumetric receiver that comprises two concentric transparent tubes (Fig. 9.4). The heat transfer fluid is passed through the inner transparent tube, and it directly absorbs concentrated solar radiation in three dimensions, which eliminates the process of heat transfer between the solid surface and the heat transfer fluid. Again, the transparent tubes must withstand high temperatures and pressure of the conduit structure.

Nonetheless, most of the heat transfer fluids weakly absorb solar radiation. Consequently, the concept of using nanofluid volume absorbers was proposed.

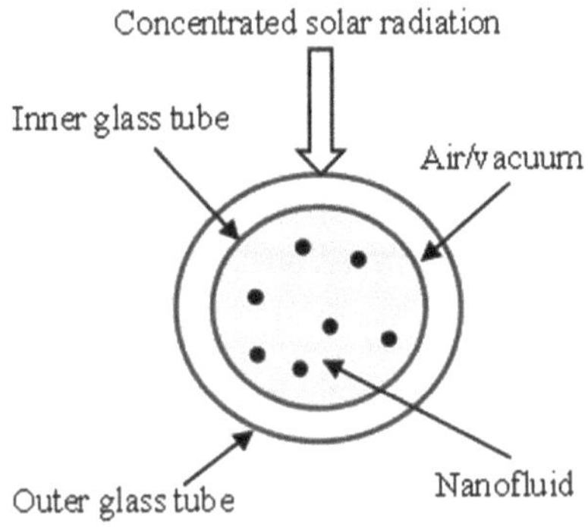

FIGURE 9.4 Example of a direct absorption volumetric receiver.

Optically, nanofluids with very high levels of absorptance can be engineered by using a small volume fraction of suitable nanoparticles. The absorption of solar radiation is a key parameter to determine the energy capture ability of fluids. Generally, the absorption coefficient increases with the volume fraction, particle size, and path length (Ahmad et al., 2017). Nevertheless, particle size becomes very insignificant to the overall improvement in optical properties in certain cases. In addition, the absorption also increased with increase in path length. A maximum achievable absorption coefficient of 22 cm^{-1} has been reported for gold nano-ellipsoids.

9.4 Temperature gradients

In general, enhancement of heat transfer contributes to the reduction of temperature gradients. Techniques for increasing the rate of heat transfer fall into passive and active categories (ASHRAE, 2017). Some of the passive techniques utilize rough surfaces, extended surfaces, displaced promoters, and vortex flow devices. Included in the active techniques are mechanical aids, surface or fluid vibration, and electrostatic fields. For example, there have been several innovations in the development of heat transfer fluids. One of these advances is the use of solid particles as a heat transfer fluid in a rotary solar receiver. In this type of receiver, solid particles are introduced into a rotating cylinder that is irradiated with concentrated solar radiation. A gas is passed through the cylinder in a cocurrent or counter current flow pattern (see Chapter 3). Rotation of the cylinder is an active technique for raising the rate of heat transmission, and for reducing temperature gradients. The efficiency of a specific enhancement technique is predominantly dependent on the mode of heat transfer or type of heat exchanger to which it is applied. It is therefore necessary to consider the major thermal resistances in the flow circuit before implementing a given technique. Some techniques may increase the friction factor, and so the benefits of their exploitation should be weighed against the additional required pumping power.

9.4.1 Temperature gradients in surface receivers

In conventional surface receivers, solar radiation is firstly captured by a thin metal plate or tube and then transferred to a heat transfer fluid (Otanicar et al., 2010). Unfortunately, a lot of energy is lost during this process and therefore the rise in outlet temperature and thermal efficiency is limited. Consequently, surface receivers are efficient at conversion of solar radiation to heat, but they are not well suited for transferring heat to the heat transfer fluid (Lenert & Wang, 2012). The implication of this is that the irradiated part of the opaque surface can attain significantly higher temperatures than the rest of the absorber element, resulting in large temperature gradients. For instance, concentrated solar radiation is incident on the bottom part of the tubular

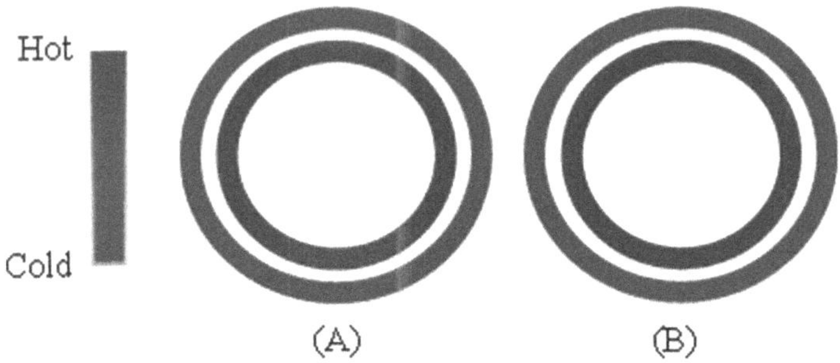

FIGURE 9.5 Simplified representation of temperature distribution in tube walls of a surface tubular receiver with concentrated solar radiation incident on: (A) on the bottom part and (B) circumferentially.

receiver of a PTC system, while the top part of the receiver is only irradiated by unconcentrated solar radiation. So, the heat flux difference between the upper and lower parts of the absorber tube creates large temperature gradients (Sandá et al., 2019). These gradients cause high thermal stresses that can bend the absorber tube and in turn lead to the breakage of the glass envelope. Fig. 9.5 illustrates the nonuniform distribution of temperature in the tubular receiver. Cyclical weather conditions and the shadows of the clouds also produce thermal gradients is due to the radiation transients. Another drawback of surface receivers is their finite absorption area that is confined to two dimensions. So, these receivers are not appropriate to effectively absorb the high radiant fluxes. Unfortunately, substantial reduction of the useful life of a system is predominantly caused by both temperature gradients and stresses (Pérez-Álvarez et al., 2020). It is therefore critical to enhance the uniform distribution of temperature in receivers, and the absorption of high intensities of concentrated solar fluxes. Strategies for solving these problems include the following: (a) the substitution of the heat transfer fluid for those with higher thermal conductivity and (b) the design of new receivers that inhibit the overheating of tube walls.

Many studies have investigated means of augmenting heat transmission in surface receivers. For instance, the use of bayonet tubes in solar receivers has been investigated (Pérez-Álvarez et al., 2020). A bayonet tube is made up of one tube inside another. In this design, the heat transfer fluid flows first through the inner tube and then back through the annular section formed between the inner and outer tubes (Fig. 9.6). Concentrated solar radiation is absorbed and converted to heat by the outer tube, thereby raising the temperature of the fluid flow in the annular section. Part of this heat is exchanged with the flow in the inner tube, which is at lower temperature, hence passively preventing overheating of the fluid in the annular section. The axis of the inner tube can be displaced from the center of the outer tube to a position opposite the angular direction where the solar radiation is maximum, at $\theta = 0°$, to create an

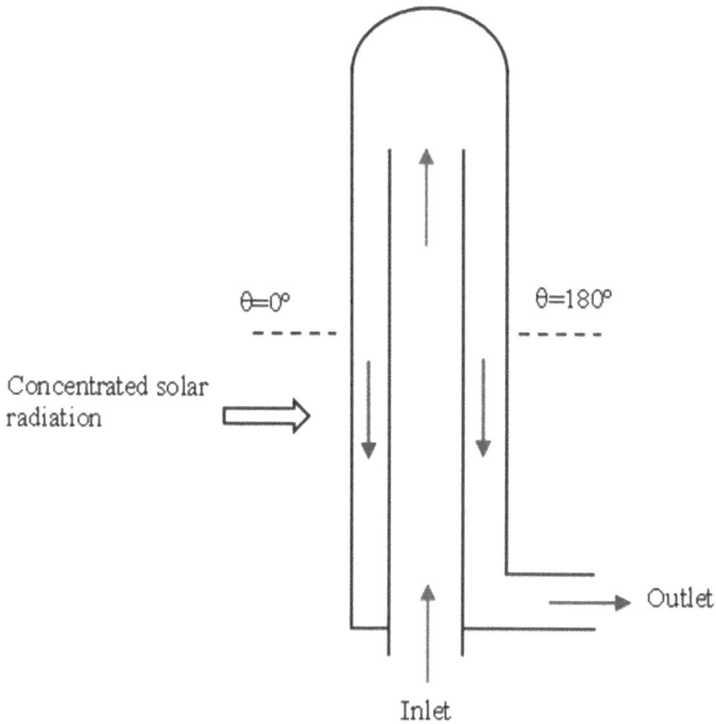

FIGURE 9.6 Schematic representation of a bayonet tube. *Adapted from Pérez-Álvarez et al. (2020).*

asymmetric structure. This geometric design affects the hydrodynamics of the bayonet tube, and it can be used to augment the heat transfer coefficient in the angular direction of $\theta = 0°$. This configuration is known as the eccentric bayonet tube, and it is a promising possibility of reducing the absorber surface temperature and the pressure drop compared to simple tubes.

The geometrical eccentricity (e) can be given by Eq. (9.1).

$$e = \frac{x}{D_i - d_e} \tag{9.1}$$

Where d_e is the outer diameter of the inner tube, D_i is the inner diameter of the outer tube, and x is the distance between the centers of the inner and outer tubes.

The performance of the bayonet receiver has been investigated numerically using molten salt and liquid sodium (Pérez-Álvarez et al., 2020). It has been found that the eccentricity of the tube decreases the temperature gradients in the outer tube wall, which results in a decline of the stresses in the absorber element. Liquid sodium yields lower temperature gradients and stresses in the

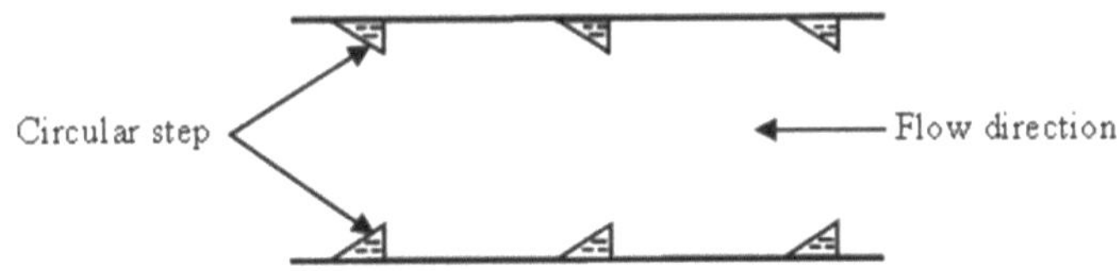

FIGURE 9.7 Linear receiver tube with perforated circular steps. *Adapted from Mahmoudi et al. (2020).*

wall, regardless of the tube configuration, due to its higher thermal conductivity. In general, temperature gradients decrease with increasing the thermal conductivity of the heat transfer fluid and the tubes.

Another milestone in the improvement of surface-based tubular receivers has been the inclusion of forward ring steps inside circular pipes (Mahmoudi et al., 2020). The rings are perforated and have a triangular cross-section (Fig. 9.7). This is a passive technique of enhancing the rate of heat transfer. The innovative receiver has been studied numerically by using governing equations of continuity, momentum, and energy in three dimensions of cylindrical coordinates for a steady-state turbulent nanofluid flow using the homogeneous single-phase model (see Chapter 8). It should also be noted that turbulent flow promotes the rate of heat transfer. Different fluids (ethylene glycol, thermal oil, water as well as CuO−water and TiO_2−water nanofluids) have been investigated, and it is found that using CuO−water as the working fluid leads to the highest thermal performance compared with that of pure water. The best thermo-hydraulic performance of the tube occurs when the distance between the steps is equal to the inner diameter of the tube.

New concepts in tubular receiver development include the use of absorber tubes with multiple layers consisting of Inconel and copper metals to enhance radial and circumferential heat transfer through the tube (Ho, 2017). Thermomechanical properties of Inconel superalloys and other selected structural materials are presented and discussed in Section 9.7.

9.4.2 Temperature gradients in volumetric receivers

It has been practically established that the aperture of a conventional volumetric receiver is the hottest region with a high level of heat loss through radiation (Ho, 2017). Nevertheless, one of the basic operating principles of volumetric receivers is that the volumetric effect causes the temperature on the irradiated side of the absorber to fall below the outlet temperature (Ávila-Marín, 2011). This temperature distribution helps to reduce radiative heat losses from the front part of the receiver. Therefore, research attention has also been given to the development of volumetric receivers with lower front temperature.

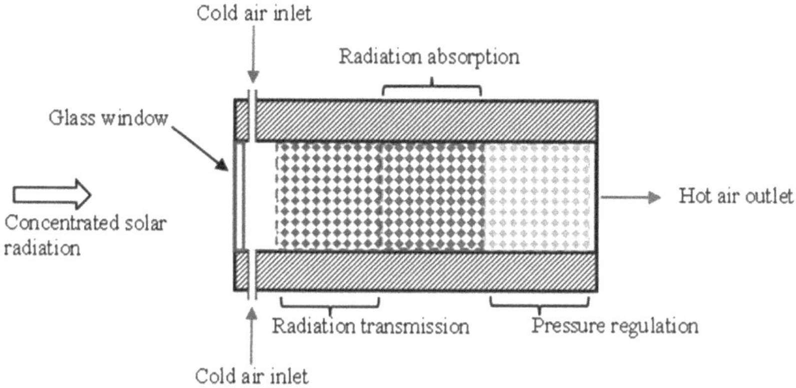

FIGURE 9.8 Schematic representation of a volumetric receiver with three layers of a porous structure. *Adapted from Zhu and Xuan (2019).*

In a conventional volumetric receiver, both the glass window and the absorber operate at high temperature, but the outer part of the window is generally exposed to ambient conditions. Thus, there are high temperature gradients in the glass window and absorber elements. To shift the highest temperature front to the inner position of the receiver, Zhu and Xuan (2019) designed a spectrally selective gradual structure receiver with swirling characteristics. In this design, a three-layer porous structure (radiation-transmitting layer, radiation-absorbing layer, and pressure-regulating layer) was used along the depth of a cylindrical volumetric receiver to augment the radiation depth and suppress the pressure drop. Four inlet flow channels were staggered along the circumference of the receiver (Fig. 9.8). This configuration of the flow channels creates rotational flow, and therefore enhances the heat exchange process inside the receiver.

Findings show that the mean temperature of the receiver frontal surface was 1158.78 K, while the average outlet temperature of the air was 1225.80 K. Consequently, the volumetric effect was achieved in this receiver design. In addition, the maximum temperature of the receiver was 1348.16 K. These levels of temperature indicate that some elements of the volumetric receiver need to be made of temperature-resistant materials.

The use of a nanofluid in a direct absorption volumetric receiver is another promising technique for improving the distribution of temperature. For instance, in a study on the performance of surface-based solar receivers with and conventional fluids and volumetric solar receivers with nanofluids, Lee et al. (2016) found that the difference between the inlet and outlet temperatures of volumetric receivers with nanofluids was larger than that of surface receivers, independent of the mass flow rate in the range of 0.75–1.5×10^{-3} kg s^{-1}. In addition, the temperature difference of volumetric

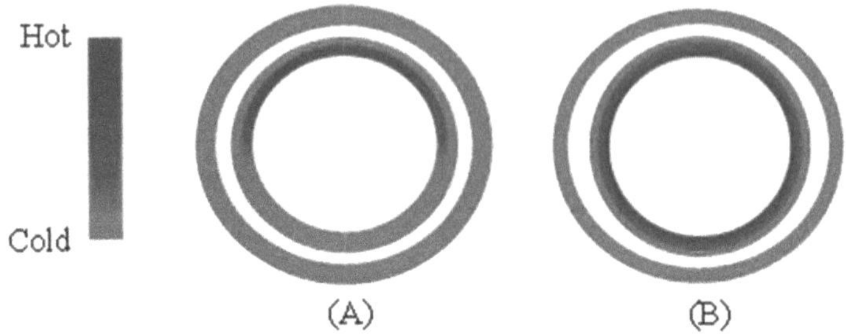

FIGURE 9.9 Simplified representation of temperature distribution in walls of the tubes in direct-absorption volumetric receiver irradiated: (A) on the top part and (B) circumferentially.

receivers increased up to a volume fraction of 0.002 vol% and then slightly decreased with the volume fraction increased. This indicates that there exists an optimal volume fraction of the nanofluid-based volumetric receiver at the prevailing conditions in this work. This observation is attributed to the solar absorptance of the nanofluid. The absorptance of the nanofluid increases with the volume fraction. However, the top-surface temperature also increases with the volume fraction. The implication of this is that the top heat loss of the receiver also rises because the heat loss is directly proportional to the temperature gradient between the top surface and ambient. So, there is an optimal volume fraction of nanofluid-based volumetric receivers at a certain combination of the magnitudes of absorbed energy and heat loss.

In case of direct absorption volumetric receivers irradiated on the top part, most of the incident radiation is absorbed by the top part of the nanofluid, and so it becomes hottest. Consequently, the inner glass tube is also hottest on the top part (Fig. 9.9). This glass tube is exposed to higher stresses than the outer one. Irradiating the receiver all around its circumference reduces the angular temperature gradient. However, heat is conducted from the inner surface of the inner tube to the outer surface of the inner tube. So, there are still radial temperature gradients within the wall of the inner tube.

9.5 Thermomechanical stresses

A structure may be simultaneously subjected to mechanical and thermal stresses. For example, in a solar receiver, a flowing heat transfer fluid exerts pressure (flow pressure) on the walls of a conduit. A high proportion of thermal power plants operating in the world use water as the working fluid in the steam Rankine cycle. In this vein, the supercritical steam cycle (s-H_2O) yields higher efficiency, and therefore more preferred. Recent research efforts are directed toward the development of ultrasupercritical steam power plants that operate at $T = 893$ K and $P = 31.0$ MPa (Boss et al., 2007). Steam can be directly generated by solar receivers, and some receivers utilize gases as heat transfer media. One main advantage of gaseous heat transfer media is their ability to attain high temperatures above 1273 K (Ho, 2017). Nevertheless, gases have low density and therefore require high operational pressure. Clearly, these magnitudes of flow pressure and temperature create high levels of mechanical and thermal stresses on the walls of conduits. It is therefore imperative to use conduits that can withstand high temperatures and pressure.

9.5.1 Thermal stress

The thermal stress created due to the temperature gradient can be expressed as follows (Panda et al., 2002):

$$\sigma = \frac{\alpha \, E \, T_c}{1 - \nu} \tag{9.2}$$

Where E is the Young's modulus (modulus of elasticity), T_c is the temperature change, α is the coefficient of thermal expansion, and ν is the Poisson's ratio. The coefficient of thermal expansion, Young's modulus, and Poisson's ratio are properties of a material and may vary with direction. More details about these parameters are presented and discussed in Section 9.6.

As seen from Eq. (9.2), thermal stress is directly proportional to the temperature gradient, for a given set of material properties (E, α, and ν). It has been established that surface-based receivers exhibit high temperature gradients. Consequently, it is necessary to avoid or at least to minimize these gradients by using appropriate designs and materials.

9.5.2 Mechanical stress

In a three-dimensional structure, stresses are created in different directions. Consider an infinitesimal volume element with six faces, presented in Fig. 9.4. There are 3 stress components per face, and therefore, a total of 18 components are associated with the element (Huston & Josephs, 2009). For this volume element to be in equilibrium, the corresponding stress components on opposite and parallel faces must balance with oppositely directed components. Consequently, only 9 out of the 18 stress components are required for consideration. A subscript notation such as σ_{ij} is used to account for the nine stress components for convenience. The subscripts i and j have the values x, y, and z, and the first subscript (i) represents the face on which the stress is applied, while the second subscript (j) represents the direction of the stress component. So, the stress components can be arranged into a matrix ($\boldsymbol{\sigma}$):

$$\boldsymbol{\sigma} = \begin{pmatrix} \sigma_{xx} & \sigma_{xy} & \sigma_{xz} \\ \sigma_{yx} & \sigma_{yy} & \sigma_{yz} \\ \sigma_{zx} & \sigma_{zy} & \sigma_{zz} \end{pmatrix} \tag{9.3}$$

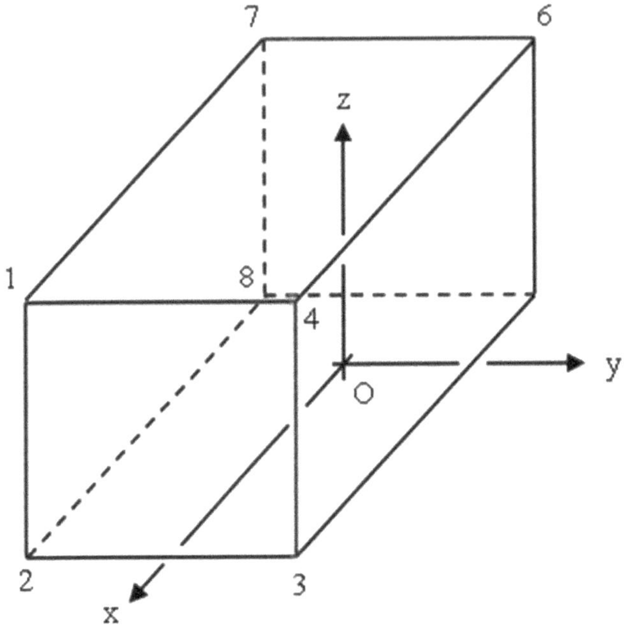

FIGURE 9.10 Vertices and surfaces of a volume element in Cartesian coordinates. *Adapted from Huston and Josephs (2009).*

Diagonal elements of the matrix in Eq. (9.3) are the normal stresses (tension = compression), while the off-diagonal elements (σ_{ij}, where $i \neq j$) are shear stresses. A stress component is considered positive if it is exerted on a positive face in a positive direction or on a negative face in a negative direction (Fig. 9.10). On the other hand, a stress component is considered negative if it is exerted on a negative face in the positive direction or a positive face in the negative direction. This analysis uses the sign convention: tension is positive and compression is negative.

Faces of the volume element can be identified by using normals because each face is normal to one of x, y, and z axes. There are two faces normal to each axis, and therefore, these faces can be seen as positive or negative depending upon their location relative to the side of the origin O. A face is positive if the movement is in the positive direction of the axis when going from the interior of the element to the exterior across a face. Otherwise, a face is negative if the movement is in the negative axis direction when traversing the face. A summary of face signs and normal axes is presented in Table 9.1.

TABLE 9.1 Signs and normal axes of faces of an infinitesimal volume element.

Face	Normal axis	Face sign
1234	X	Positive
4356	Y	Positive
6714	Z	Positive
7658	-x	Negative
2178	-y	Negative
2853	-z	Negative

9.6 Thermomechanical strains

Most materials tend to expand when heated and contract when cooled. Thus, a structural member may be deformed due to heating or cooling. The strength in tension, hardness, and resistance to fatigue for most solid materials (that do not undergo transitions) increases with a decline in service temperature (Huston and Josephs, 2009). This rule applies without exception to such metals as aluminum, copper, nickel alloys, and austenitic stainless steels. For ordinary carbon steels, nevertheless, the advantages of improved properties are significantly compromised by the tendency of materials to become brittle.

9.6.1 Thermal strain

Consider a bar with fixed ends undergoing a temperature increase from T_0 to T_1. If not constrained, the one-dimensional thermal strain (ε) in the bar can be defined by the following:

$$\varepsilon = \frac{L_1 - L_0}{L_0} = \frac{\alpha\,(T_1 - T_0)}{1 - \nu} = \frac{\alpha\,T_c}{1 - \nu} \tag{9.4}$$

Where L_0 is the initial length of the bar at temperature T_0, L_1 is the length of the bar at final temperature T_1, T_c is the temperature change, α is the coefficient of thermal expansion, and ν is the Poisson's ratio.

The difference $(L_1 - L_0)$ is the extension (or contraction) of the bar. In case of cooling, the difference between the initial and final length gives the level of contraction. Elastic strains are produced that compensate the thermal strains because the thermal expansion in the length direction is completely suppressed (sum of thermal and elastic strains is equal to zero). It is also possible for the bar to undergo strain in two or three dimensions.

9.6.2 Mechanical strain

When a force is applied to a structure, the structure may undergo deformation (such as extension, compression, or shearing). The magnitude of deformation depends on the applied force and characteristic strength of the material.

9.6.2.1 *Material strength*

Consider a circular bar with an original length (L_0) and cross-sectional area (A) that is fixed at one end (Fig. 9.11). A force (F) is applied at the free end and causes some linear displacement along the axis of the rode. The applied force and displacement are related through Hooke's law. This law states that the force is directly proportional to the displacement. Alternatively, the law states that stress is directly proportional to strain.

In one dimension, Hooke's law can be expressed as follows:

$$F = k\delta \tag{9.5}$$

Where F is the applied force, k is the constant of proportionality, and δ is the displacement in one dimension.

However, stress is related to the applied force and the cross-sectional area, while strain is related to the displacement and original length as follows:

$$\sigma = F/A \text{ and } \varepsilon = \delta/L_0 \tag{9.6}$$

Therefore, Hooke's law can also be expressed as:

$$\sigma = E\varepsilon \tag{9.7}$$

Where $E = L_0k/A$ is the Young's modulus.

Hooke's law is valid only within the elastic range up to yield point (Fig. 9.12). Plastic deformation starts at this point and the structure no longer obeys Hooke's law. As the stress increases, the structural member eventually attains the fracture point. The stress just prior to fracture (rupture) is called the breaking strength. In brittle materials, such as glass, yield and fracture points almost coincide, and so these materials break immediately after the yield point. Table 9.2 shows the values of modulus of elasticity and maximum strength (σ_{max}) for selected materials. Steel is one of the commonly-used structural materials in solar receivers.

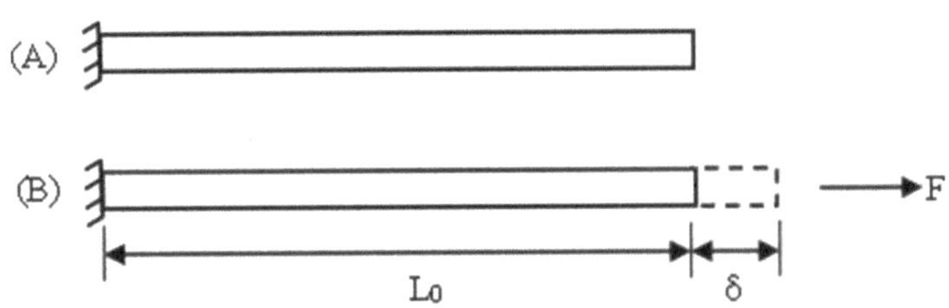

FIGURE 9.11 A rod extended by an axial force (F): (A) before extension and (B) after extension.

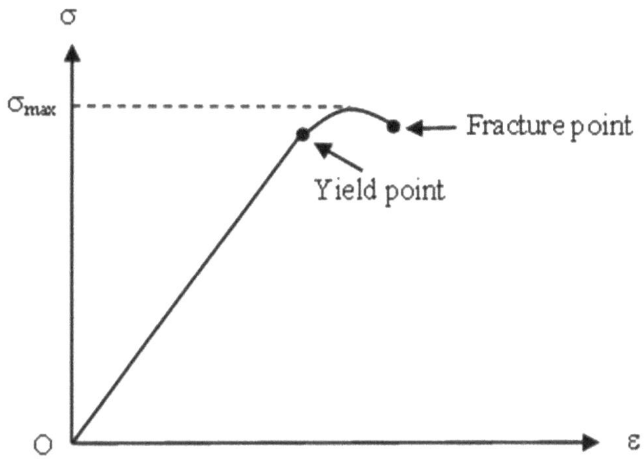

FIGURE 9.12 Relationship between stress and strain.

TABLE 9.2 Values of the Young's modulus and maximum strength of selected materials.

Material	Young's modulus (GPa)	Maximum strength (MPa)
Steel	207	410–820
Aluminum	69	275–550
Copper	117	200–400
Concrete	28	28–40

9.6.2.2 Hooke's law in two and three dimensions

The elementary concept of Hooke's law can be extended to two and three dimensions. In this case, the displacement occurs in multiple directions when a force is applied. In one dimension, it was assumed that the cross-sectional area perpendicular to the direction of the applied force remains constant. In multiple dimensions, there is induced strain in directions perpendicular to an applied strain. For instance, a rod may contract laterally as it is elongated by the applied strain. Thus, there is need to quantify the induced strain.

Poisson's ratio is a parameter that quantifies induced strain in directions perpendicular to an applied strain. Consider a rod with a square cross-section submitted to an axial load (Fig. 9.13). As the rod extends, the cross-section area decreases.

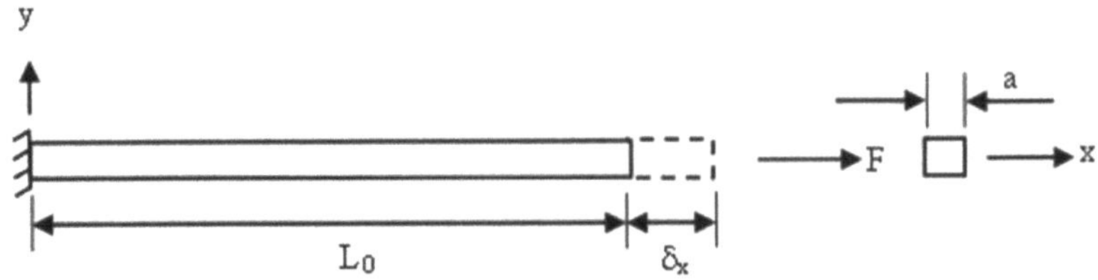

FIGURE 9.13 A stretched road with square cross-section.

Poisson's ratio (ν) is a measure of this effect, and it is given by Eq. (9.8). Values of ν for selected materials are presented in Table 9.3.

$$\nu = \frac{-\varepsilon_{yy}}{\varepsilon_{xx}} \tag{9.8}$$

Where ε_{xx} is the strain in the x-direction and ε_{yy} is the strain in the y-direction.

Based on the geometry of the rod, the longitudinal (ε_{xx}) and transverse (ε_{yy}) strains can be expressed as follows:

$$\varepsilon_{xx} = \frac{\delta_x}{L_0} \text{ and } \varepsilon_{yy} = \frac{-\delta_y}{a} \tag{9.9}$$

Where δ_x is the elongation of the rod in the x-direction, δ_y is the shrinking of the cross-section, L_0 is the original rod length, and a is the side length of the cross-section.

TABLE 9.3 Values of Poisson's ratio for selected materials.

Material	Poisson's ratio (dimensionless)
Upper theoretical limit (perfectly deformable material)	0.5
Lead	0.43
Gold	0.42
Platinum	0.39
Aluminum (pure)	0.36
Copper	0.34
Aluminum (wrought)	0.33
Stainless steel	0.31
Structural steel	0.3
Cast iron (gray)	0.26
Glass	0.24
Lower theoretical limit (perfectly brittle material)	0

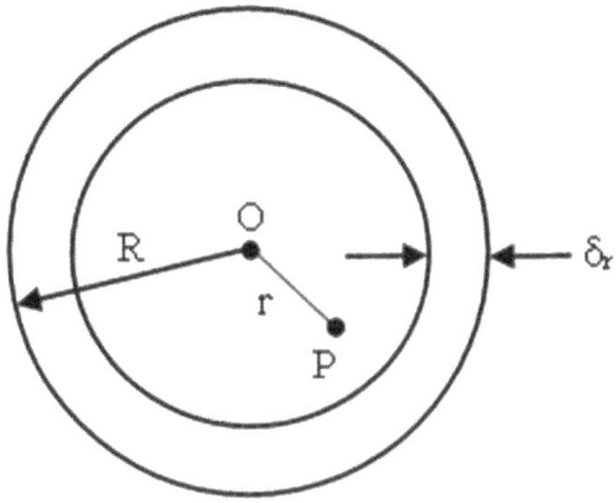

FIGURE 9.14 Cross-section of a circular rod with typical point P.

Using Eqs. (9.8) and (9.9), the Poisson's ratio for a rod with square cross-section can be defined by Eq. (9.10).

$$\nu = \frac{-\varepsilon_{yy}}{\varepsilon_{xx}} = \left(\frac{\delta_y}{\delta_x}\right)\left(\frac{L_0}{a}\right) \tag{9.10}$$

Similar expressions can also be derived for a rod with circular cross-section. As the rod extends, the circular cross-section shrinks where R is the undeformed cross-section radius and δ_r is the decrease in the radius. During extension, the radial displacement u_r of a point P in the cross-section is proportional to the distance of P from the axis (Fig. 9.14).

The radial displacement u_r is the following:

$$u_r = -\delta_r \frac{r}{R} \tag{9.11}$$

Where the negative sign shows that P is displaced toward the center O as the cross-section contracts.

The radial strain ε_{rr} is then given by:

$$\varepsilon_{rr} = \frac{\partial u_r}{\partial r} = \frac{-\delta_r}{R} \tag{9.12}$$

If the circular rod is stretched by a distance δ_x along its axis (x direction), then the longitudinal strain is expressed as follows:

$$\varepsilon_{xx} = \frac{\delta_x}{L_0} \tag{9.13}$$

Therefore, the Poisson's ratio for a circular rod is defined as:

$$\nu = -\varepsilon_{rr}/\varepsilon_{xx} = \left(\frac{\delta_r}{\delta_x}\right)\left(\frac{L_0}{R}\right) \tag{9.14}$$

Example 9.1

A stainless steel bar with a length of 0.20 m and square cross-section is subjected to a tensile stress resulting in a longitudinal extension of 4×10^{-4} m, while the cross-section displacement is 1×10^{-6} m. Calculate the area of the cross-section.

Solution

Area of cross-section $= a^2$
From Eq. (9.10),

$$\nu = \left(\frac{\delta_y}{\delta_x}\right)\left(\frac{L_0}{a}\right)$$

Therefore,

$$a = \frac{\delta_y \, L_0}{\delta_x \, \nu}$$

From Table 9.3, $\nu = 0.31$.
This implies that

$$\text{Area} = a^2 = \left(\frac{\delta_y L_0}{\delta_x}\right)^2 = \left(\frac{1 \times 10^{-6} \times 0.20}{4 \times 10^{-4} \times 0.31}\right)^2 = 260.1 \text{ mm}^2$$

Poisson's ratio is a measure of the shrinkage of a loaded rod, and so it also measures the volumetric change of a loaded structural member. In this vein, the elongated round bar can also be analyzed in three dimensions, and the initial volume (V_0) of the rod is as follows:

$$V_0 = \pi R^2 L_0 \tag{9.15}$$

From Fig. 9.8, the deformed volume (V_1) of the road is given by Eq. (9.16).

$$V_1 = \pi(R - \delta_r)^2 (L_0 + \delta_x) \tag{9.16}$$

The displacements δ_r and δ_x are relatively small. Consequently, Eq. (9.16) can be expressed as:

$$V_1 = \pi\left(R^2 - 2R\delta_r\right)(L_0 + \delta_x) = \pi\left(R^2 L_0 - 2R\, L_0 \delta_r + R^2\, \delta_x\right) \tag{9.17}$$

The change in volume is given by the following equation:

$$\Delta V = V_1 - V_0 = \pi\left(-\,2R\, L_0 \delta_r + R^2 \delta_x\right) \tag{9.18}$$

From Eq. (9.14), δ_r is expressed as follows:

$$\delta_r = \nu\, \delta_x \left(\frac{R}{L_0}\right) \tag{9.19}$$

So, Eq. (9.18) becomes Eq. (9.20).

$$\Delta V = \pi(-2\nu + 1)R^2\delta_x \quad (9.20)$$

Consequently, the volumetric strain ($\Delta V/V_0$) can be expressed as follows:

$$\Delta V/V_0 = (1 - 2\nu)(\delta_x / L_0) = (1 - 2\nu)\varepsilon_{xx} \quad (9.21)$$

The concept of volumetric strain can also be extended to a small rectangular parallelepiped block with edges a, b, and c before loading (Fig. 9.15). After loading, the edges will become the following: $(1 + \varepsilon_{xx})a$, $(1 + \varepsilon_{yy})b$, and $(1 + \varepsilon_{zz})c$.

From the foregoing discussion, the volumes of the block before (V_0) and after (V_1) deformation are respectively defined by Eqs. (9.22) and (9.23).

$$V_0 = abc \quad (9.22)$$

$$V_1 = abc(1 + \varepsilon_{xx})(1 + \varepsilon_{yy})(1 + \varepsilon_{zz}) = abc(1 + \varepsilon_{xx} + \varepsilon_{yy} + \varepsilon_{zz}) \quad (9.23)$$

Therefore, the volumetric strain of a structural member can be expressed as follows:

$$\Delta V/V_0 = \varepsilon_{xx} + \varepsilon_{yy} + \varepsilon_{zz} \quad (9.24)$$

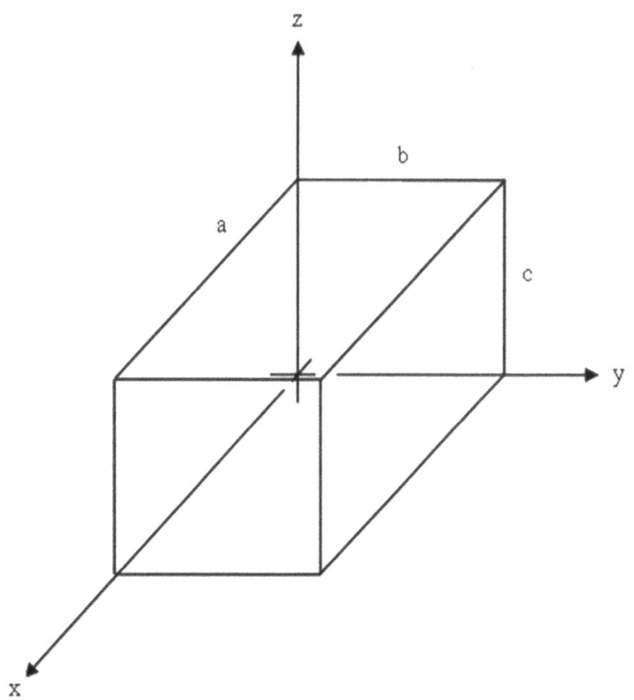

FIGURE 9.15 Uniform x-direction load and deformation of a rectangular block.

But from Eq. (9.8),

$$\varepsilon_{yy} = -\nu\varepsilon_{xx} \text{ and } \varepsilon_{zz} = -\nu\varepsilon_{xx} \tag{9.25}$$

This shows that the volumetric strain can be expressed as follows:

$$\Delta V/V_0 = (1 - 2\nu)\varepsilon_{xx} \tag{9.26}$$

A material is said to be brittle if it does not contract or shrink transversely when loaded. Alternatively, a material is described as incompressible or fully compliant if it undergoes shrinkage without the volume change of the loaded body (Huston and Josephs, 2009). In this case, the Poisson's ratio is zero, and the thermal strain (Eq. 9.4) can be simplified to (αT_c). Generally, brittle materials (such as glass) have low values of ν, and so their thermal strain can be described by a simple equation. For a fully compliant material, the change of volume is zero during loading. Thus, from Eq. (9.26), we obtain the following:

$$1 - 2\nu = 0 \text{ or } \nu = 0.5 \tag{9.27}$$

Consequently, the range of Poisson's ratio is 0–0.5. Similar to stress, strain can be generalized in three dimensions. In this regard, consider the displacement of a point P relative to a convenient set of x, y, and z axes to be, respectively, u, v, and w (Fig. 9.16).

It can be shown that the strains along the edges of a cubic element intersecting at P are as follows:

$$\varepsilon_{xx} = \frac{\partial u}{\partial x},\ \varepsilon_{yy} = \frac{\partial v}{\partial y} \text{ and } \varepsilon_{zz} = \frac{\partial w}{\partial z} \tag{9.28}$$

$$\varepsilon_{xy} = \frac{1}{2}\left(\frac{\partial u}{\partial y} + \frac{\partial v}{\partial x}\right),\ \varepsilon_{yz} = \frac{1}{2}\left(\frac{\partial v}{\partial z} + \frac{\partial w}{\partial y}\right) \text{ and } \varepsilon_{zx} = \frac{1}{2}\left(\frac{\partial w}{\partial x} + \frac{\partial u}{\partial z}\right) \tag{9.29}$$

It should be noted that $\varepsilon_{ij} = \varepsilon_{ji}$, and the components of the mechanical strain can be represented by a matrix:

$$\boldsymbol{\varepsilon} = \begin{pmatrix} \varepsilon_{xx} & \varepsilon_{xy} & \varepsilon_{xz} \\ \varepsilon_{yx} & \varepsilon_{yy} & \varepsilon_{yz} \\ \varepsilon_{zx} & \varepsilon_{zy} & \varepsilon_{zz} \end{pmatrix} \tag{9.30}$$

Structural members may be subjected to multiple loadings. In this context, the principle of superposition states that multiple loadings on an elastic structural member may be considered individually and in any order for analyzing the stresses and strains caused by the loadings (Faupel & Fisher, 1981). The stress (or strain) of an elastic body subjected to multiple loads is equal to the sum of stresses (or strains) caused by individual loadings. The principle of superposition is useful in stress–strain analysis, but it is not always applicable, particularly in heavily loaded structures with large

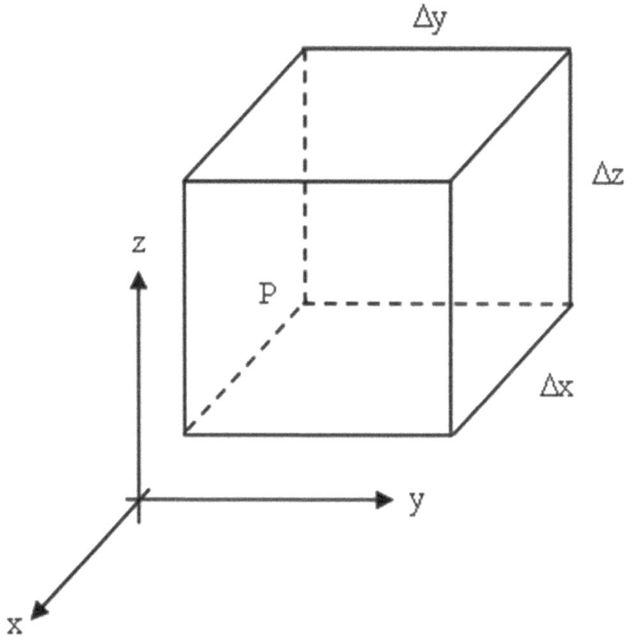

FIGURE 9.16 Displacement of point P in a cubical element.

deformation. Nevertheless, this principle holds if the deformation is small and if linear stress–strain equations are applicable.

The Poisson's ratio and the principle of superposition can be used to extend the fundamental relations of stress and strain in one dimension to two or three dimensions to obtain combined stresses and strains in two or more directions. In this vein, consider a small rectangular elastic element subjected to tension (Fig. 9.17).

According to Hooke's law, the strain is directly proportional to the stress for sufficiently small deformations (Kittel, 2005). Consequently, strain components can be expressed as linear functions of stress components.

$$\begin{pmatrix} \varepsilon_{xx} \\ \varepsilon_{yy} \\ \varepsilon_{zz} \\ \varepsilon_{yz} \\ \varepsilon_{zx} \\ \varepsilon_{xy} \end{pmatrix} = \begin{pmatrix} S_{11} & S_{12} & S_{13} & S_{14} & S_{15} & S_{16} \\ S_{21} & S_{22} & S_{23} & S_{24} & S_{25} & S_{26} \\ S_{31} & S_{32} & S_{33} & S_{34} & S_{35} & S_{36} \\ S_{41} & S_{42} & S_{43} & S_{44} & S_{45} & S_{46} \\ S_{51} & S_{52} & S_{53} & S_{54} & S_{55} & S_{56} \\ S_{61} & S_{62} & S_{63} & S_{64} & S_{65} & S_{66} \end{pmatrix} \begin{pmatrix} \sigma_{xx} \\ \sigma_{yy} \\ \sigma_{zz} \\ \sigma_{yz} \\ \sigma_{zx} \\ \sigma_{xy} \end{pmatrix} \tag{9.31}$$

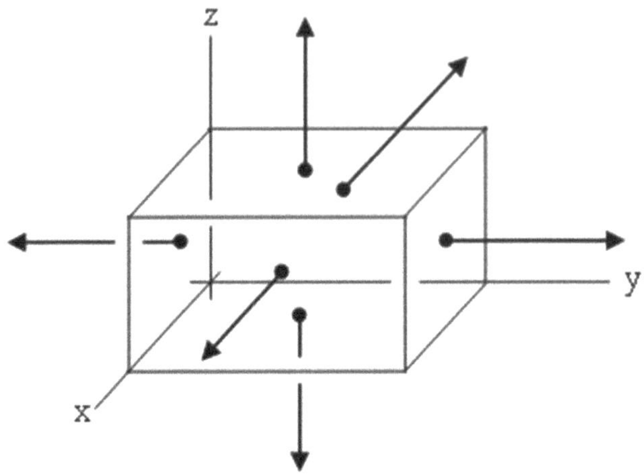

FIGURE 9.17 A rectangular elastic element subjected to tension loading.

The quantities S_{ij} are known as elastic compliance constants or elastic constants. These constants are influenced by the properties of materials. In this vein, properties of anisotropic materials depend on the direction considered. For instance, a material can have different values of strength in different directions. If properties vary with direction, then the material is said to be anisotropic. A material is orthotropic if its properties are independent in three mutually perpendicular directions. For example, the Poisson's ratio may be different for transversal deformations in y and z directions ($\nu_1 = -\varepsilon_{yy}/\varepsilon_{xx}$ and $\nu_2 = -\varepsilon_{zz}/\varepsilon_{xx}$). Similarly, the Young's modulus may also significantly vary with direction ($\sigma_{xx} = E_1\varepsilon_{xx}$, $\sigma_{yy} = E_2\varepsilon_{yy}$, and $\sigma_{zz} = E_3\varepsilon_{zz}$). On the other hand, properties of isotropic materials do not depend on the direction. In this case, $\nu_1 = \nu_2 = \nu$, $E_1 = E_2 = E_3 = E$, and strain and stress components can be related by the following equation:

$$\begin{pmatrix} \varepsilon_{xx} \\ \varepsilon_{yy} \\ \varepsilon_{zz} \\ \varepsilon_{yz} \\ \varepsilon_{zx} \\ \varepsilon_{xy} \end{pmatrix} = \frac{1}{E} \begin{pmatrix} 1 & -\nu & -\nu & 0 & 0 & 0 \\ -\nu & 1 & -\nu & 0 & 0 & 0 \\ -\nu & -\nu & 1 & 0 & 0 & 0 \\ 0 & 0 & 0 & 2(1+\nu) & 0 & 0 \\ 0 & 0 & 0 & 0 & 2(1+\nu) & 0 \\ 0 & 0 & 0 & 0 & 0 & 2(1+\nu) \end{pmatrix} \begin{pmatrix} \sigma_{xx} \\ \sigma_{yy} \\ \sigma_{zz} \\ \sigma_{yz} \\ \sigma_{zx} \\ \sigma_{xy} \end{pmatrix} \tag{9.32}$$

Similarly, based on Hooke's law, stress components can be expressed as linear functions of strains:

$$\begin{pmatrix} \sigma_{xx} \\ \sigma_{yy} \\ \sigma_{zz} \\ \sigma_{yz} \\ \sigma_{zx} \\ \sigma_{xy} \end{pmatrix} = \begin{pmatrix} C_{11} & C_{12} & C_{13} & C_{14} & C_{15} & C_{16} \\ C_{21} & C_{22} & C_{23} & C_{24} & C_{25} & C_{26} \\ C_{31} & C_{32} & C_{33} & C_{34} & C_{35} & C_{36} \\ C_{41} & C_{42} & C_{43} & C_{44} & C_{45} & C_{46} \\ C_{51} & C_{52} & C_{53} & C_{54} & C_{55} & C_{56} \\ C_{61} & C_{62} & C_{63} & C_{64} & C_{65} & C_{66} \end{pmatrix} \begin{pmatrix} \varepsilon_{xx} \\ \varepsilon_{yy} \\ \varepsilon_{zz} \\ \varepsilon_{yz} \\ \varepsilon_{zx} \\ \varepsilon_{xy} \end{pmatrix} \tag{9.33}$$

The quantities C_{ij} are known as stiffness constants or moduli of elasticity. Again, these constants depend on the properties of a material. It should be noted that C_{ij} is an inverse matrix of S_{ij}. For isotropic materials, the stiffness matrix C can be given by the following:

$$C = \begin{pmatrix} \psi & \lambda & \lambda & 0 & 0 & 0 \\ \lambda & \psi & \lambda & 0 & 0 & 0 \\ \lambda & \lambda & \psi & 0 & 0 & 0 \\ 0 & 0 & 0 & \mu & 0 & 0 \\ 0 & 0 & 0 & 0 & \mu & 0 \\ 0 & 0 & 0 & 0 & 0 & \mu \end{pmatrix} \tag{9.34}$$

Where $\lambda = \frac{\nu E}{(1+\nu)(1-2\nu)}$, $\mu = \frac{E}{2(1+\nu)}$ and $\psi = \lambda + 2\mu$.

If we let the resulting tensile stresses be σ_{xx}, σ_{yy}, and σ_{zz}, and then use the Poisson's ratio and the principle of superposition, the mechanical strains on the elemental structure of an isotropic material in Cartesian coordinates are as follows (Faupel & Fisher, 1981):

$$\begin{pmatrix} \varepsilon_{xx} \\ \varepsilon_{yy} \\ \varepsilon_{zz} \end{pmatrix} = \frac{1}{E} \begin{pmatrix} 1 & -\nu & -\nu \\ -\nu & 1 & -\nu \\ -\nu & -\nu & 1 \end{pmatrix} \begin{pmatrix} \sigma_{xx} \\ \sigma_{yy} \\ \sigma_{zz} \end{pmatrix} \tag{9.35}$$

Eq. (9.35) is derived by assuming that all the stresses are positive (tension). Nevertheless, if some or all of the stresses are negative (compression), the expressions are still valid. We then just insert negative values for those negative stresses on the right side of this equation. It should also be noted that Eq. (9.35) is linear in both the stresses and the strains. Consequently, it can be solved for the stresses in terms of the strains (by inverting the matrix of elastic compliance constants) to obtain the following:

$$\begin{pmatrix} \sigma_{xx} \\ \sigma_{yy} \\ \sigma_{zz} \end{pmatrix} = \frac{E}{(1+\nu)(1-2\nu)} \begin{pmatrix} 1-\nu & \nu & \nu \\ \nu & 1-\nu & \nu \\ \nu & \nu & 1-\nu \end{pmatrix} \begin{pmatrix} \varepsilon_{xx} \\ \varepsilon_{yy} \\ \varepsilon_{zz} \end{pmatrix} \tag{9.36}$$

The effective stress (σ_{eff}) can be computed by using von Mises theory:

$$\sigma_{eff} = \sqrt{\left[\sigma_{xx}^2 + \sigma_{yy}^2 + \sigma_{zz}^2 - \left(\sigma_{xx}\sigma_{yy} + \sigma_{xx}\sigma_{zz} + \sigma_{yy}\sigma_{zz}\right)\right]} \tag{9.37}$$

For a circular structural member, three principal stresses are hoop stress (σ_t), radial stress (σ_r), and longitudinal stress (σ_z). For an isotropic material, components of mechanical strain and stress can be related by the following:

$$\begin{pmatrix} \varepsilon_t \\ \varepsilon_r \\ \varepsilon_z \end{pmatrix} = \frac{1}{E} \begin{pmatrix} 1 & -\nu & -\nu \\ -\nu & 1 & -\nu \\ -\nu & -\nu & 1 \end{pmatrix} \begin{pmatrix} \sigma_t \\ \sigma_r \\ \sigma_z \end{pmatrix} \tag{9.38}$$

It should be noted that Eq. (9.38) is similar to Eq. (9.35). As expected, the two equations have the same matrix of elastic compliance constants. Consequently, their corresponding elastic stiffness matrices are also equal.

Tubular conduits are exploited in many types of solar receivers. For a thick-walled cylinder or tube with internal and external radii r_i and r_o, respectively, and internal and external pressure P_i and P_o, respectively, the radial and circumferential stresses can be given by the following (Kanlıkama et al., 2013):

$$\sigma_t = P_i \frac{r_i^2\left(r^2 + r_o^2\right)}{r^2\left(r_o^2 - r_i^2\right)} - P_o \frac{r_o^2\left(r^2 + r_i^2\right)}{r^2\left(r_o^2 - r_i^2\right)} \tag{9.39}$$

$$\sigma_r = P_i \frac{r_i^2\left(r^2 - r_o^2\right)}{r^2\left(r_o^2 - r_i^2\right)} - P_o \frac{r_o^2\left(r^2 - r_i^2\right)}{r^2\left(r_o^2 - r_i^2\right)} \tag{9.40}$$

$$\sigma_z = \frac{P_i r_i^2 - r_o^2 P_o}{r_o^2 - r_i^2} \tag{9.41}$$

Where r is the radial distance from the axis of the tube.

Sometimes, only the inner pressure is assumed on thick-walled cylinders ($P_o = 0$). In this case, Eqs. (9.39)–(9.41) can be simplified for an external pressure or pressure gradient over the whole cylinder or tube.

9.6.3 Relationships between thermomechanical stress and strain

In general, anisotropic materials have different coefficients of linear expansion in different directions. In such cases, it important to consider the coefficient of thermal expansion as a tensor ($\boldsymbol{\alpha}$) with up to six independent elements: α_{11}, α_{22}, α_{33}, α_{12}, α_{13}, and α_{23}. This shows that thermal gradient can cause tensile and shear strains. In practical engineering materials, this is not the case. Thermal gradient causes extensional strains, which leads to three possibilities of coefficient of linear expansion: α_{11}, α_{22}, and α_{33} for orthotropic materials.

For an isotropic material, the sum of mechanical and thermal strains along the principal axes can be expressed in terms of stresses:

$$\varepsilon_{xx} = \frac{1}{E}\left[\sigma_{xx} - \nu\left(\sigma_{yy} + \sigma_{zz}\right)\right] + \frac{\alpha T_c}{1-\nu} \tag{9.42}$$

$$\varepsilon_{yy} = \frac{1}{E}\left[\sigma_{yy} - \nu\left(\sigma_{xx} + \sigma_{zz}\right)\right] + \frac{\alpha T_c}{1-\nu} \tag{9.43}$$

$$\varepsilon_{zz} = \frac{1}{E}\left[\sigma_{zz} - \nu\left(\sigma_{xx} + \sigma_{yy}\right)\right] + \frac{\alpha T_c}{1-\nu} \tag{9.44}$$

Where T_c is the temperature change above the initial value.

Eqs. (9.42)–(9.44) are linear in thermomechanical stresses and strains. Consequently, for a given temperature change, stresses can be expressed in terms of strains:

$$\sigma_{xx} = E\frac{(1-\nu)\varepsilon_{xx} + \nu\left(\varepsilon_{yy} + \varepsilon_{zz}\right) - \left(\frac{1+\nu}{1-\nu}\right)\alpha T_c}{(1+\nu)(1-2\nu)} \tag{9.45}$$

$$\sigma_{yy} = E\frac{(1-\nu)\varepsilon_{yy} + \nu\left(\varepsilon_{xx} + \varepsilon_{zz}\right) - \left(\frac{1+\nu}{1-\nu}\right)\alpha T_c}{(1+\nu)(1-2\nu)} \tag{9.46}$$

$$\sigma_{zz} = E\frac{(1-\nu)\varepsilon_{zz} + \nu\left(\varepsilon_{xx} + \varepsilon_{yy}\right) - \left(\frac{1+\nu}{1-\nu}\right)\alpha T_c}{(1+\nu)(1-2\nu)} \tag{9.47}$$

For a circular structural member, thermomechanical stresses and strains can be related as follows:

$$\varepsilon_t = \frac{1}{E}\left[\sigma_t - \nu(\sigma_r + \sigma_z)\right] + \frac{\alpha T_c}{1-\nu} \tag{9.48}$$

$$\varepsilon_r = \frac{1}{E}\left[\sigma_r - \nu(\sigma_t + \sigma_z)\right] + \frac{\alpha T_c}{1-\nu} \tag{9.49}$$

$$\varepsilon_z = \frac{1}{E}[\sigma_z - v(\sigma_r + \sigma_t)] + \frac{\alpha T_c}{1 - v} \quad (9.50)$$

The stresses for a circular structural member can also be expressed in terms of strains, using Eqs. (9.48)–(9.50).

For a tubular structural member, the thermomechanical stress and strain can be related as follows (Kandil et al., 1995; Salifu et al., 2019):

$$\sigma_t = \frac{\alpha E}{(1 - v)r^2}\left[\frac{r^2 + r_i^2}{r_o^2 - r_i^2}\int_{r_i}^{r_o} T_c r\, dr - \int_{r_i}^{r} T_c r\, dr - T_c r^2\right] \quad (9.51)$$

$$\sigma_r = \frac{\alpha E}{(1 - v)r^2}\left[\frac{r^2 - r_i^2}{r_o^2 - r_i^2}\int_{r_i}^{r_o} T_c r\, dr - \int_{r_i}^{r} T_c r\, dr\right] \quad (9.52)$$

$$\sigma_z = \frac{\alpha E}{(1 - v)}\left[\frac{2}{r_o^2 - r_i^2}\int_{r_i}^{r_o} T_c r\, dr - T_c\right] \quad (9.53)$$

Salifu et al. (2019) studied the thermomechanical behavior of an unconstrained X20 steam pipe by using numerical and analytical methods. The pipe was insulated with pyrogel material (Fig. 9.18A), and the operating temperature and pressure of the steam were, respectively, 823 K and $P_i = 18$ MPa. Thus, the internal surface of the pipe was heated by steam, and heat was conducted from the internal surface to the outer surface of the steel pipe. The values of design parameters are presented in Table 9.4. It was found that thermomechanical stress was highest on the inner side of the pipe, declining gradually outwards (Fig. 9.18B). The maximum developed stress was 91.71 MPa, which is significantly lower than the maximum stress of steel (see Table 9.2).

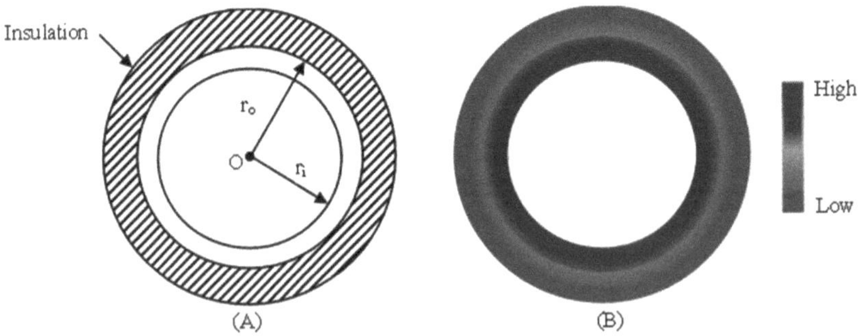

FIGURE 9.18 Schematic representation of an insulated pipe showing: (A) cross-section and (B) distribution of thermomechanical stress in the structural member.

TABLE 9.4 Design parameters of an insulated X20 steel pipe.

Parameter	Unit	X20 steel pipe	Insulation (pyro gel)
Density	kgm^{-3}	7800	171
Elasticity	GPa	200	1×10^{-2}
Poisson's ratio	Dimensionless	0.28	0.2
Thermal expansion	K^{-1}	1×10^{-5}	4.0×10^{-6}
Thermal conductivity	$W\ m^{-1}K^{-1}$	28	6.4×10^{-6}
Specific heat capacity	$J\ kg^{-1}K^{-1}$	460	2300
Internal diameter	m	0.26	0.32
External diameter	m	0.32	0.48
Thickness	m	0.03	0.05

Solar towers commonly use cavity or external tubular receivers at very high temperatures (see Chapter 3). In this connection, the most important constraint for designing these receivers is to maintain the intercepted solar flux within the tube mechanical safety limits. By using a fine grid model (FGM), Rodríguez-Sánchez et al. (2018) studied the effect of incident solar flux resolution on the thermomechanical performance of an external solar receiver for a solar tower plant. The FGM model was compared with the coarse grid model (CGM). Some aiming strategies were used to establish differences between the two models. These strategies are characterized by an aiming factor (f) that varies between 0 and 3, with 0 being an aiming of all the heliostats to the upper and lower edges of the receiver, and 3 being an aiming equivalent to an equatorial aiming strategy. Diminished values of f reduce the solar peak flux but augment the spillage losses and therefore decrease the average incident solar flux and the optical efficiency of the solar field.

The heat transfer fluid used in this study was molten salt. The maximum tube wall temperature is one of the major limitations of the receivers, because it is responsible of degradation of molten salt and the tube corrosion. There are two types of corrosion: general corrosion and pitting corrosion (Liang et al., 2019). General corrosion is defined as the uniform reduction in the effective cross-sectional area of steel elements, and it usually occurs in steel structures without a surface protection layer. Pitting corrosion is a very localized corrosive attack that is relatively smaller than the overall exposed surface and it usually happens in the surface protective layer of elements, such as locations of where antirust thin layer of paint is damaged. It accounts for more than 70% of the total cases of corrosion. Pitting corrosion is more dangerous because of the difficulty to detect it, and it causes very sudden failures in a structural component.

The effective thermomechanical stress in the tubes was computed from the equation reported by Rodríguez-Sánchez et al. (2018):

$$\sigma_{eff} \approx \gamma E\left[\left(\frac{T_w + T_{fi}}{2}\right) - T_m\right] + \frac{T_w - T_{fi}}{2(1-\nu)} \tag{9.54}$$

$$T_m = T_{salt} + \frac{1}{\pi}\left(\frac{T_{wa} + T_{fi}}{2} - T_{salt}\right) \tag{9.55}$$

Where T_{fi} is the film temperature (inner surface of tube), T_w is the temperature of tube wall, T_m is the mean temperature, E is the Young's modulus, ν is the Poisson's ratio, and γ ($\approx 3\alpha$) is the volumetric coefficient of thermal expansion of the tube material. All the properties of the tube (E, ν, and γ) depend on the tube wall temperature.

Numerical results show that the effective thermo-mechanical stress slightly decreases to a minimum level before increasing as the aiming factor increases from $f = 1$ to $f = 1.5$. Thereafter, the stress increases with f. Another notable finding is that the CGM underestimates the effective thermo-mechanical stress. The implication of this is that the CGM leads to an overestimation of the solar flux distribution on the receiver (Faupel & Fisher, 1981) that is required for its safe operation.

The thermal stress experienced by a material depends on the rate of heating or cooling, for a given set of the thermomechanical properties of the material (Panda et al., 2002). High thermal stresses are generated in rapidly cooled or heated components. In metals, the temperature gradients only create small plastic deformations, while as in the case of ceramics, due to its linear elastic material behavior, high stresses are generated.

9.6.4 Thermal stress index

The strength in tension, hardness, and resistance to fatigue of most solid materials, that do not undergo transitions, decreases with an increase in service temperature (Faupel & Fisher, 1981). As the temperature increases, the strength of materials decreases. A measure of the ability of a material to resist thermal gradients and therefore thermally induced stresses is called thermal stress index (I_{ts}), defined by the following:

$$I_{ts} = \frac{\sigma_u k}{\alpha E} \tag{9.56}$$

Where E is the Young's modulus, k is the thermal conductivity, α is the coefficient of linear thermal expansion, and σ_u is the ultimate stress.

The ability of a material to resist temperature gradients increases with the thermal stress index. Thus, materials with higher values of the thermal stress index are more preferred.

9.6.5 Thermal shock and fatigue

The thermal stress index is useful in correlating the properties of a material with reference to the resistance to fracture by thermal shock (Faupel & Fisher, 1981; Kelble & Bernados, 1963). The occurrence of only one single severe thermal cycle is called a thermal shock. This phenomenon may cause serious damage, such as cracking of a structural member. In addition, the occurrence of many cycles with moderate stresses may result in an incremental increase in damage. This effect is important for materials that exhibit subcritical crack growth and it is known as thermal fatigue.

Materials can resist certain magnitudes of sudden changes in temperature. The maximum temperature (T_{max}) that a material can withstand under thermal shock conditions depends on the mechanical properties of the material. It can be expressed by Eq. (9.57).

$$T_{max} = \frac{\sigma_u(1 - \nu)}{\alpha E} \tag{9.57}$$

It should be noted that T_{max} is independent of the thermal conductivity of the material. Both the thermal stress index and maximum thermal shock temperature are useful parameters for determining the thermal shock resistance of a given material.

Example 9.2

Thermomechanical properties of some metallic materials A, B, and C are given in Table 9.5. Determine the thermal stress index of the materials and comment on their suitability for possible application in solar absorber elements.

Solution

From Eq. (9.56),

$$I_{ts} = \frac{\sigma_u k}{\alpha E}$$

Material A, $I_{ts} = \frac{450\times10^{6}\times25}{13\times10^{-6}\times72\times10^{9}} = 12019\ Wm^{-1}$

TABLE 9.5 Thermomechanical properties of materials A, B and C.

Material	Young's modulus (GPa)	Ultimate stress (MPa)	Coefficient of linear thermal expansion (x $10^{-6}K^{-1}$)	Thermal conductivity ($Wm^{-1}K^{-1}$)
A	72	450	13	25
B	100	689	10	12
C	80	550	18	30

Material B, $I_{ts} = \frac{689\times10^6\times12}{10\times10^{-6}*100\times10^9} = 8268\ Wm^{-1}$

Material C, $I_{ts} = \frac{550\times10^6\times30}{18\times10^{-6}\times80\times10^9} = 11458\ Wm^{-1}$

Based on the thermal stress index, material A would be most suitable for application in a solar receiver even though material C has the highest thermal conductivity.

9.7 Thermomechanical properties of materials

Many materials have been developed for thermal and other applications. However, there are still limitations on the range of desirable characteristics. Consequently, the search for advanced concepts of materials is continuing. In this section, a selected number of materials for absorber, transparent, and insulation elements are presented and discussed.

9.7.1 Absorber materials

Central to a solar receiver is the absorber element, which is the hottest element of the receiver. Consequently, materials for this element must be resistant to high temperature. Copper and steel are common materials for tube walls. However, approximately above 813 K, ordinary steels and titanium alloys are not strong enough for application (ASM International, 2002). Moreover, steels may suffer from enhanced corrosion attack at elevated temperatures. In high temperature solar applications, values of temperature above 1073 K are attained (Capuano et al., 2016). In view of this, superalloys and ceramic materials have also been considered for fabrication of absorber tubes.

A conventional volumetric receiver consists of a porous absorber structure. Air is passed through the irradiated porous structure and it can be heated to temperatures of up to 1773 K (Ávila-Marín, 2011). Heat is transferred from the porous element to the air through convection. Consequently, absorber element must be made of materials with large thermal conductivity, high strength, and temperature resistance. These materials may be metallic or ceramic. On the other hand, advanced volumetric receivers use nanofluids to directly absorb concentrated solar radiation (Wang et al., 2019). The fluid is passed through a transparent conduit, often made of glass. So, the glass structure must have high strength and temperature resistance.

9.7.1.1 Copper

Copper is one of the common structural materials in conventional tubular and other absorbers. It has a very high thermal conductivity (391 $Wm^{-1}K^{-1}$ at 293 K), and so it is widely exploited where high thermal conductivity is required. Pure copper is defined as having a minimum copper content of 99.3% (Li & Zinkle, 2012). Usually, the application of unalloyed copper is limited by its low strength. The strength of copper can be improved by using various

TABLE 9.6 Thermomechanical properties of selected structural materials for absorber tubes.

Material	Melting point (K)	Yield strength at 0.2% offset and T = 1033 K (MPa)
Copper (Cu)	1358	<50
CuCrZr	1348	100–150
Stainless steel (Grade 316)	1673	146
Inconel 617	1605–1650	389
Inconel 625	1563–1623	381

processes such as cold working, grain refinement, solid solution hardening, precipitation hardening, and dispersion strengthening. Nevertheless, these approaches can also lead to a significant reduction in thermal conductivity. Thus, the challenge is to design a material with the best combination of strength and thermal conductivity.

An alternative approach is alloying copper with other metals. Alloying can significantly improve mechanical strength and increase the softening temperature (Table 9.6). Addition of alloying elements also decreases electrical and thermal conductivity, but some of the alloying techniques yield copper alloys of relatively high strength and conductivity. For instance, an alloy of copper with chromium and zirconium (CuCrZr) has a thermal conductivity range of 314–335 $Wm^{-1}K^{-1}$ and yield strength of 100–150 MPa at 1033 K. This alloy is commercially available under different trade names such Elbrodur CuCrZr from KME Germany AG, Zollen CuCrZr, C18150, and Trefimetaux CuCrZr with slight variations in their compositions.

9.7.1.2 Stainless steel

Stainless steel is one of the most important structural materials in engineering. It is an iron alloy with a high proportion of chromium that prevents the iron from rusting and provides heat-resistant, and favorable mechanical properties. There are five different classes of stainless steels: austenitic, ferritic, duplex, martensitic, and precipitation hardening steels.

Austenitic steels are iron based with a face-centered-cubic crystal structure and a high alloying content (16–25 mass %) of chromium (Cr); molybdenum (Mo) that enhances resistance against pitting; and nickel (Ni) that produces the austenite (Samaras, 2011). These steels are widely used in both conventional and nuclear power plants due to their excellent corrosion resistance and mechanical properties at elevated temperatures. Grades of austenitic steel include 316, 316H, and 316L, and some of them have been applied in solar receiver absorber tubes. Ferritic stainless steel is characterized by a higher chromium content than the martensitic stainless steel. Usually, the chromium content of the ferritic stainless steel is in the range of 14–27 mass % (Banerjee, 2017). Ferritic stainless steel has a microstructure of ferrite, and can be hardened by only cold working. Some of the grades of this class of stainless steel are 409, 430, and 430F. Duplex stainless steel has a mixed microstructure of about equal proportions of austenite and ferrite. It covers ferritic/austenitic Fe–Cr–Ni alloy with ferrite content in the range of 30–70 mass % (Kahar, 2017). Grades of duplex steels include 2101, 2205, and 2304. Martensitic stainless steel contains 10.5–18 mass % of Cr (Hara et al., 2022, pp. 140–152). Although there are some cases that carbon content exceeds 1.2 mass %, carbon content is often equal to or less than 0.02 mass %. Grades of martensitic steel include 410, 416, and 420. Precipitation hardening is a heat treatment technique used to raise the yield strength of malleable materials, including alloys of stainless steels. Precipitation hardening stainless contains chromium and nickel that can develop very high tensile strengths. The commonest grade in this class is Grade 630, also known as 17–4 PH, with the composition of 17% chromium, 4% nickel, 4% copper, and 0.3% niobium (Kazior et al., 2013).

9.7.1.3 Superalloys

Superalloys are nickel-, iron-nickel-, and cobalt-base alloys that are generally used at temperatures approximately above 811 K (ASM International, 2002). The iron-nickel-base superalloys are an extension of the stainless steel technology and are generally wrought. Depending on the intended application/composition, cobalt-base and nickel-base superalloys may be wrought or cast. Over the years, many alloys have been invented and investigated, and the search for better alloys is ongoing. Nickel-base superalloys include the Haynes (214, 230), Hastelloy (B, N, S, W, X, CR-276), Inconel (600, 601, 617, 625) and Nimonic (75, 86) groups, while the iron-nickel-base alloys include Incoloy (800, 801, 802, 800H, 800HT, 907, 909, 925) and Pyromet CTX-1 groups. Included under the cobalt-base superalloys are Haynes (25, 188), Alloy S-816, MP35-N, MP159, and Stellite B.

Super alloys retain the highest proportion of strength (over 75%), followed by refractory metal alloys when their temperature increases from 394 to about 1158 K (Faupel & Fisher, 1981). Thereafter, there is a switch over of the trend. At about 1366 K, refractory metals retain more than 50% of strength, while superalloys stand at about 20%. Superalloys have lower melting points than refractory metals, but they have different desirable characteristics from refractory metals. Consequently, superalloys are much more widely exploited in various industries including concentrating solar power.

Despite their higher costs relative to stainless steels, nickel-based alloys may be required for high-temperature receivers operating above 973 K to maintain the desired strength and corrosion resistance (Ho, 2017). It has been found that nickel-based alloys such as Inconel 625 and Hastelloy C-276 are the most corrosion-resistant to high-temperature salts compared to stainless steels. Haynes 230 has also been used in high-temperature solar receivers. Olivares et al. (2013) evaluated three nickel-based superalloys (Haynes HR160, Hastelloy X, and Haynes 230) at temperatures of up to 1323 K. They found that Haynes 230 showed the greatest resistance to oxidation in CO_2 and air. All three alloys exhibited less corrosion in air than CO_2 due to the formation of a more protective oxide layer that formed in air.

9.7.1.4 Ceramic materials

Silicon carbide (SiC) is an attractive material for high-temperature solar receivers due to its high strength, high thermal conductivity, low thermal expansion, high solar absorptance, and chemical inertness at elevated temperatures (Ho, 2017). Nevertheless, one drawback is about its brittleness that can cause failure from mechanical shock.

Fabrication of SiC fiber-reinforced SiC matrix composites (SiCf/SiC) for toughened high-temperature tubular receivers has been investigated (Ortona, Fend, et al., 2015; Ortona, Yoon, et al., 2015). Test results of the SiCf/SiC tube samples show excellent heat transfer characteristics. Ultrahigh-temperature ceramics (hafnium and tantalum carbides) have been proposed for use in high temperature receiver applications (Sani et al., 2011, 2012). These carbides have melting points above 3200 K with good thermophysical properties and are exploited in aerospace and high-temperature energy systems.

9.7.1.5 Selective coatings

A solar selective absorbing coating is applied on the outer surface of the absorber tube wall to enhance the optical efficiency of the receiver (Xu et al., 2020). High-temperature coatings have been developed for application in solar receivers. These coatings have high absorptance for solar radiation but low thermal emittance. The efficiency of the absorber ($\eta_{abs)}$ is defined as follows (Ho, 2017):

$$\eta_{abs} = \alpha_s - \frac{\sigma_{sb} T^4}{G} \in \qquad (9.58)$$

Where G is the irradiance on the receiver, T is the surface temperature of the receiver, α_s is the solar absorptivity, $\in$ is the emittance, and σ_{st} is the Stefan–Boltzmann constant.

It is evident from Eq. (9.58) that relative importance of the thermal emissivity to the solar absorptivity is given by the ratio of the thermal emission to the solar irradiance ($\sigma_{sb}\ T^4\ G^{-1}$). Increasing the absorptivity and reducing the thermal emission can enhance the efficiency of the absorber. The relative importance of the thermal emissivity increases with the surface temperature. Nevertheless, a high solar absorptance is more critical than a low thermal emissivity for efficient receivers under practical receiver temperatures and irradiances. Consequently, many studies have investigated coatings and methods to develop selective absorbers with high solar absorptivity and low thermal emissivity (Atkinson et al., 2015; Hall et al., 2012; Moon et al., 2015). Some of the challenges are high costs, scalability of synthesis, and durability of these specialized coatings at elevated temperatures in oxidizing environments. Other studies have investigated laser sintering, texturing, and the use of metamaterials to enhance solar absorptivity while reducing thermal emissivity (Shah et al., 2015; Shah & Gupta, 2013). A commercial standard absorbing coating that has been used on solar tower receivers for many years is high-temperature Pyromark 2500 paint (Ho et al., 2014). This coating is not a selective absorber, but it has a very high solar absorptivity (0.96). Moreover, it can degrade if not properly applied or when subjected to temperatures above 973−1073 K for a period. Instead of applying coatings and textures, researchers have suggested the concept of using oxidized substrates (such as the superalloy Haynes 230) that can give a solar absorptivity higher than 0.90. These and other efforts aim at increasing the optical efficiency of the absorber while also considering the thermomechanical suitability of the materials at elevated temperatures.

9.7.2 Glazing materials

Absorber elements are often covered with a transparent structure to reduce heat loss. In this vein, fused silica (quartz glass) is commonly used as a glazing material for tubular and volumetric receivers. Fused silica is an amorphous synthetic material which contains pure silicon dioxide (SiO_2). It has a melting point of 1858 K, and an abraded modulus of rapture of 52.4 MPa and a Poisson's ratio of 0.16 at 298 K (Corning Incorporated, 2003). Some modifier atoms can be introduced into the basic microstructure of fused silica to obtain different types and grades of glass. For instance, sodium (Na) and calcium (Ca) atoms can be substitutes for the Si atom in fused silica to obtain soda-lime glass. Similarly, borosilicate glass is produced by substituting boron (B) for the Si atom in fused silica.

A typical receiver tube of the PTC comprises an inner steel pipe containing the heat transfer fluid and an outer glass tube surrounding the steel pipe. The glass tube is made of low-iron borosilicate glass to enhance its transmittance for solar radiation (Moya, 2012). In windowed volumetric receivers, the transparent cover is usually made of quartz glass.

9.7.3 Insulation materials

Insulation is one of the strategies for reducing heat loss from a thermal system. Usually, one surface of the insulation wall is exposed to a high temperature environment, while the other one is exposed to a low-temperature

TABLE 9.7 Selected industrial high-temperature insulation materials.

Material	Maximum service temperature (K)	Density (kg m $^{-3}$)	Thermal conductivity at 530K (W m^{-1})
Alumina-silica fiber blanket	1530	48	0.071
Calcium silicate	920	190	0.075
Diatomaceous silica	1145	345	0.092
Insulating cement mineral fiber (with clay binder)	1255	430	0.105
Insulating cement mineral fiber (with hydraulic setting binder)	922		0.137
Pyrogel XT	923	180	0.034

environment. For instance, the inner surface of an insulation wall may be in contact with a hot absorber, while the outer surface is exposed to ambient temperature. Consequently, materials with a low thermal conductivity are required to suppress heat conduction across the insulation wall. In addition, these materials must have high resistance to extreme temperature environments to maintain the desired thermal and structural characteristics. In this vein, some of the important properties of an insulation material are its maximum service temperature and thermal conductivity of the material. Generally, the latter parameter increases with temperature. Selected industrial insulation materials for use at elevated temperature are presented in Table 9.7.

Heat is conducted through the insulation wall, from the hot surface ($T = T_h$) to the cold surface ($T = T_c$) of the wall because the conductivity of a practical insulation materials is greater than zero. In addition, the cold surface loses heat to the ambient environment via convective (h_c) and radiative (h_r) modes of heat transfer. So, there is resistance to heat flow between the hot surface of the insulation layer and the environment (Fig. 9.19).

The effective resistance (R_{eff}) in the flow network shown in Fig. 9.19 is as follows:

$$R_{eff} = R_{ins} + \frac{1}{\frac{1}{R_c} + \frac{1}{R_r}} \tag{9.59}$$

Where R_{ins} is the resistance due to heat conduction through the insulation layer, R_c is the resistance due to convection from the cold surface of the

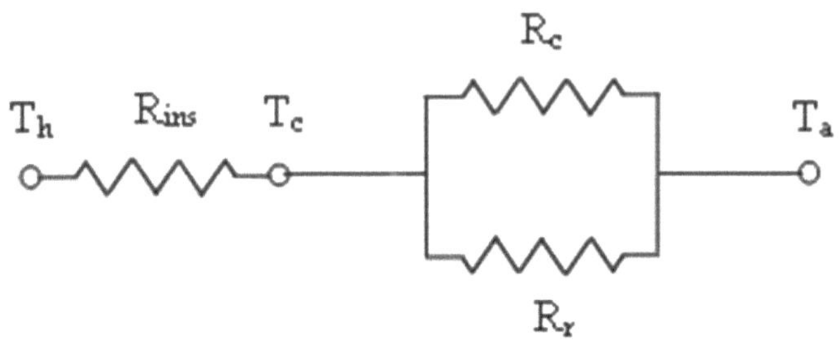

FIGURE 9.19 Thermal network of correspond to flow of heat through an insulation wall with hot (T_h) and cold (T_c) surfaces to the ambient environment (T_a).

insulation wall, and R_r is the resistance due radiation from the cold surface of the insulation wall.

$$R_{ins} = \frac{L}{k} \tag{9.60}$$

Where k and L are the thermal conductivity and the thickness of the wall, respectively.

Eq. (9.60) shows that flow resistance increases with the thickness of the wall, but it declines with increasing the thermal conductivity of wall. In view of this, thicker insulation with low thermal conductivity is more effective in reducing heat loss from a system. Resistances due to convection and radiation are as follows:

$$R_c = \frac{1}{h_c} \tag{9.61}$$

$$R_r = \frac{1}{h_r} \tag{9.62}$$

Therefore, the effective resistance can be expressed as follows:

$$R_{eff} = \frac{L}{k} + \frac{1}{h_c + h_r} \tag{9.63}$$

The coefficient of heat loss (U) from the wall to the ambient environment can be computed from the equation reported by Duffie & Beckman (2013):

$$U = \frac{1}{R_{eff}} \tag{9.64}$$

The rate of heat loss (Q_{loss}) via the insulator can be computed from the following:

$$Q_{loss} = AU(T_h - T_c) \tag{9.65}$$

Where A is the surface area, T_c is the cold surface temperature of the insulation wall, and T_h is the hot surface temperature of the insulation wall.

Example 9.3

The inner and out surfaces of a diatomaceous silica insulation wall of thickness 0.05 m are exposed to thermal environments of a solar absorber at 650 K and ambient temperature of 290 K, respectively. Assuming the thermal conductivity of the insulation material is 0.092 Wm^{-1} at the mean temperature, and the coefficients of convective and radiative heat transfers are, respectively, 5.0 and 7.5 $Wm^{-2}K^{-1}$, calculate the rate of heat loss per unit surface area of the insulation layer.

Solution

From Eq. (9.65), the rate of heat loss per unit area is the following:

$$\frac{Q_{loss}}{A} = U(T_h - T_c)$$

From Eq. (9.64),

$$U = \frac{1}{R_{eff}} = \frac{1}{\left(\frac{L}{k} + \frac{1}{h_c + h_r}\right)} = \frac{1}{\left(\frac{0.05}{0.092} + \frac{1}{5.0 + 7.5}\right)} = 1.604\ Wm^{-2}K^{-1}$$

Therefore,

$$\frac{Q_{loss}}{A} = 1.604 \times (650 - 290) = 577\ Wm^{-2}$$

Nomenclature

A area (m^2)
D diameter (m)
e eccentricity (dimensionless)
E Young's modulus (Pa)
f aiming factor (dimensionless)
G irradiance (Wm^{-2})
$\mathbf{I_{ts}}$ thermal stress index ($W\ m^{-1}$)
k thermal conductivity ($Wm^{-1}K^{-1}$)
L length (m)
P pressure (Pa)
Q heat rate (W)
R, r radius (m)
T temperature (K)
U heat loss (Wm^{-2})
V volume (m^3)
x distance in x direction (m)
y distance in y direction (m)
z distance in z direction (m)

Greek symbols

α coefficient of thermal expansion (K^{-1})
δ displacement (m)
ε strain (dimensionless)
$\in$ emissivity (dimensionless)
ϕ inclination angle (degree)
ν Poisson's ratio (dimensionless)
θ angular direction of solar radiation (degree)
σ stress (Pa)
α_s solar absorptivity (dimensionless)
σ_{sb} Stefan–Boltzmann constant ($Wm^{-2}K^{-4}$)

References

Ahmad, S. H. A., Saidur, R., Mahbubul, I. M., & Al-Sulaiman, F. A. (2017). Optical properties of various nanofluids used in solar collector: A review. *Renewable and Sustainable Energy Reviews, 73*, 1014–1030. https://doi.org/10.1016/j.rser.2017.01.173

Anwar, M. K., Yilbas, B. S., & Shuja, S. Z. (2016). A thermal battery mimicking a concentrated volumetric solar receiver. *Applied Energy, 175*, 16–30. https://doi.org/10.1016/j.apenergy.2016.04.110

ASHRAE. (2017). *Handbook - fundamentals*. American Society of HeatiHandbook - Fundamentalsng, Refrigerating and Air-Conditioning Engineers (ASHRAE).

ASM International. (2002). *Superalloys for high temperatures - a primer*. American Society for Metals (ASM) International. http://www.asminternational.org.

Atkinson, C., Sansom, C. L., Almond, H. J., & Shaw, C. P. (2015). Coatings for concentrating solar systems - a review. *Renewable and Sustainable Energy Reviews, 45*, 113–122. https://doi.org/10.1016/j.rser.2015.01.015

Ávila-Marín, A. L. (2011). Volumetric receivers in solar thermal power plants with central receiver system technology: A review. *Solar Energy, 85*(5), 891–910. https://doi.org/10.1016/j.solener.2011.02.002

Banerjee, M. K. (2017). Heat treatment of commercial steels for engineering applications. In M. S. J. Hashmi (Ed.), *Comprehensive materials finishing* (Vols. 2–3, pp. 180–213). Elsevier Inc. https://doi.org/10.1016/B978-0-12-803581-8.09190-6

Barron, R. F., & Barron, B. R. (2012). *Design for thermal stresses*. John Wiley & Sons, Inc.

Bellos, E., Bousi, E., Tzivanidis, C., & Pavlovic, S. (2019). Optical and thermal analysis of different cavity receiver designs for solar dish concentrators. *Energy Conversion and Management: X, 2*, 100013. https://doi.org/10.1016/j.ecmx.2019.100013

Boss, M., Gadoury, T., Feeny, S., Montgomery, M., Nah, U. H., Cho, S. I., & Yang, S. H. (2007). *Recent Advances in turbine technology*. GE Energy.

Capuano, R., Fend, T., Schwarzbözl, P., Smirnova, O., Stadler, H., Hoffschmidt, B., & Pitz-Paal, R. (2016). Numerical models of advanced ceramic absorbers for volumetric solar receivers. *Renewable and Sustainable Energy Reviews, 58*, 656–665. https://doi.org/10.1016/j.rser.2015.12.068

Corning Incorporated. (2003). *HPFS® fused silica standard grade*. Corning Incorporated.

Cui, Y., Geng, Z., Zhu, Q., & Han, Y. (2017). Review: Multi-objective optimization methods and application in energy saving. *Energy, 125*, 681–704. https://doi.org/10.1016/j.energy.2017.02.174

Davidge. (1991). Thermal shock. In R. J. Brook (Ed.), *Concise encyclopaedia of advanced ceramic materials*. Elsevier Ltd. https://doi.org/10.1016/C2009-1-28294-3

Duffie, J. A., & Beckman, W. A. (2013). Solar engineering of thermal processes: Fourth edition. In *Solar engineering of thermal processes* (4th ed.). John Wiley and Sons. https://doi.org/10.1002/9781118671603

Faupel, J. H., & Fisher, F. E. (1981). *Engineering design: A synthesis of stress analysis and materials engineering*. John Wiley & Sons Inc.

Hall, A., Ambrosini, A., & Ho, C. (2012). Solar selective coatings for concentrating. *Advanced Materials and Processes, 170*(1), 28–32.

Hara, T., Semba, H., & Amaya, H. (2022). *Pipe and tube steels for oil and gas industry and thermal power plant*. Elsevier BV. https://doi.org/10.1016/b978-0-12-803581-8.12107-1

Ho, C. K. (2017). Advances in central receivers for concentrating solar applications. *Solar Energy, 152*, 38–56. https://doi.org/10.1016/j.solener.2017.03.048

Ho, C. K., Mahoney, A. R., Ambrosini, A., Bencomo, M., Hall, A., & Lambert, T. N. (2014). Characterization of pyromark 2500 paint for high-temperature solar receivers. *Journal of Solar Energy Engineering, Transactions of the ASME, 136*(1). https://doi.org/10.1115/1.4024031

Huston, R., & Josephs, H. (2009). *Practical stress analysis in engineering design* (3rd ed.). Taylor & Francis Group, LLC.

Kahar, Dr S. D. (2017). Duplex stainless steels-an overview. *International Journal of Engineering Research and Applications*, 27–36. https://doi.org/10.9790/9622-0704042736

Kandil, A., El-Kady, A. A., & El-Kafrawy, A. (1995). Transient thermal stress analysis of thick-walled cylinders. *International Journal of Mechanical Sciences, 37*(7), 721–732. https://doi.org/10.1016/0020-7403(94)00105-S

Kanlıkama, B. (2013). Coupled thermoelastic analysis of thick-walled pressurized cylinders. *International Journal of Energy and Power Engineering, 60*. https://doi.org/10.11648/j.ijepe.20130202.15

Kazior, J., Szewczyk-Nykiel, A., Pieczonka, T., Hebda, M., & Nykiel, M. (2013). Properties of precipitation hardening 17-4 PH stainless steel manufactured by powder metallurgy technology. *Advanced Materials Research, 811*, 87–92. https://doi.org/10.4028/www.scientific.net/amr.811.87

Kelble, J. M., & Bernados, J. E. (1963). High-temperature nonmetallic materials. *Aerospace Engineering, 22*(1), 56–75.

Kittel, C. (2005). *Introduction to solid state physics*. New Jersey: John Wiley & Sons, Inc.

Lee, S. H., Choi, T. J., & Jang, S. P. (2016). Thermal efficiency comparison: Surface-based solar receivers with conventional fluids and volumetric solar receivers with nanofluids. *Energy, 115*, 404–417. https://doi.org/10.1016/j.energy.2016.09.024

Lenert, A., & Wang, E. N. (2012). Optimization of nanofluid volumetric receivers for solar thermal energy conversion. *Solar Energy, 86*(1), 253–265. https://doi.org/10.1016/j.solener.2011.09.029

Liang, X., Sheng, J., & Wang, K. (2019). Investigation of the mechanical properties of steel plates with artificial pitting and the effects of mutual pitting on the stress concentration factor. *Results in Physics, 14*, 102520. https://doi.org/10.1016/j.rinp.2019.102520

Li, M., & Zinkle, S. J. (2012). Physical and mechanical properties of copper and copper alloys. In *Comprehensive nuclear materials* (Vol. 4, pp. 667–690). Elsevier Ltd. https://doi.org/10.1016/B978-0-08-056033-5.00122-1

Mahmoudi, A., Fazli, M., Morad, M. R., & Gholamalizadeh, E. (2020). Thermo-hydraulic performance enhancement of nanofluid-based linear solar receiver tubes with forward perforated ring steps and triangular cross section; a numerical investigation. *Applied Thermal Engineering, 169*, 114909. https://doi.org/10.1016/j.applthermaleng.2020.114909

Moon, J., Kyoung Kim, T., VanSaders, B., Choi, C., Liu, Z., Jin, S., & Chen, R. (2015). Black oxide nanoparticles as durable solar absorbing material for high-temperature concentrating solar power system. *Solar Energy Materials and Solar Cells, 134*, 417–424. https://doi.org/10.1016/j.solmat.2014.12.004

Moya, E. Z. (2012). Parabolic-trough concentrating solar power (CSP) systems. In *Concentrating solar power technology: Principles, developments and applications* (pp. 197–239). Elsevier Ltd. https://doi.org/10.1016/B978-1-84569-769-3.50007-8

Msaddak, A., Sediki, E., & Ben Salah, M. (2018). Assessment of thermal heat loss from solar cavity receiver with Lattice Boltzmann method. *Solar Energy, 173*, 1115–1125. https://doi.org/10.1016/j.solener.2018.08.059

Olivares, R. I., Stein, W., & Marvig, P. (2013). Thermogravimetric study of oxidation-resistant alloys for high-temperature solar receivers. *JOM, 65*(12), 1660–1669. https://doi.org/10.1007/s11837-013-0706-0

Ortona, A., Fend, T., Yu, H. W., Raju, K., Fitriani, P., & Yoon, D. H. (2015). Tubular Si-infiltrated SiCf/SiC composites for solar receiver application - Part 1: Fabrication by replica and electrophoretic deposition. *Solar Energy Materials and Solar Cells, 132*, 123–130. https://doi.org/10.1016/j.solmat.2014.08.029

Ortona, A., Yoon, D. H., Fend, T., Feckler, G., & Smirnova, O. (2015). Tubular Si-infiltrated SiCf/SiC composites for solar receiver application – Part 2: Thermal performance analysis and prediction. *Solar Energy Materials and Solar Cells, 140*, 382–387. https://doi.org/10.1016/j.solmat.2015.04.034

Otanicar, T. P., Phelan, P. E., Prasher, R. S., Rosengarten, G., & Taylor, R. A. (2010). Nanofluid-based direct absorption solar collector. *Journal of Renewable and Sustainable Energy, 2*(3). https://doi.org/10.1063/1.3429737

Panda, P. K., Kannan, T. S., Dubois, J., Olagnon, C., & Fantozzi, G. (2002). Thermal shock and thermal fatigue study of ceramic materials on a newly developed ascending thermal shock test equipment. *Science and Technology of Advanced Materials, 3*(4), 327–334. https://doi.org/10.1016/S1468-6996(02)00029-3

Pérez-Álvarez, R., Acosta-Iborra, A., & Santana, D. (2020). Thermal and mechanical stresses in bayonet tubes of solar central receivers working with molten salt and liquid sodium. *Results in Engineering, 5*, 100073. https://doi.org/10.1016/j.rineng.2019.100073

Rodríguez-Sánchez, M. R., Marugán-Cruz, C., Acosta-Iborra, A., & Santana, D. (2018). Thermo-mechanical modelling of solar central receivers: Effect of incident solar flux resolution. *Solar Energy, 165*, 43–54. https://doi.org/10.1016/j.solener.2018.03.005

Salifu, S. A., Desai, D. A., Kok, S., & Ogunbiyi, O. F. (2019). Thermo-mechanical stress simulation of unconstrained region of straight X20 steam pipe. In *Procedia manufacturing* (Vol. 35, pp. 1330–1336). Elsevier B.V. https://doi.org/10.1016/j.promfg.2019.05.021

Samaras, M. (2011). In K. H. J. Buschow, M. C. Flemings, E. J. Kramer, P. Veyssière, R. W. Cahn, B. Ilschner, & S. Mahajan (Eds.), *Nuclear energy systems: Advanced alloys and cladding*. Elsevier Ltd. https://doi.org/10.1016/B978-0-08-043152-9.02263-6

Sandá, A., Moya, S. L., & Valenzuela, L. (2019). Modelling and simulation tools for direct steam generation in parabolic trough solar collectors: a review. *Renewable and Sustainable Energy Reviews, 113*, 109226. https://doi.org/10.1016/j.rser.2019.06.033

Sani, E., Mercatelli, L., Fontani, D., Sans, J. L., & Sciti, D. (2011). Hafnium and tantalum carbides for high temperature solar receivers. *Journal of Renewable and Sustainable Energy, 3*(6). https://doi.org/10.1063/1.3662099

Sani, E., Mercatelli, L., Sansoni, P., Silvestroni, L., & Sciti, D. (2012). Spectrally selective ultra-high temperature ceramic absorbers for high-temperature solar plants. *Journal of Renewable and Sustainable Energy, 4*(3). https://doi.org/10.1063/1.4717515

Shah, A. A., & Gupta, M. C. (2013). Spectral selective surfaces for concentrated solar power receivers by laser sintering of tungsten micro and nano particles. *Solar Energy Materials and Solar Cells, 117*, 489–493. https://doi.org/10.1016/j.solmat.2013.07.013

Shah, A. A., Ungaro, C., & Gupta, M. C. (2015). High temperature spectral selective coatings for solar thermal systems by laser sintering. *Solar Energy Materials and Solar Cells, 134*, 209–214. https://doi.org/10.1016/j.solmat.2014.11.009

Sumelka, W., & Łodygowski, T. (2013). Thermal stresses in metallic materials due to extreme loading conditions. *Journal of Engineering Materials and Technology, 135*(2). https://doi.org/10.1115/1.4023777

Wang, Y., Li, Q., & Xuan, Y. (2019). Thermal and chemical reaction performance analyses of solar thermochemical volumetric receiver/reactor with nanofluid. *Energy, 189*, 116123. https://doi.org/10.1016/j.energy.2019.116123

Xu, K., Du, M., Hao, L., Mi, J., Yu, Q., & Li, S. (2020). A review of high-temperature selective absorbing coatings for solar thermal applications. *Journal of Materiomics, 6*(1), 167–182. https://doi.org/10.1016/j.jmat.2019.12.012

Zaversky, F., Les, I., Sorbet, P., Sánchez, M., Valentin, B., Brau, J. F., & Siros, F. (2020). The challenge of solar powered combined cycles – providing dispatchability and increasing efficiency by integrating the open volumetric air receiver technology. *Energy, 194*. https://doi.org/10.1016/j.energy.2019.116796

Zhu, Q., & Xuan, Y. (2019). Improving the performance of volumetric solar receivers with a spectrally selective gradual structure and swirling characteristics. *Energy, 172*, 467–476. https://doi.org/10.1016/j.energy.2019.01.166

Chapter 10

Modeling and optimization of solar receivers

10.1 Introduction

A solar receiver converts solar radiation into heat that is transferred to different components of the concentrating solar power (CSP) system. Advances in the development of solar receivers are resulting in complex configurations and processes that require suitable computational tools for system design. It takes more time and financial resources to conduct experimental investigations for every new concept of receiver design. In this vein, simulation of receivers has gained popularity as a suitable design tool. A simulation is a numerical experiment in which the processes are quickly and inexpensively conducted on computers (Duffie & Beckman, 2013). It involves four major tasks: (a) building a physical model of the actual system, (b) formulating a mathematical model from the physical model, (c) applying a numerical method to obtain a solution, and (d) putting into operation the numerical model to obtain results (Houbak, 1995). Each step in the simulation process is vital to find accurate results.

10.1.1 Concentrating solar collector configurations

A solar collector is a special kind of heat exchanger that converts solar radiation into heat (Duffie & Beckman, 2013). Solar collectors are classified into two main categories: flat-plate and concentrating collectors. Flat-plate collectors are designed for low-temperature applications and harness both beam and diffuse radiation. A concentrating solar collector is made up of a concentrator and a receiver as distinct components. In this vein, imaging concentrating solar collectors capture only beam radiation and they can attain very high temperatures.

In building a physical model, it is necessary to draw the diagram of a system that is being modeled. Details of the major elements of the system need to be captured in the drawing, with clearly defined boundaries among the elements. The central element of a solar receiver is the absorber that absorbs solar radiation and converts it to heat. In addition, some receivers have

Solar Receivers for Thermal Power Generation. https://doi.org/10.1016/B978-0-323-85271-5.00010-0

transparent and insulation elements. A heat transfer fluid is passed through the absorber element to evacuate heat from the receiver to other components of the CSP system.

Common types of CSP technology are the: parabolic trough concentrator (PTC), linear Fresnel reflector (LFR), solar tower (ST) and parabolic dish concentrator (PDC). PTC and LFR systems focus solar radiation on a line while ST and PDC systems focus radiation on a point. Concentrators of all these CSP systems form an image of the sun on the receiver, and so they are known as imaging concentrators. In addition, they have relatively high concentration ratios. In contrast, non-imaging concentrators do not form such an image, they have low concentration ratio and are unsuitable for high-temperature industrial application such as in thermal power plants. The focus of this book is on solar receivers for thermal power generation, and so, only imaging concentrators are presented and discussed in this section.

Line focusing receivers are used on the PTC and LFR systems, and they are commonly made of evacuated or non-evacuated tubes. Cavity, external and volumetric receivers collect solar radiation concentrated at a point. However, the PDC is more compatible with cavity receivers. Details about the classification of solar receivers have been presented in Chapter 3, and examples of collector configurations are shown in Fig. 10.1.

The CSP technology utilizes direct normal irradiance which is the amount of solar radiation intercepted per unit area by a surface held at right angles to the rays that propagate in a straight line from the sun at its instantaneous position in the sky. Nevertheless, the position of the sun changes with time and so, the surface must be mounted on a solar tracking system. In this case, the surface follows the sun on daily (one-axis) or daily and seasonal (two-axis) bases to maintain normal incidence of the beam radiation. Therefore, concentrators are often mounted on sun-tracking systems to optimize solar collection. Depending on the configuration of the system, both the concentrator and the receiver may move together during the sun tracking process. Receivers of the parabolic trough and parabolic dish concentrators are mobile, while those of the linear Fresnel reflector and solar tower are fixed (Behar et al., 2013). More details about sun tracking are reported in Chapter 2.

10.1.2 Building mathematical models

A mathematical model is a description of the way in which variables of a problem are related and how the independent variables affect the performance criterion (Bejan et al., 1996). Variables are features/factors that are liable to change, and they are classified into dependent and independent categories. Values of a dependent variable rely on those of other variables while the values of an independent variable are not influenced by those of other variables.

The performance of a solar receiver is significantly affected by optical and thermal parameters. Optical processes involve the reflection, transmission, and

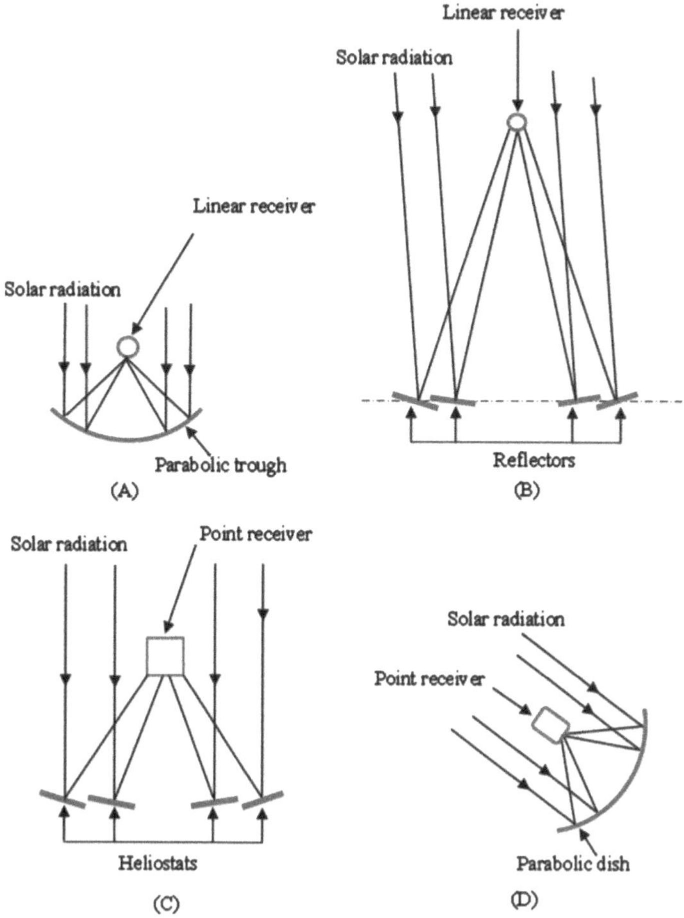

FIGURE 10.1 Selected configurations of concentrating solar collectors: (A) linear receiver for parabolic trough concentrator (PTC), (B) linear receiver for the linear Fresnel reflector, (C) point receiver for the solar tower, and (D) point receiver for the parabolic dish concentrator.

absorption of solar radiation. The amount of solar radiation absorbed by the receiver increases with the level of optical efficiency. Consequently, the optical efficiency of a receiver is a key indicator of the system performance. The absorbed solar radiation is converted to heat, and energy and exergy are important thermodynamic parameters that affect the performance of receivers. In addition, the thermal performance of receivers is influenced by the hydrodynamics of heat transfer/working fluids and modes of heat transfer. The flow of heat causes temperature gradients that produce thermal stresses in the system. Consequently, it is necessary to model the various important processes that occur in the receiver.

There are many designs of receivers, and new designs are still being developed to improve the performance of the CSP technology. Some important design considerations are geometry and materials. A wide range of receiver shapes is available, and different materials are used in receivers. Moreover, receivers operate at elevated temperatures, and therefore, structural materials and heat transfer media must have a high resistance to temperature. Considering the wide variation in receiver designs, it is rather difficult to formulate a single mathematical model that can accurately describe the performance of all receivers. However, basic principles (that include natural laws) can be applied in modeling specific receiver designs. This approach is also used in the development of new designs. It is usual to make assumptions about processes. For example, the density of solids can be assumed constant in a wide range of operational temperatures.

Processes may be steady or transient, and some of them may be neglected depending on reality. In steady processes, parameters at a given point in the system do not change with time. For instance, the temperatures of hot and cold surfaces of an insulation wall remain constant as time changes, in a steady flow of heat across the wall (Fig. 10.2). On the other hand, transient processes entail changes of variables with time. Most real processes involve changes of variables (such as mass and energy) with time within a control volume. While the mass flow rate can be kept constant (except during start up or shut down), some processes in solar receivers are inherently unsteady, and one of them is the availability of solar radiation. The solar resource is highly intermittent. Thus, it is important to establish the nature of the process when attempting to model it. Basic principles are applied in formulating appropriate mathematical models. Design, operational, and environmental factors that influence the

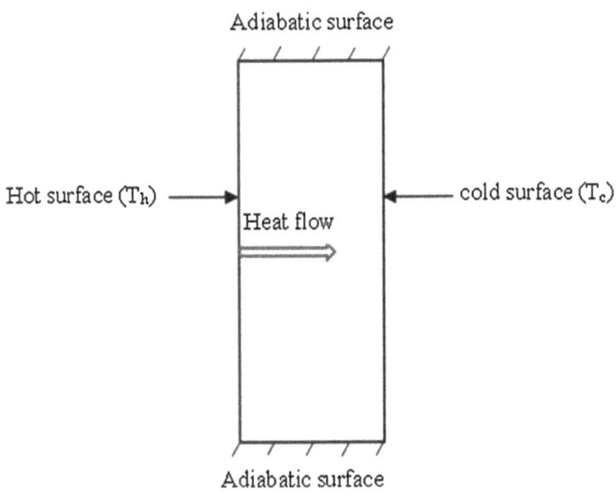

FIGURE 10.2 Steady heat conduction through a slab with adiabatic, cold, and hot surfaces.

performance of a solar receiver are included in the mathematical models. Usually, this leads to a system of differential equations that can be solved analytically or numerically. Computer programs are utilized to find numerical solutions.

10.2 Optical performance

There are many designs of concentrating solar collectors, and so, their optical performance depends on the geometric and surface properties of the concentrator and receiver. An important geometric parameter is the concentration ratio (C_r). Generally, point focus receivers have higher values of C_r than linear systems. The operational temperature of the receiver increases with C_r. Consequently, point focus receivers are more suitable for driving advanced thermodynamic cycles that operate at high temperatures.

Surface properties are the specular reflectance (ρ) except when the reflector is diffuse (ρ is replaced by transmittance when the concentrator is a refractor), absorptance of the absorber on the solar receiver (α), and transmittance of the transparent cover on the solar receiver (τ). For unglazed solar receivers, the transmittance $\tau = 1.0$. Despite the variations in the geometry of solar concentrators and receivers, it is possible to apply a generic mathematical model in describing the optical performance of all of them. Details about optical properties of materials, and various optical concepts have been discussed in Chapter 4. The optical efficiency of a CSP system is one of the key performance parameters in addition to the thermal efficiency. It is the ratio of solar energy absorbed by the absorber element to the solar radiation incident on the aperture of the concentrator. It influences the amount of absorbed radiation per unit area of unshaded aperture (S) that can be expressed as (Duffie & Beckman, 2013):

$$S = \rho(\alpha\tau\gamma)_n K_\theta\, G_{ap} = \eta_{opt} G_{ap} \tag{10.1}$$

Where G_{ap} is the effective incident radiation measured on the plane of the aperture that includes only beam radiation for all concentrators except those of low concentration ratio (10 or below), K_θ is the incidence angle modifier (established from empirical data), α is the absorptance of the absorber on the receiver, the subscript n represents normal incidence, ρ is the specular reflectance except when the reflector is diffuse (ρ is replaced by transmittance when the concentrator is a refractor), τ is the transmittance of the transparent cover on the solar receiver ($\tau = 1.0$ for unglazed receivers), γ is the proportion of the reflected solar radiation that is incident on the absorbing surface of the receiver (values of γ may be greater than 0.9), and $\eta_{opt} = \rho(\alpha\tau\gamma)_n K_\theta$ is the optical efficiency of the system.

The concept of the intercept factor (γ) is mainly useful in describing imaging concentrators (see Section 1.2.1). For instance, Fig. 10.3 shows an image

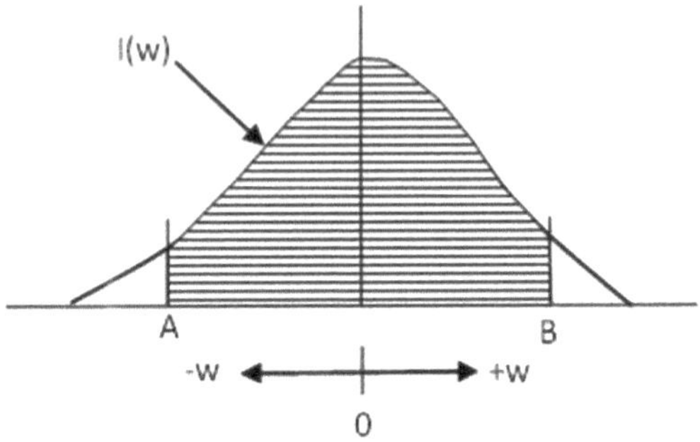

FIGURE 10.3 Schematic representation of a possible flux distribution in the focal plane of a linear imaging concentrator. *Adapted from Duffie and Beckman (2013).*

formed in the focal plane of a linear concentrator, with the receiver extending from A to B.

In Fig. 10.3, if a receiver extends from A to B, the intercept factor will be given by (Duffie and Beckman, 2013):

$$\gamma = \frac{\int_A^B I(w)dw}{\int_{-\infty}^{+\infty} I(w)dw} \tag{10.2}$$

Concentrating systems have lower heat losses from the absorber due to its reduced area. Most of the imaging collectors are constructed with receivers large enough area to intercept a high proportion of the radiation reflected from the concentrator but not large enough to intercept the low-intensity fringes of the images. The calculation of the transmittance (τ) may be difficult because of the possible uncertainty about the angles of incidence of the solar radiation from the reflector on the cover. Similarly, the absorptance of the absorber for the reflected (and transmitted) solar radiation (α) may be rather difficult to compute. More details about computation of various optical properties are presented in Chapter 4. The use of the incidence angle modifier ($K_{\gamma\tau\alpha}$) is to account for deviations from the normal angle of incidence of the radiation on the system aperture. Biaxial incidence modifiers are required for cylindrical receivers, with distinct treatments for the longitudinal and transverse planes. Nonetheless, the transverse incidence angle modifier would only be utilized for tracking errors and may not be practically required. One modifier is enough for circular systems that are completely symmetric. If a given system is not completely symmetric, then biaxial or possibly more complex modifiers are required. In Eq. (10.1), K_θ includes the influence of angle of incidence on the intercept factor. Generally, the size of the image increases, while the fraction of the reflected radiation intercepted by the receiver decreases as the angle of incidence of radiation increases. For cylindrical (linear) systems, end effects may arise if the collector is not long enough compared to its focal length. These effects can be incorporated into the angular modifier.

10.2.1 Linear parabolic collectors

Most of the CSP power plants operating around the world are based on the parabolic trough concentrator with tubular linear receivers that fall into two categories: evacuated and nonevacuated (see Chapter 9). Several tubular receivers can be assembled into a panel, and panels can be assembled into an array. Fig. 10.4 shows a solar collector with evacuated tubular receiver.

Like in the generic model, the optical efficiency of the PTC can be decomposed into several parameters such as the reflectance of the reflector, transmittance of the glass envelope, absorptance of the absorber tube, intercept factor, cleanliness factor, and the incidence angle modifier (Yılmaz & Mwesigye, 2018). In addition, an end loss factor that considers the end section of the collector where the radiation cannot reach is also included in optical efficiency. Nevertheless, most studies do not take this factor into consideration. In this vein, the optical efficiency of a PTC collector (η_{opt}) is defined by Sanda et al. (2019) as:

$$\eta_{opt} = \rho\ \alpha\ \tau\ \gamma\ F_c K_\theta \tag{10.3}$$

Where F_c is the cleanliness factor and K_θ is incidence angle modifier.

Eq. (10.3) is like the generic optical Eq. (10.1) except for the additional factor of cleanliness in the former equation. This factor is important because covers are exposed to dust and other dirty that reduce transmittance.

Linear concentrators with parabolic cross section are most prevalent and have received a lot of research attention. Knowledge about the characteristics of the images produced by reflectors is necessary for computation of the amount of solar radiation absorbed by the receiver. Perfect images are theoretically produced by perfect concentrators that are thoroughly aligned and depend on the geometry of the concentrator. Figs. 10.5 and 10.6 show cross sections of a linear parabolic concentrator.

Using the coordinate system shown in Fig. 10.5, the equation of the parabola is:

$$y^2 = 4fx \tag{10.4}$$

Where f is the focal length (distance between the vertex A and the focal point F).

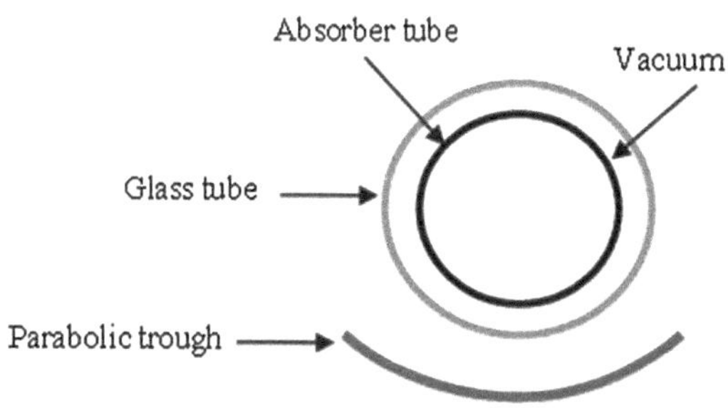

FIGURE 10.4 Schematic of a solar collector with evacuated tubular receiver for a parabolic trough concentrator.

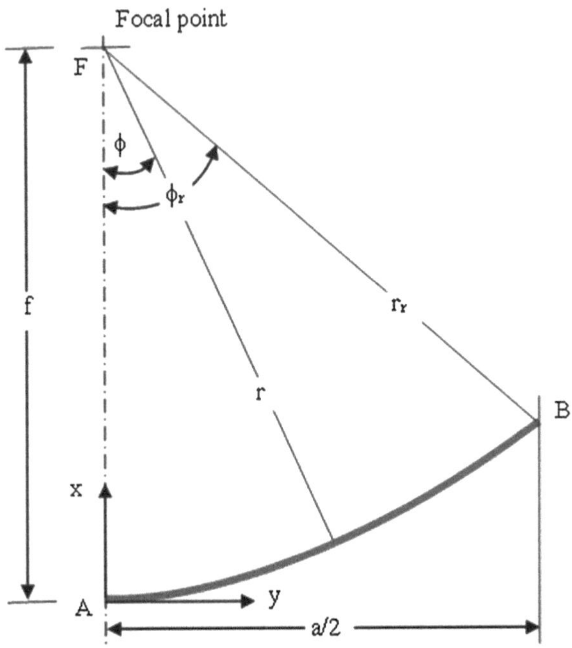

FIGURE 10.5 Cross section of a linear parabolic concentrator. The aperture of the parabolic trough is a. *Adapted from Duffie and Beckman (2013).*

In Fig. 10.5, the radiation beam is incident on the reflector at point B at the rim where the mirror radius is a maximum at r_r. The rim angle is ϕ_r and it is given by:

$$\phi_r = \tan^{-1}\left[\frac{8(f/a)}{16(f/a)^2 - 1}\right] = \sin^{-1}\left(\frac{a}{2r_r}\right) \tag{10.5}$$

Where a is the width of the aperture (Fig. 10.6).

The local mirror radius for any point on the parabolic reflector is give n by:

$$r = \frac{2f}{1 + \cos\phi} \tag{10.6}$$

The incident beam radiation is a cone with an angular width of 0.53° (half angle of 0.267 degrees or 16′). Let us assume that the concentrator is symmetrical and that the beam radiation is normal to the aperture. This implies that, in a perfect concentrator, the beam radiation is incident on the concentrator in a direction parallel to the central plane of the parabolic trough (the x-z plane described by the axis and focus of the parabola). Nevertheless, the effects of alignment errors arise in practical concentrators. The width of the solar

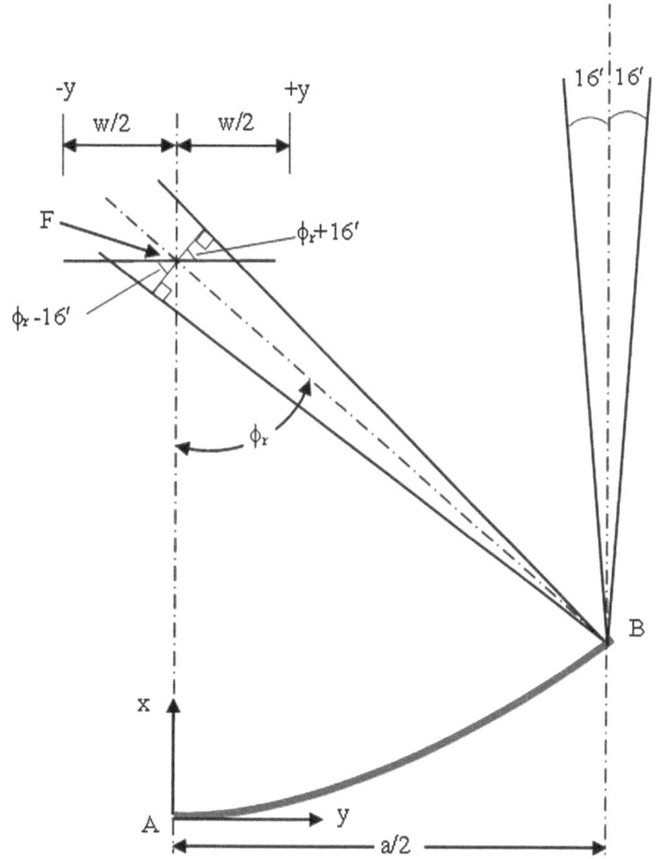

FIGURE 10.6 Schematic representation of image formation by a linear concentrator. *Adapted from Duffie and Beckman (2013).*

image in the focal plane increases with the rim angle. The minimum sizes of flat, circular, and semi-circular receivers centered at the focal point to intercept all of the reflected radiation are shown in Fig. 10.7. It is evident from this schematic that the angle of incidence of radiation on the surface of any of these absorber shape is variable.

For specular parabolic reflectors of perfect shape and alignment, the size of the absorber to intercept the entire solar image can be computed. The diameter (D) of a circular absorber is given by:

$$D = 2r_r \sin 0.267^\circ = \frac{a \sin 0.267^\circ}{\sin\phi_r} \tag{10.7}$$

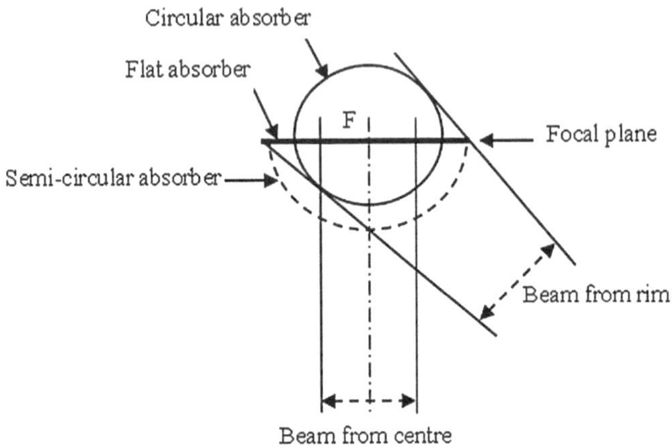

FIGURE 10.7 Schematic of reflected beam radiation from the center and rim of half of a parabolic reflector with minimum plane, circular, and semi-circular absorbers to intercept all the reflected radiation from a full parabola. *Adapted from Duffie and Beckman (2013).*

For a flat absorber in the focal plane of the parabola shown in Fig. 10.6, the width W is also the diameter of the semi-circular absorber, and it is expressed as:

$$W = \frac{2r_r \sin 0.267^o}{\cos(\phi_r + 0.267^o)} = \frac{a \sin 0.267^o}{\sin\phi_r \cos(\phi_r + 0.267^o)} \tag{10.8}$$

For a flat receiver, r increases from f to r_r, and the theoretical image size in the focal plane increases from D (evaluated with r_r equal to f) to W when ϕ rises from zero to ϕ_r. The focal length is a determinant of the image size, while the aperture is the determinant of the total energy. Consequently, the image brightness or energy flux concentration at the receiver of a focusing system is a function of the ratio a/f.

10.2.2 Images formed by perfect linear concentrators

Consider a flat absorber at right angles to the axis of a perfect parabola at its focal point, with beam radiation normal to the aperture. In this case, the intersection of the focal plane and the cone of the reflected beam radiation from an element is an ellipse, with lengths of minor (L_x) and major (L_y) axes given by:

$$L_x = 2r \sin 16' \tag{10.9}$$

$$L_y = r\sin 16' \left[\frac{1}{\cos(\phi - 16')} - \frac{1}{\cos(\phi + 16')}\right] \tag{10.10}$$

The sum of the ellipses for all the elements of the reflector is the total image. Evans (1977) determined the distribution of energy flux in the integrated images on the focal plane for three models of the sun, including a nonuniform disk of the sun that considers the fact that the sun radiates more from its central portion than it does from the edge. The local concentration ratio $C_l = I(y)/I_{b,ap}$ is the ratio of the intensity at any position y in the image to the intensity on the aperture of the concentrator. Both the local concentration ratio and the size of the image increase with the rim angle. For a uniform disk of the sun, lower concentration ratios are observed at the center of the image and higher ratios in the outer parts of the images than for the nonuniform model.

Linear parabolic concentrators may be oriented in different ways to track the disk of the sun. Generally, the beam radiation is incident on the principal plane (x-z plane) of the collector as described by the focus and vertex of the parabola but it will not be perpendicular to the aperture. Under these conditions, a parabolic trough collector can be analyzed by projections in the x-y plane. Common modes of orientation are adjustments about the horizontal axes, aligned either east-west or north-south while rotation about a polar axis is occasionally used.

The apparent half-angle subtended by the sun increases with the angle of incidence in the x-z plane (the projection on the x-y plane of the 0.267° half-angle). The effect on image width (W) is derived from Eq. (10.8) by substituting 0.267/cos θ for 0.267 to obtain:

$$W = \frac{2r_r \sin(0.267^\circ/\cos\theta)}{\cos(\phi_r + 0.267^\circ/cos\ \theta)} \tag{10.11}$$

The effect of θ on image width is examined by considering the ratio of image width (W) at an incidence angle (θ) to the width (W_0) at $\theta = 0$:

$$W/W_0 = \frac{\sin(0.267^\circ/\cos\theta)}{\sin 0.267^\circ}\ \frac{\cos(\phi_r + 0.267^\circ)}{\cos(\phi_r + 0.267^\circ/\cos\theta)} \approx \frac{1}{\cos\theta} \tag{10.12}$$

The W/W_0 ratio varies by a small margin with θ until the rim angle is greater than about 80 degrees. Until θ becomes large, the second fraction does not significantly change with θ and the dominant effect is produced by the first fraction. In addition, the sine of a small angle is approximately equal to the angle, and therefore, W/W_0 is almost equal to one/cos θ. The widening of the image causes a linear parabolic concentrator oriented on a horizontal east-west axis to have an over-enlarged image on the receiver in the early and late hours of a day and of minimum size at noon. On the other hand, a collector oriented on a north-south horizontal axis would have an image size that remains constant through the day. Thus, θ does not only affect the image size but also the energy in the image. The energy in the image is reduced by the factor of cos θ.

Example

The aperture and focal length of a perfect cylindrical parabolic reflector are respectively 8 and 4.82 m. Its length is along the z direction such that end effects are insignificant. The incident beam radiation normal to the aperture is 800 W m^{-2}. The specular reflectance of the concentrator is 0.90.

Calculate the image width. What is the radiation intensity at the center of the image for the nonuniform solar disk if the local concentration ratio is 202?

Solution:

The ratio f/a of this concentrator is 4.82/8.00 = 0.603. Using Eq. (10.5) the rim angle for a concentrator of these dimensions is

$$\phi_r = \tan^{-1}\left[\frac{8(f/a)}{16(f/a)^2 - 1}\right] = \tan^{-1}\left[\frac{8 * 0.603}{16 * (0.603)^2 - 1}\right] = 45^o$$

Using Eq. (10.8), the width of the image is

$$W = \frac{a \sin 0.267^o}{\sin \ \phi_r \ \text{cosin}(\ \phi_r + 0.267^o)} = \frac{8.00 * \sin 0.267^o}{\sin 45^o * \ \text{cosin}(\ 45^o + 0.267^o)} = 0.075 \text{ m}$$

The flux in the image at the axis (y = 0) is:

$$I(y=0) = I_{b,ap}\ \rho\ C_r = 800 * 0.90 * 202 = 145 \text{ kWm}^{-2}$$

10.2.3 Images formed by imperfect linear concentrators

In a perfect concentrator, the intensity of flux is highest on the image at the axis (y = 0). Real concentrators are often not perfect, and so they may create imperfect images on the receivers. Imperfect concentrators produce larger images than the theoretical images. It is possible to use image distributions for different standard deviations (σ) of normally distributed reflector slope errors to consider imperfections in reflectors of a given rim angle (Fig. 10.8).

Another method of accounting for imperfections in the geometry of a parabola is to examine the reflected beam as having an angular width of (0.53 + δ) degrees, where δ is a dispersion angle (Fig. 10.9). In this case, δ is a measure of the angular error limit of the reflector surface.

Using δ, equations can be derived for the size of images created on cylinders or planes at the focus. The diameter of a cylindrical receiver that can intercept the whole image is:

$$D = 2r_r \sin\left(\frac{\delta}{2} + 0.267^o\right) = \frac{a \sin\left(\frac{\delta}{2} + 0.267^o\right)}{\sin\phi_r} \tag{10.13}$$

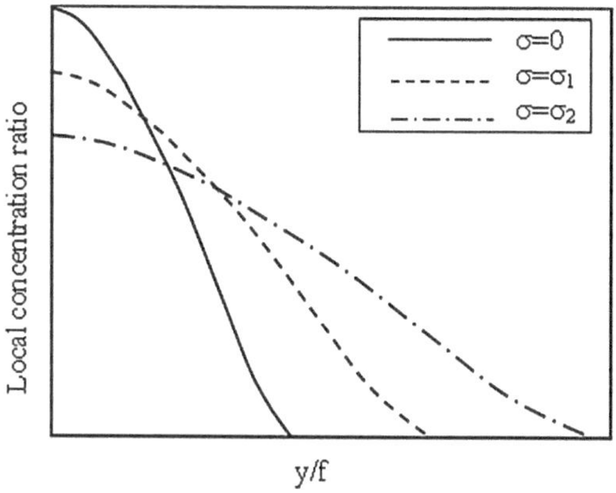

FIGURE 10.8 Illustration of image distributions for imperfect reflectors for a given rim angle (ϕ_r) for different standard deviations (s) of normally distributed reflector slope errors, where $0 < \sigma_1 < \sigma_2$. *Adapted from Duffie and Beckman (2013).*

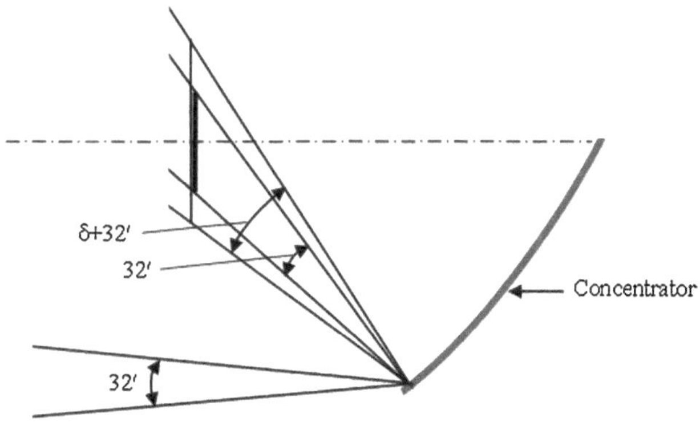

FIGURE 10.9 Schematic representation of part of a concentrator with a dispersion angle δ added to the solar intercept angle (0.53 degrees). *Adapted from Duffie and Beckman (2013).*

If the incident radiation is normal to the aperture, then the image width on the focal plane will be:

$$W = \frac{2r_r \sin\left(\frac{\delta}{2} + 0.267^\circ\right)}{\cos\left(\phi_r + \frac{\delta}{2} + 0.267^\circ\right)} = \frac{a\sin\left(\frac{\delta}{2} + 0.267^\circ\right)}{\sin\phi_r \cos\left(\phi_r + \frac{\delta}{2} + 0.267^\circ\right)} \tag{10.14}$$

The dispersion angle δ can be estimated from measurements of flux distribution, by ray-tracing methods where the distribution of angular errors is known, or by knowledge of the distribution of angular errors anticipated from the process that is used in manufacturing a given reflector. An assumption of normal distribution of angular errors may be more accurate.

The diameter and width of images of imperfect parabolic troughs, that can be described by the dispersion angle (δ), are given by Eqs. (10.13) and (10.14). The ratio $(a/W)_{\gamma=1}$ is defined as the ratio of the area of the aperture to the total area of the receiver when the receiver is just large enough to intercept all the radiation specularly reflected from the concentrator (when $\gamma = 1$). Using the ratio $(a/W)_{\gamma=1}$, we can derive the maximum concentration ratio (C_{max}) that enables the interception of the whole image. For a concentrator producing an image with clearly defined boundaries and without pointing errors or mispositioning of the receiver, for a flat receiver of width (W) with a shadow W wide, the maximum area concentration ratio for $\gamma = 1$ is $[(a/W)_{\gamma=1} - 1]$. Consequently,

$$C_{max} = \frac{\sin\phi_r \cos\left(\phi_r + \frac{\delta}{2} + 0.267^{\circ}\right)}{a\sin\left(\frac{\delta}{2} + 0.267^{\circ}\right)} - 1 \tag{10.15}$$

10.2.4 Point focusing collectors

The parabolic dish and solar tower systems focus solar radiation on a point. These systems have high concentration ratios and can attain higher temperatures than the linear systems.

10.2.4.1 Paraboloidal collectors

Paraboloidal collectors use surfaces of revolution to concentrate solar radiation in three dimensions to a point (Duffie and Beckman, 2013). A section through the axis of the paraboloid of these collectors is like that of the linear parabolic collector as shown in Figs. 10.4 and 10.5. The rim angle ϕ_r and mirror radius r correspond to those for the linear concentrator. Dispersion of also takes place in paraboloidal collectors, and equations analogous to Eq. (10.4) can be written for paraboloidal collectors without tracking errors. For spherical receivers (assuming minimum shading by the receiver):

$$C_{max} = \frac{\sin^2\phi_r}{4\sin^2\left(\frac{\delta}{2} + 0.267^{\circ}\right)} - 1 \tag{10.16}$$

For flat receivers,

$$C_{max} = \frac{\sin^2\phi_r \cos^2\left(\phi_r + \frac{\delta}{2} + 0.267^{\circ}\right)}{4\sin^2\left(\frac{\delta}{2} + 0.267^{\circ}\right)} - 1 \tag{10.17}$$

Where C_{max} is the maximum local concentration ratio that can be achieved, based on interception of all the specularly reflected radiation that falls within the cone of angular width $(0.534 + \delta)$.

Cavity receivers can be utilized with paraboloidal concentrators to augment absorptance and to reduce convective losses from the absorbing surface (Duffie and Beckman, 2013). There are different designs of this receiver type (see Section 3.3.5). Tubular conduits for the heat transfer fluid are often used in cavity receivers. An example of the design of a cavity receiver is shown in Fig. 10.10.

A cavity receiver absorbs a high proportion of solar radiation approximately close to a blackbody. The effective absorptance of the cavity (α_{eff}) is the fraction of incident radiation absorbed by the cavity. It is a function of the absorptance of the inner surfaces of the cavity and the ratio of the areas of the cavity aperture and inner surfaces (see Section 4.3.8). So, the effective absorptance (Eq. 4.67) must be used in the computation of the optical efficiency of cavity receivers.

Equations for optical properties of systems with flat receivers will be applicable to cavity receivers provided absorptances of the cavities are used. The solar radiation absorbed by the absorber of a paraboloidal collector is influenced by properties in the same manner as linear parabolic collectors. Nevertheless, these collectors attain higher concentration ratios and therefore, operate at higher temperatures. Any absorption of solar radiation in the transparent cover material would lead to heating of the cover, and so, these

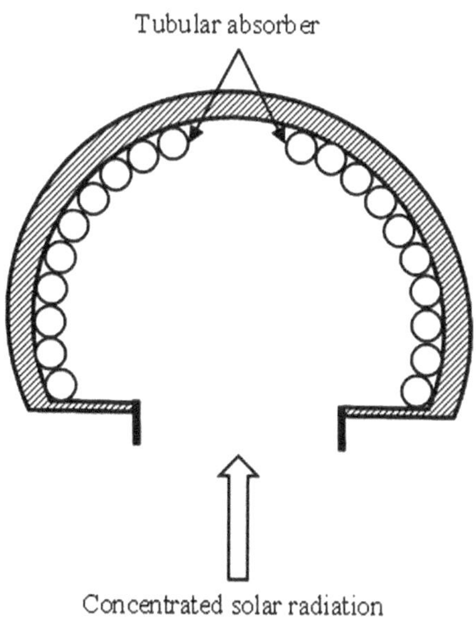

FIGURE 10.10 Cross-section of a spherical cavity solar receiver with a tubular absorber in the shape of a coil.

receivers are often unglazed. In the absence of a cover, the absorbed energy for the unshaded aperture area of the collector can be given by:

$$S = I_{b,ap}\rho(\alpha\gamma)_n K_\theta \tag{10.18}$$

The incidence angle modifier (K_θ) accounts for pointing errors in the alignment of the collector. It may be circumferentially symmetrical. Biaxial or other modifiers may be applied if the optical system is not symmetrical.

10.2.4.2 Solar tower collector

A solar tower (ST) utilizes many heliostats that reflect beam radiation onto a central receiver. These heliostats form a parabolic reflector broken up into small segments that are key to the optical performance of the solar tower system. Several additional optical phenomena must be considered, such as shading and blocking. Heliostats can shade one another from incident beam radiation and block reflected radiation from a heliostat by another, which prevents that radiation from reaching the receiver. Some researchers have investigated heliostat generation models. For example, Sánchez and Romero (2006) developed a method for generating the heliostat field layout of a solar tower power plant using the tower height and the yearly normalized energy surface. A method for the design of heliostat field layout for solar tower power plants was proposed by Wei et al. (2010). Using this method, it was possible to design heliostats of higher optical efficiency and at a faster speed. The method also enables assessment of the feasibility of the land for crop growth under the heliostats to optimize land use. The performance of an ST power plant is significantly affected by the solar input subsystem which in turn depends on the heliostat field. The annual performance of the heliostat system, according to Collado and Guallar (2012), depends on the time of the year (seasonal variations of solar irradiance), its position relative to the solar receiver and the relative position of the adjacent heliostats. This is ascribed to the shadowing and blocking challenges that regularly occur in the field. Collado and Guallar (2012) developed an accurate and detailed procedure for optimization of the heliostat positioning in the field by computing the blocking and shadowing factor. This method is useful in proper design, optimization, and analysis of heliostat fields to improve the overall performance of the solar field of the ST technology. The effect of the gap size between adjacent heliostats on the wind load on the heliostats must reduce the structural requirement of the heliostat field members (Wu et al., 2010). The gap size has minimal effect on the wind load on the heliostat but the drag force increases with the gap size. The heliostats are spaced apart, and only a fraction of the ground area is covered by mirrors. Suggested practical values of ψ fall in the range 0.3−0.5 (Duffie & Beckman, 2013).

For a three-dimensional concentrator system with radiation incident at an angle (θ_i) on the plane of the heliostat array (θ_i is equal to the zenith angle θ_z

for a horizontal array), a rim angle of ϕ_r, and a dispersion angle of δ, if all reflected beam radiation is to be intercepted by a spherical receiver, the maximum concentration ratio is:

$$C_{max} = \frac{\psi \sin^2\phi_r}{4\sin^2\left(\frac{\delta}{2} + 0.267^\circ\right)} - 1 \tag{10.19}$$

Where ψ is the fraction of the ground area covered by mirrors.

For a flat receiver, the maximum concentration ratio is:

$$C_{max} = \psi\left[\frac{\sin\phi_r \cos\left(\phi_r + \frac{\delta}{2} + 0.267^\circ\right)}{\sin\left(\frac{\delta}{2} + 0.267^\circ\right)}\right]^2 - 1 \tag{10.20}$$

Like line focusing concentrators, the optimum performance may be found with intercept factors below one. The designer of these collectors must also consider other factors: The heliostat field does not need to be symmetrical, the ground cover factor ψ does not have to be uniform, and the heliostat array need not necessarily be all in one plane.

10.2.4.3 External receivers

An external receiver comprises panels of tubes arranged in a cylindrical or polygon shape, with the former design being most prevalent (Fig. 10.11). Usually, the exterior surface of the receiver is circumferentially irradiated by concentrated solar radiation from the heliostats of a solar tower system. This type of solar receivers has high operational temperatures and absorber areas that tend to raise heat losses to the environment, compared to the cavity receiver.

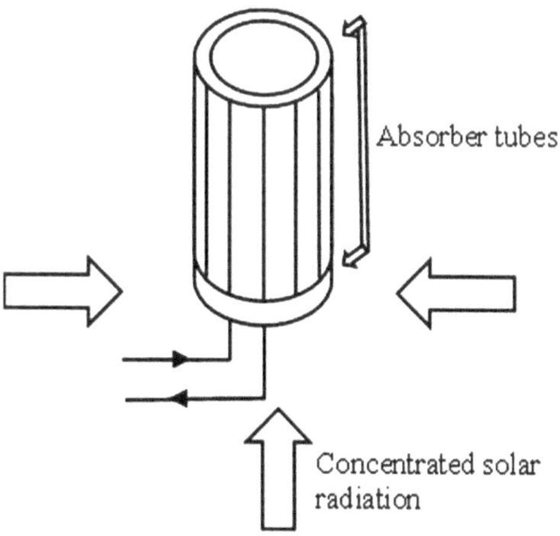

FIGURE 10.11 Example of an external receiver design.

The concentrator of a solar tower is made up of many heliostats. In this vein, the total flux incident on the heliostat field (G_{ap}) and the heliostat efficiency (η_h) can be defined according to Rodríguez-Sánchez et al. (2018):

$$G_{ap} = \sum_{i=1}^{N_h} C_i . G_{b,n} \tag{10.21}$$

Where C_i is the concentration ratio of the ith heliostat, $G_{b,n}$ is the direct normal irradiance, and N_h is the number of heliostats.

$$\eta_h = \rho \cos\omega_h f_{atm} f_{sb} f_{sp} \tag{10.22}$$

Where f_{atm} is the atmospheric attenuation factor, f_{sb} is the shadowing and blocking factor, f_{sp} is the spillage factor, and ω_h corresponds to the angle between the sun and the heliostats normal.

Thus, solar radiation absorbed by an unglazed external receiver can be given by:

$$S = \rho \alpha_t \cos\omega_h f_{atm} f_{sb} f_{sp} G_{ap} \tag{10.23}$$

Where α_t is the absorptance of the tubes.

10.2.4.4 Volumetric receivers

Absorption of solar radiation by the absorber element of the receiver is a critical optical process. In surface receivers (such as the evacuated tube), solar radiation is absorbed in two dimensions by a solid surface, and then heat is transferred from the surface to a heat transfer fluid through convection. This process entails heat losses and is therefore less efficient in heat removal from the absorber element. This observation led to the concept of volumetric receivers that use a volumetric structure or heat transfer fluid to absorb concentrated solar radiation in three dimensions. There has been significant advancement in the development of volumetric receivers.

10.2.4.5 Direct-absorption volumetric receivers

For direct solar absorption, the fraction of absorbed radiation depends on the coefficient of absorption and depth of the fluid. Concentrated solar radiation is incident on the transparent cover with high transmittance (Fig. 10.12). So, a high proportion of the radiation reaches the top part of the fluid. The fluid absorbs most of the radiation before it reaches the back plate that has a high specular reflectance to promote direct absorption. Most ordinary fluids have low absorptance, and so research attention has been given to advanced concepts of using nanofluids and suspensions of solid particles (see Chapter 5). It is possible to tune the absorptance of a nanofluid by changing the volume fraction (f_v) of nanoparticles and the depth of the nanofluid. There are optimal levels of both the volume fraction and the pathlength. The magnitude of absorption is also dependent on the wavelength of the incident radiation.

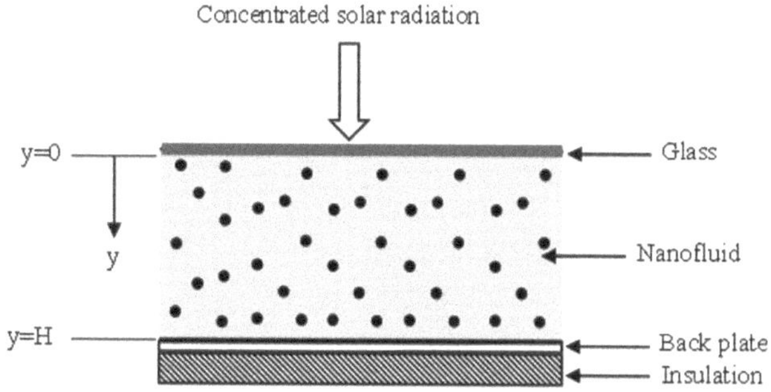

FIGURE 10.12 Schematic of model for a 1-D volumetric solar receiver with a transparent top (y = 0) with high transmittance, and back plate (y = H) with high specular reflectance.

Consider a direct-absorption volumetric receiver in which a nanofluid flows between two parallel plates separated by a variable height (H), as shown in Fig. 10.12. Let us assume that the length of the receiver in the horizontal direction is significantly larger than the height. The incident solar heat flux (G_{tr}) is transmitted through the transparent cover and absorbed volumetrically by the suspended nanoparticles, thereby causing volumetric heat generation. The back of the receiver is insulated and so it is assumed to be adiabatic. The receiver has a specularly reflective back plate at the bottom of the flow channel. Radiative heat loss can be modeled separately for two spectral Regions I ($\lambda < 2$ μm) and II ($\lambda > 2$ μm) (Lenert & Wang, 2012). The nanofluid is approximated as a gray medium with a tuneable optical thickness and a refractive index equal to one ($n_I = 1$). A gray body is a medium with a spectral emittance (ε_λ) that is independent of the wavelength (λ) of the radiation and with equal medium (ε) and spectral (ε_λ) emittances (ASHRAE, 2017). In practical calculations, surfaces are often assumed to be gray because of the unavailability of information about the relationship between the spectral emittance and wavelength. In spectral Region II ($\lambda > 2$ μm), the nanofluid is modeled as fully absorptive ($\varepsilon_{II} = 1$) because of the assumption that the broad vibrational and rotational absorption bands of the liquid overlap and dominate the infrared properties. It is also assumed that that convective and conductive heat losses are negligible in the high-temperature regime where radiative loss is more significant.

Generally, base fluids have low absorptance of solar radiation. The absorbing properties of these fluids are augmented by the addition of nanoparticles that absorb solar radiation in a predictable and tuneable way. Radiative properties of a nanoparticle of diameter (D) can be well described by the Rayleigh scattering principle provided that the particle diameter is small ($x = \pi D n_f/\lambda << 1$) compared to the wavelength of the incident light (Bohren & Huffman, 1998). The Rayleigh scattering concept is also applicable to a suspension of polydisperse particles in the limit of low volume fraction ($f_v << 0.006$), where the volume fraction of a polydisperse suspension is equal to the relative volume occupied by the particles over the total volume of the suspension:

$$f_v = \int_0^\infty \frac{\pi D^3}{6} N(D)dD \tag{10.24}$$

Where N is the number density of that particle size.

The fraction of absorbed solar radiation (F_s) can be given by:

$$F_s = \frac{\int I_{s,\lambda} \exp(-K_{eff,\lambda} H) d\lambda}{\int I_{s,\lambda} d\lambda} \tag{10.25}$$

Where H is the depth of the nanofluid layer, $I_{s,\lambda}$ is spectral intensity, $K_{eff,\lambda}$ is the coefficient of monochromatic radiation absorption, and λ is the wavelength of the monochromatic radiation.

10.2.4.6 Indirect-absorption volumetric receivers

The absorber element of an indirect-absorption volumetric receiver is a three-dimensional porous structure with a certain volume on which solar radiation is concentrated and then absorbed throughout the depth of this volume (Fig. 10.13). Heat is then transferred from the hot solid to the fluid through volumetric convection. The porous structure can be interlocking shapes, knit-wire packs, foam, or foil arrangements, made of metal, ceramic, or other materials with an adequate specific porosity. The absorber is painted black to enhance solar absorption. The aperture of the volumetric receiver is often covered with a quartz glass, while the sides are insulated to reduce heat loss from the system.

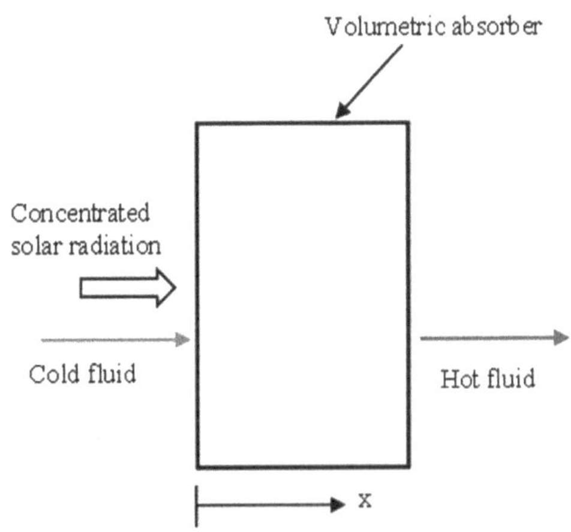

FIGURE 10.13 Schematic representation of an indirect-absorption volumetric receiver irradiated with concentrated solar radiation.

Part of the incident concentrated solar radiation is absorbed as it propagates through the porous structure. The absorbed radiation can be considered as a volumetric heat source described by an exponential law (Capuano et al., 2016; Fend et al., 2005):

$$S = G_o \exp(-K\ x) \tag{10.26}$$

Where G_0 is the intensity of concentrated solar radiation incident on the front part of the porous structure, and x is the depth into the absorber structure.

For glazed volumetric receivers, concentrated solar radiation is incident on the transparent cover. Consequently, a factor for the transmittance of the glass is required to obtain the correct amount of solar radiation absorbed by the absorber element.

10.3 Thermodynamic models

Scientists and engineers often use energy and exergy balances to formulate mathematical models of energy devices. In this context, the first and second laws of thermodynamics deal with energy conservations respectively (see Section 7.2), and so they are applied in modeling energy flows in solar receivers.

A solar receiver absorbs concentrated solar radiation and converts it to heat. The solar flux absorbed per unit area of the aperture (S) must be estimated from the radiation and the optical characteristics of the concentrator and receiver. Part of the generated heat is lost to the environment, and thermal losses from the receiver can be estimated in terms of a loss coefficient U_L, which is based on the area of the receiver (Duffie & Beckman, 2013). A flow factor (F_R) can be used to account for temperature gradients along the receiver and to permit the use of the inlet fluid temperatures in the energy balance calculations.

Unlike flat-plate collectors, it is difficult to provide a single generic method for calculating thermal losses from CSP receivers. There are wide variations in shapes and designs, high operating temperatures, more significant edge effects, higher conduction terms, and the problems may be complicated by the nonuniformity of radiation flux on receivers that can create substantial temperature gradients across the energy-absorbing surfaces. Consequently, each receiver geometry must be analyzed as a special case.

The most complete mathematical model to describe fluids motion with heat transfer from a solid structure to fluid is a system of partial differential equations (PDEs) constituted by mass, momentum, and energy conservation equations, that must be satisfied at each time step in the whole spatial domain (Sandá et al., 2019). For instance, thermal, hydrodynamic, and thermomechanical processes may simultaneously interact with one another. Consequently, it may be necessary to solve the governing equations simultaneously. The use of a system of PDEs allows coupling of equations for these and other processes. The level of complexity of these equations and associated numerical solution algorithms depend on the assumptions made for the independent variables, time, and space.

Transient processes are described by time dependent systems of equations. Otherwise, time-independent variables are said to be steady. Even though weather conditions (such as solar radiation, temperature, and wind speed) are intermittent, most authors consider steady-state models when weather and operation parameters are relatively stable. These conditions are also known as quasi-steady state. The system of partial differential equations can capture spatial variations in one or more dimensions. Generally, the complexity of the system increases with the number of dimensions. In general, the useful heat from a receiver (Q_u) can be given by:

$$Q_u = A_{ab}S - Q_{loss} = \dot{m}\, C_{p,htf}\left(T_{htf,out} - T_{htf,in}\right) \tag{10.27}$$

Where A_{ab} is the area of the absorber, $C_{p,htf}$ is the specific heat capacity of the heat transfer fluid, $\dot{m}$ is the mass flowrate of the heat transfer fluid, S is the absorbed solar radiation, $T_{htf,in}$ is the inlet temperature of the heat transfer fluid, and $T_{htf,out}$ is the outlet temperature of the heat transfer fluid.

Eq. (10.27) can be used to compute the useful gain by using empirical data. It is a critical equation in the process of collector testing (see Chapter 11). In the absence of empirical data, mathematical models are useful in performance evaluation of solar collectors. These models usually comprise coupled nonlinear algebraic and/or differential equations. The system of equations can be expressed in a simple form through linearization of nonlinear equations and discretization of differential equations. Simplified forms of equations are convenient for finding solutions by hand or computer programs. Also currently available are computer programs that can handle solar collector equations in natural complex forms. In this section, both simplified and natural forms of collector equations are presented.

10.3.1 Equations of flat-plate collectors

For a flat-plate collector operating under steady state conditions, the useful heat can be expressed as (Duffie & Beckman, 2013):

$$Q_u = A_c F_R [S - U_L (T_i - T_a)] \tag{10.28}$$

$$F_R = \frac{\dot{m}\, C_p}{A_c U_L}\left[\frac{T_{fo} - T_{fi}}{S/U_L - \left(T_{fi} - T_a\right)}\right] = \frac{\dot{m}\, C_p}{A_c U_L}\left[1 - \frac{S/U_L - \left(T_{fo} - T_a\right)}{S/U_L - \left(T_{fi} - T_a\right)}\right] \tag{10.29}$$

Where A_c is the collector area, C_p is the specific heat capacity of the fluid, F_R is the collector heat removal factor, S is the absorbed solar radiation, T_a is the ambient temperature, T_{fi} is the fluid inlet temperature, T_{fo} is the fluid outlet temperature, and U_L is the overall heat loss coefficient from the collector.

Combining Eq. (10.28) and the definition of instantaneous efficiency (η_i) yields:

$$\eta_i = \frac{Q_u}{A_c G_T} = F_R(\tau\alpha)_m - \frac{F_R U_L (T_i - T_a)}{G_T} \tag{10.30}$$

Where G_T is the solar flux intercepted by the collector, and $(\tau\alpha)_m$ is mean product of the transmittance (τ) of the transparent cover and the absorptance (α) of the absorber plate.

$F_R(\tau\alpha)_n$ and $F_R U_L$ are the two parameters that describe how the collector works if most of the radiation is beam radiation that is approximately normal to the collector and if F_R and U_L vary by a small margin within the range of operation of the collector. $F_R(\tau\alpha)_n$ is an indication of how energy is absorbed, while the product $F_R U_L$ shows how energy is lost from the collector. The third parameter is the incidence angle modifier (K_θ) for beam radiation that is defined as the ratio $(\tau\alpha)_b/(\tau\alpha)_n$. It is a function of the incidence angle of the beam radiation (θ). The incidence angle modifier can be approximately represented in terms of a single coefficient. These three parameters constitute the simplest model of a practical flat-plate collector.

$$K_\theta = 1 - b_o\left(\frac{1}{\cos\theta} - 1\right) \tag{10.31}$$

Where b_o is the incidence angle modifier coefficient which is generally a positive number.

Other models can be formulated to take into consideration the temperature dependence of U_L and F_R and complex angular relationships for $(\tau\alpha)_{av}$. Consequently, there is a wide range of collector thermal performance models that vary in their level of detail. The purpose at hand determines the choice of a model. The simplest model is suitable if the collector operates over restricted ranges of conditions, otherwise more complex models may be needed if operating conditions are highly variable.

10.3.2 Equations of concentrating collectors

In general, the thermal analysis of a concentrating collector is like that of a flat-plate collector. However, one significant difference between the two types of collectors is the high temperatures encountered in the concentrating collector. This implies that thermal radiation is significant, resulting in the temperature dependence of the heat loss coefficient. Even though not necessary, it is convenient to derive suitable expressions for the collector efficiency factor (F'), the collector heat removal factor (F_R) and the heat loss coefficient (U_L). Once F_R and U_L are known, the concentrating collector useful gain can be calculated from an expression that is like that for a flat-plate collector.

10.3.2.1 Line focusing collectors

Consider thermal losses from an unglazed absorber tube used as a receiver with a linear concentrator. Here, we assume that the temperature is uniform around the absorber tube. The heat loss and loss coefficient, considering convection and radiation from the surface and conduction through the support structure, are:

$$\begin{aligned}\frac{Q_{loss}}{A_{rec}} &= h_w(T_{rec} - T_a) + \varepsilon\sigma\left(T_{rec}^4 - T_{sky}^4\right) + U_{cond}(T_{rec} - T_a) \\ &= (h_w + h_r + U_{cond})(T_{rec} - T_a) \\ &= U_L(T_{rec} - T_a)\end{aligned} \tag{10.32}$$

Where A_{rec} is the area of the receiver, h_r is the coefficient of radiative heat transfer, h_w is the wind induced coefficient of convective heat transfer, T_a is the ambient temperature, T_{rec} is the temperature of the receiver, T_{sky} is the sky temperature, U_L is the overall coefficient of heat loss, and U_{cond} is the coefficient of conductive heat loss.

For an unglazed receiver, heat is lost from the absorber tube of the receiver to the environment. For glazed systems, heat is lost from the outer surface of the glazing element to the environment.

The linearized coefficient of radiative heat transfer (h_r) can be computed from:

$$h_r = \frac{\left(T_{rec}^4 - T_{sky}^4\right)}{(T_{rec} - T_a)} \tag{10.33}$$

The collector can be divided into segments, each with constant U_L, if one value of U_L is unacceptable because of large temperature gradients in the flow direction (Stuetzle et al., 2004). The coefficient of wind-induced convective heat transfer (h_w) for cylinders can be estimated from:

$$h_w = \frac{Nu\ k}{D} \tag{10.34}$$

Where D is the diameter of the tube, k is the thermal conductivity of the fluid, and Nu is the Nusselt number.

$$Nu = \begin{cases} 0.40 + 0.54\,Re^{0.52}, & 0.1 < Re < 1000 \\ 0.30\,Re^{0.6}, & 1000 < Re < 50000 \end{cases} \tag{10.35}$$

$$Re = \frac{\rho V D_h}{\mu} \tag{10.36}$$

Where D_h is the hydraulic diameter of the tube, V is the velocity of the fluid, ρ is the density of the fluid, and μ is the dynamic viscosity of the fluid.

The coefficient of conductive losses is computed from the knowledge of the construction details or from measurements on a particular collector. For a glazed tubular receiver of length L, heat is transferred from the absorber at a temperature (T_{ab}) to the inner surface of the cover at T_{ci}, to the outer surface of cover at T_{co} and then to the surroundings at T_a and T_{sky}, and it is given by (Duffie and Beckman, 2013):

$$Q_{loss} = \frac{2\pi k_{eff} L}{\ln(D_{ci}/D_{ab})}(T_{ab} - T_{ci}) + \frac{\pi D_{ab} L \sigma \left(T_{ab}^4 - T_{ci}^4\right)}{\frac{1}{\varepsilon_{ab}} + \frac{1-\varepsilon_c}{\varepsilon_c}\left(\frac{D_{ab}}{D_{ci}}\right)} \quad (10.37)$$

$$Q_{loss} = \frac{2\pi k_c \, L(T_{ci} - T_{co})}{\ln(D_{co}/D_{ci})} \quad (10.38)$$

$$Q_{loss} = 2\pi D_{co} \, L h_w(T_{co} - T_a) + \varepsilon_c \pi D_{co} L \sigma \left(T_{co}^4 - T_{sky}^4\right) \quad (10.39)$$

$$\frac{k_{eff}}{k} = \max\left[1, \; 0.386\left(\frac{Pr \, Ra^*}{0.861 + Pr}\right)\right] \quad (10.40)$$

$$Ra^* = \frac{[\ln(D_o/D_i)]^4}{L_s^3\left(D_i^{-3/5} + D_o^{-3/5}\right)^5} Ra_L \quad (10.41)$$

$$Pr = \frac{\nu}{\xi} \quad (10.42)$$

$$Ra_L = \frac{g\beta \, \Delta T L_s^3}{\nu \, \xi} \quad (10.43)$$

Where g is the acceleration due to gravity, L_s is the space between the plates, ξ is the thermal diffusivity, ΔT is the temperature difference between the plates, β is the volumetric coefficient of expansion (for an ideal gas, $\beta = T^{-1}$), ν is the kinematic viscosity, c represents cover, ci and co respectively represent the inner and outer surfaces of the cover, and eff stands for effective.

If the space between the absorber and glass tubes is evacuated, convection is suppressed and k_{eff} can be zero at very low pressures. The outside convective coefficient h_w is calculated by using Eq. (10.35) with $Re > 1000$. Eqs. (10.37)–(10.39) can be solved by using different equation solvers on a computer. Alternatively, the procedure to solve them by hand is to estimate T_{co} (which is nearer to T_a than T_{ab}). Calculate Q_{loss} from Eq. (10.39), and substitute these values into Eq. (10.38) to find an estimate of T_{ci}. Eq. (10.37) is used to check the initial guess of T_{co} by comparing Q_{loss} from Eq. (10.39) with that calculated from Eq. (10.37).

Example

An evacuated tubular receiver (1.0 m long) has an absorber tube of diameter 5×10^{-2} m that is operated at 573 K with a surface emittance of 0.30. The absorber is covered with a tubular borosilicate glass 8×10^{-2} m in the outer diameter and 4×10^{-3} m thick. The wind speed is 4 ms^{-1} and the air and sky temperatures are 300 and 275 K, respectively. Calculate the overall heat loss coefficient (U_L) from the receiver. Assume the thermal conductivity of glass is 1.40 $Wm^{-1}K^{-1}$.

Solution:

We can solve heat loss equations by using computer-based programs for solving equations. Alternatively, these equations can be solved using an iterative hand calculation. For the first guess, we assume that the outer surface of the transparent cover is at a temperature of 315 K. We then find the Reynolds number for air at the average temperature of the cover and ambient to estimate the coefficient of convective heat transfer induced by wind.

$$Re = \frac{\rho V D_h}{\mu} = \frac{1.1398 * 4 * 0.08}{1.8767 * 10^{-5}} = 21864$$

The coefficient of heat transfer induced by wind is found from Eq. (10.35) with Re > 1000

$$Nu = 0.30 * (21864)^{0.6} = 120$$

$$h_w = \frac{Nu\ k_{air}}{D_{co}} = \frac{120 * 0.0268}{0.08} = 40.3\ Wm^{-2}K^{-1}$$

This value of h_w is substituted in Eq. (10.39) to estimate heat loss:

$$Q_{loss} = 2\pi * 0.08 * \ 1.0 * 40.3 * (315 - 300) + 0.88 * \pi * 0.08 * 1 * 5.67 * 10^{-8}\left(315^4 - 275^4\right) = 224.1\ W$$

Eq. (10.38) is used to find the temperature of the inner surface of the glass cover:

$$T_{ci} = T_{co} + \frac{Q_{loss}\ \ln(D_{co}/D_{ci})}{2\pi\ L\ k_c} = 315 + \frac{224.1 * \ \ln(0.08/0.072)}{2 * \pi * 1.0 * \ 1.40} = 317.7$$

Eq. (10.37) is then used to find the heat loss for comparison with the original value. Another guess is required if the two values are significantly different. From Eq. (10.37) the heat loss is:

$$Q_{loss} = \frac{\pi * 0.05 * 1.0 * 5.67 * 10^{-8}\left(573^4 - 275^4\right)}{\frac{1}{0.30} + \frac{1 - 0.88}{0.88}\left(\frac{0.05}{0.072}\right)} = 253.6\ W$$

There is a difference of +29.5 W between the values of Q_{loss}, which indicates that the initial guess of outer cover temperature was too low. A second guess of the outer cover temperature of 320 K gives an error of −10.4 W. A linear interpolation to determine the temperature where the error is zero yields a cover temperature of 319.14 K. Using this new cover temperature, the heat loss computed from Eqs. (10.37) and (10.39) is almost the same and equal to 252.1 W. The heat loss coefficient (based on receiver area) is computed by using Eq. (10.32):

$$U_L = \frac{Q_{loss}}{A_{rec}(T_{rec} - T_a)} = \frac{Q_{loss}}{\pi D_{ab} L (T_{rec} - T_a)}$$
$$= \frac{252.1}{\pi * 0.08 * 1 * (573 - 300)} = 3.67\ \mathrm{Wm^{-2}K^{-1}}$$

It is also important to consider the factors that account for temperature gradients in the flow direction. Heat is transferred from the outer surface of the absorber tube to the glass cover and from the inner surface of the absorber tube to the heat transfer fluid inside the tube. The heat transfer resistance from the outer surface of the absorber tube to the fluid in the tube must include the tube wall because the heat flux in a concentrating system may be high. Consequently, the overall heat transfer coefficient (based on the outer absorber tube diameter) between the surroundings and the fluid is:

$$U_o = \left[\frac{1}{U_L} + \frac{D_o}{h_{fi} D_i} + \frac{D_o \ln(D_o/D_i)}{2k}\right]^{-1} \tag{10.44}$$

Where D_i and D_o are respectively the inner and outer absorber tube diameters, h_{fi} is the heat transfer coefficient inside the absorber tube, and k is the thermal conductivity of the absorber tube.

The useful energy gain per unit length of the collector (q'_u), expressed in terms of the local absorber temperature (T_{ab}) and the absorbed solar radiation per unit area of the aperture (S) is:

$$q'_u = \frac{A_{ap} S}{L} - \frac{A_{rec} U_L}{L}(T_{rec} - T_a) \tag{10.45}$$

where A_{ap} is the unshaded area of the concentrator aperture and A_{rec} is the area of the receiver ($A_{rec} = \pi D_o L$ for the circular absorber).

The useful energy gain per unit length of the collector can also be expressed in terms of the energy transfer to the fluid at local fluid temperature T_f:

$$q'_u = \frac{A_{rec}/L (T_{rec} - T_f)}{\frac{D_o}{h_{fi} D_i} + \left(\frac{D_o}{2k} \ln \frac{D_o}{D_i}\right)} \tag{10.46}$$

Elimination of T_{rec} from Eqs. (10.45) and (10.46), yields:

$$q'_u = F' \frac{A_{ap}}{L}\left[S - \frac{A_{rec}}{A_{ap}} U_L (T_f - T_a)\right] \tag{10.47}$$

Where F′ is the collector efficiency factor.

The collector efficiency factor can be expresses as:

$$F' = \frac{U_o}{U_L} \tag{10.48}$$

Like Eq. (10.28), the useful energy from the concentrating collector operating under steady conditions can be expressed as:

$$Q_u = F_R A_{ap}\left[S - \frac{A_{rec}}{A_{ap}} U_L (T_i - T_a)\right] \tag{10.49}$$

The collector flow factor (F'') is given by:

$$F'' = \frac{F_R}{F'} = \frac{\dot{m}C_p}{A_{rec}\, U_L F'}\left[1 - \exp\left(-\frac{A_{rec}\, U_L F'}{\dot{m}C_p}\right)\right] \tag{10.50}$$

So, the heat removal factor can also be expressed as:

$$F_R = \frac{\dot{m}C_p}{A_{rec}\, U_L}\left[1 - \exp\left(-\frac{A_{rec}\, U_L F'}{\dot{m}C_p}\right)\right] \tag{10.51}$$

The differences between glazed and unglazed receivers are in the computation of S and U_L. This receiver type can be used as a boiler, and in this case, F' is given by Eq. (10.36), but F_R and F are then equal because there is no temperature gradient in the flow direction. The two or three sections of the receiver must be treated separately if a part of the receiver functions as a boiler and other parts as heaters of the fluid.

Example

If the absorber tube of the receiver for Example 10.2 is made of stainless steel ($k = 16\ Wm^{-1}K^{-1}$) of thickness 5×10^{-3} m, and the coefficient of convective heat transfer from the inner surface of the tube to the fluid is 250 $Wm^{-2}K^{-1}$, calculate the overall heat loss coefficient of the receiver.

Solution:

From Eq. (10.44), the overall heat loss coefficient (U_o) is

$$U_o = \left[\frac{1}{U_L} + \frac{D_o}{h_{fi} D_i} + \frac{D_o \ln(D_o/D_i)}{2k}\right]^{-1}$$

From Example 10.2, $U_L = 3.67\ Wm^{-2}K^{-1}$
Therefore,

$$U_o = \left[\frac{1}{3.67} + \frac{0.05}{250 * 0.04} + \frac{0.05 \ln(0.05/0.04)}{2 * 16}\right]^{-1} = 3.60\ Wm^{-2}K^{-1}$$

Some collectors are very long in the flow direction, and control problems can arise for such collector systems. In this vein, parabolic trough collectors can be longer than 1000 m so that the heat transfer fluid may take half an hour or more to flow from inlet to outlet. To maintain a uniform outlet temperature, the only available control variable is the fluid flow rate. Adjustment of the flow rate continues to influence the outlet temperature for the duration of the fluid transit time.

10.3.2.2 Point focusing collectors

Cavity, external, and volumetric receivers are commonly used on point focusing collectors. In cavity and external receivers, concentrated solar radiation is absorbed by tubular conduits, and heat is transferred from the tube surface to the heat transfer fluid (Fig. 10.14). In volumetric receivers, solar radiation is absorbed in three dimensions. Based on the Carnot principle, the efficiency of the CSP cycle can be raised by operating from a heat source at the highest temperature. Consequently, highest temperatures of the outlet fluid from a solar receiver are desirable. To achieve this, heat losses from the receiver must be minimized.

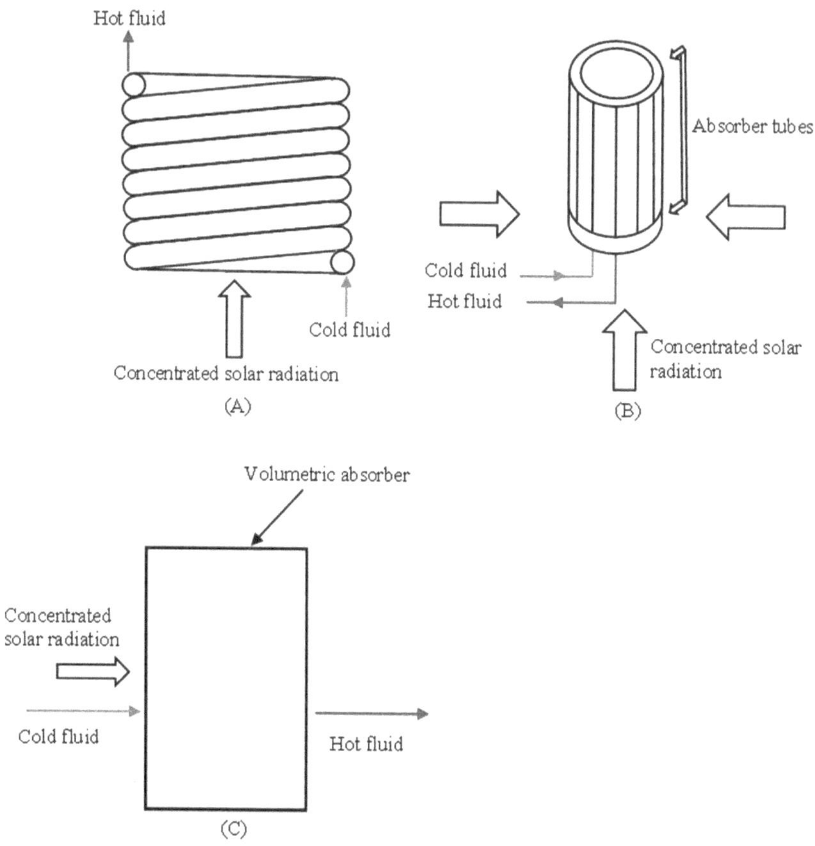

FIGURE 10.14 Schematic representation of selected designs of: (A) cylindrical cavity receiver with a helical tube, (B) cylindrical external receiver, and (C) volumetric receiver.

The thermal performance of point focusing receivers can generally also be described by Eq. (10.49). However, there are different designs of each receiver type. Consequently, the overall heat loss coefficient (U_L) and heat removal factor (F_R) must be computed from a given set of design, environmental, and operational parameters. Some assumptions can be made about the design and operation of the receiver for purposes of estimating these parameters. For example, consider a cylindrical external receiver comprising panels of absorber tubes (Fig. 10.10). The inner part of the cylinder is insulated, and therefore heat is lost from the external side by convection and radiation. The cylinder is mounted on a tower where it can exchange radiation with the ground as well as the sky.

We assume that the temperature is uniform around each absorber tube. Only half of the circumference of a given tube is irradiated with concentrated solar radiation. Turbulent natural convective heat transfer is assumed. Conductive heat loss through the support structure is negligible. Radiative heat exchange between adjacent tubes is neglected. With these assumptions, the total heat loss from the walls of the receiver can be given by:

$$\frac{Q_{loss}}{A_{rec}} = (h_c + h_r)(T_{rec} - T_a) = U_L\ (T_{rec} - T_a) \tag{10.52}$$

Where A_{rec} is the circumferential area of the receiver (πDH), D is the diameter of the receiver, H is the height of the receiver, h_c is the coefficient of convective heat loss, h_r is the coefficient of radiative heat loss, T_a is the ambient temperature, and T_{ab} is the temperature of the absorber.

For turbulent natural heat convection, the coefficient of heat loss can be computed from Siebers and Kraabel (1984):

$$h_c = Nu \frac{k}{L_{ch}} \tag{10.53}$$

Where L_{ch} is the characteristic length of the receiver (for cylindrical receiver, $L_{ch} = H$, the height of the cylinder), k is the thermal conductivity of the fluid at the mean temperature of the receiver and ambient temperature, and Nu is the Nusselt number (the ratio of convective to conductive heat transfer coefficients within a fluid).

For this type of receiver geometry, Nu can be calculated by using the following correlation:

$$Nu = 0.098\ Gr^{1/3} \left(\frac{T_{rec}}{T_a}\right)^{-0.14} \tag{10.54}$$

$$Gr = gL_{ch}^3 \frac{\beta\phi^2(\Delta T)}{\mu^2} \tag{10.55}$$

Where g is the acceleration due to gravity, L_{ch} is the characteristic length, β is the coefficient of thermal expansion of the fluid ($=T^{-1}$, for gases), ΔT is the temperature difference between the surface and fluid, ϕ is the density of the fluid, and μ is the dynamic viscosity of the fluid.

Radiation losses can be determined as follows (Too et al., 2019):

$$\frac{Q_{loss,\ rad}}{A_{rec}} = \varepsilon_{rec}\sigma\left(T_{rec}^4 - \frac{\varepsilon_{sky}T_{sky}^4 + \varepsilon_g T_a^4}{\varepsilon_{sky} + \varepsilon_g}\right) \tag{10.56}$$

Where ε_{rec} is the emittance of the receiver, ε_g is the emittance of the ground, and ε_{sky} is the emittance of the sky.

Consequently, the linearized coefficient of radiative heat transfer is:

$$h_r = \frac{\varepsilon_{rec}\sigma}{T_{rec} - T_a}\left(T_{rec}^4 - \frac{\varepsilon_{sky}T_{sky}^4 + \varepsilon_g T_a^4}{\varepsilon_{sky} + \varepsilon_g}\right) \tag{10.57}$$

Example

A cylindrical external solar receiver 9.5 m high is mounted on a tower. The receiver is made of panels of tubes, and it is insulated on the inner part. During a certain time of the day, the temperature of the outer surface of the absorber tubes is 750 K when the ambient and sky temperatures are respectively 300 and 293 K. Assuming the emittances of the absorber, ground, and sky are 0.89, 0.90, and 0.96, respectively, calculate overall heat loss from the receiver.

Solution:

From Eq. (10.53), the coefficient of convective heat transfer is:

$$h_c = Nu\frac{k}{L_{ch}}$$

$$Gr = gL_{ch}^3\frac{\beta\phi^2(\Delta T)}{\mu^2} = 10 * 9.5^3 * \frac{0.0019 * 0.6647^2(750 - 300)}{0.00002793^2} = 4.1519 * 10^{12}$$

$$Nu = 0.098\left(4.1519 * 10^{12}\right)^{1/3}\left(\frac{750}{300}\right)^{-0.14} = 1385$$

Therefore,

$$h_c = 1385 * \frac{0.0423}{9.5} = 6.2\ Wm^{-2}K^{-1}$$

Using Eq. (10.57), the coefficient of radiative heat transfer is:

$$h_r = \frac{0.89 * 5.67 * 10^{-8}}{750 - 300}\left(750^4 - \frac{0.96 * 293^4 + 0.90 * 300^4}{0.96 + 0.90}\right) = 34.6\ Wm^{-2}K^{-1}$$

$$U_L = h_c + h_r = 6.2 + 34.6 = 40.8\ W\ m^{-2}K^{-1}$$

10.3.3 Exergetic performance

Exergy is an important thermodynamic concept in the development of solar receivers. It is the maximum useful work that can be obtained when the system is taken from its given state to the thermodynamic equilibrium with the environment. This concept has been presented in Chapter 7. The exergetic efficiency of a solar receiver (η_{ex}) can be computed from:

$$\eta_{ex} = \frac{\hat{E}_{out}}{\hat{E}_{in} + \hat{E}_s} \tag{10.58}$$

Where $\hat{E}_{in}$ is the physical exergy flow into the receiver, $\hat{E}_{out}$ is the physical exergy flow out of the receiver, and $\hat{E}_s$ is the solar radiation exergy. Models for computing physical and solar radiation exergy flows are reported in Chapter 7.

Example

A solar tower receiver uses pressurized air as a heat transfer fluid. The heliostat field has an area of 5000 m^2. At a certain time of the day, the intensity of the beam radiation intercepted by the concentrator is 110 Wm^{-2}, the inlet and outlet temperatures of the air are respectively 550 and 900 K at a pressure $P = 3$ MPa and a flow rate of 4.5 kgs^{-1}. Calculate the exergy efficiency of the receiver, assuming the reference temperature (T_0) and pressure (P_o) are respectively 273.15 K and 0.101,325 MPa.

Solution:

From Eq. (10.58),

$$\eta_{ex} = \frac{\hat{E}_{out}}{\hat{E}_{in} + \hat{E}_s}$$

But from Eq. (7.34), physical exergy ($\hat{E}_{ph}$) is given by:

$$\hat{E}_{ph} = \dot{m}\left[C_p(T - T_0) - C_p T_0 \ln\left(\frac{T}{T_0}\right) + R\,T_0 \ln\left(\frac{P}{P_0}\right)\right]$$

Where C_p is the specific heat capacity of the fluid, $\dot{m}$ is the mass flowrate, P is pressure at temperature T, P_0 is the pressure at reference temperature T_0, and R is universal gas (8.314 J $mol^{-1}K^{-1}$).

Therefore,

$$\hat{E}_{in} = 4.5\left[1080 * (550 - 273.15) - 1080 * 273.15 * \ln\left(\frac{550}{273.15}\right) + 8.314 * 273.15 \ln\left(\frac{3}{0.101325}\right)\right] = 450995 \text{ W}$$

$$\hat{E}_{out} = 4.5\left[1080 * (900 - 273.15) - 1080 * 273.15 * \ln\left(\frac{900}{273.15}\right) + 8.314 * 273.15 \ln\left(\frac{3}{0.101325}\right) = 1498228 \text{ W}\right]$$

From Eq. (7.35),

$$\hat{E}_s = \left(1 - \frac{T_0}{T_{out}}\right) Q_s$$

$$Q_s = A_{ap} G_b = 5000 * 1100 = 5500000 \text{ W}$$

$$\hat{E}_s = \left(1 - \frac{273.15}{900}\right) * 5500000 = 3830750 \text{ W}$$

Therefore, the exergy efficience is:

$$\eta_{ex} = \frac{1498228}{450995 + 3830750} = 0.35$$

10.4 Hydrodynamic models

It is also possible to express the solar collector equations in natural forms. The governing equations are often complex, and so they require computer programs to organize them into a suitable form and to find solutions. The system of equations may vary with collector design and operational parameters.

Both heat and mass transfers occur in a solar receiver. Heat transfer fluids flow through absorber conduits or porous structures to evacuate heat from the absorber element to other components of the CSP system. Due to the elevated operational temperatures, it is also necessary to consider the thermomechanical performance of receivers. For instance, one of the assumptions in the simplified form of modeling the tubular receiver is that the temperature of the tube is circumferentially uniform. Concentrated solar radiation may not be uniformly incident on the absorber element (such as a tubular receiver), resulting in temperature gradients and thermal stresses in the element. Complex hydrodynamic processes can more closely be described by coupled algebraic and partial differential equations constituted by mass, momentum, and energy conservations that must be satisfied at each time step in the whole spatial in the entire domain (Bonilla et al., 2013; Sandá et al., 2019). The fluids may be considered as viscous or inviscid, thereby requiring different models. Single- and two-phase flows also occur in solar receivers, and governing equations for these and other flow types are reported in Section 8.3.

10.4.1 Single-phase flows

Single-phase flow can be modeled by using equations of mass, momentum, and energy conservations. Assuming the heat transfer fluid is viscous and compressible, then the flow field can be described by the following equations:

Mass conservation

$$\frac{\partial \phi}{\partial t} + \nabla \cdot (\phi \mathbf{V}) = 0 \tag{10.59}$$

The physical meanings of the mass conservation terms in Eq. (10.59) are:

Where **V** is the velocity vector, ∇ is the vector differential operator, and ϕ is the density of the fluid.

The density of fluids is affected by pressure and temperature. However, the density of incompressible fluids does not vary with pressure. For an ideal gas, the density of the gas can be expressed as a function of pressure (P) and temperature (T):

$$\phi = \frac{P}{RT} \tag{10.60}$$

$$R = \frac{R_U}{M} \tag{10.61}$$

Where M is the molecular mass of a gas (kg $kmol^{-1}$), R is the gas constant ($Jkg^{-1}K^{-1}$), and R_u is the universal gas constant (J $kmol^{-1}K^{-1}$).

Momentum conservation

$$\frac{\partial(\phi \mathbf{V})}{\partial t} + \nabla \cdot \left(\phi \mathbf{V} \mathbf{V}^T\right) = -\nabla P + \nabla \cdot \boldsymbol{\sigma} + \phi \mathbf{g} \tag{10.62}$$

Where **g** is the acceleration due to gravity, $\mathbf{V}^T$ is the transpose of the velocity vector **V**, and $\boldsymbol{\sigma}$ is the viscous stress tensor.

Energy conservation

$$\frac{\partial\left[\phi\left(e + \frac{V^2}{2}\right)\right]}{\partial t} + \nabla \cdot \left[\phi\left(e + \frac{V^2}{2}\right)\mathbf{V}\right] = -\nabla \cdot (P\mathbf{V}) + \nabla \cdot (\boldsymbol{\sigma}\mathbf{V}) - \nabla \cdot \mathbf{Q} + \phi\, \mathbf{g} \cdot \mathbf{V} + q \tag{10.63}$$

Where e is the internal energy per unit volume, **Q** is the heat flux, and q is the rate of heat addition per unit volume.

A high proportion of natural and industrial applications involving multiphase flows are turbulent. Consequently, modeling of two-phase flows must consider turbulent phenomena when constructing the conservation equations of the fluid. In this vein, deterministic approaches to turbulence modeling fall into four major categories. In order of decreasing complexity and computational cost, these approaches are (Versteeg & Malalasekera, 2007):

(a) Direct numerical simulation (DNS): Adequately small spatial mesh and time steps are considered to solve different turbulence scales. This approach entails a high computational cost, and therefore, its use is limited, but it neglects any specific turbulence model.

(b) Large eddy simulation (LES): This approach models large-scale eddies with time-dependent simulation, while smaller eddies are solved by using sub-grid scale models.

(c) Turbulence models of the Reynolds-Averaged-Navier-Stokes (RANS) equations: These models consider time or statistically averaged with time-dependent boundary conditions. The modeling approach centers on the mean flow and the effect of turbulence (fluctuating component) on it and its properties. The averaging process requires the addition of extra terms to the equations to model the fluid behavior. These terms vary with the selected turbulence model, with the commonest being the standard κ-ε (turbulent kinetic energy and energy dissipation rate) and the Reynolds stress model. This is the most widely used approach in multidimensional homogenous flows due to its diminished computational cost.

(d) Empirical correlations: The use of empirical correlations is a common technique to model simple turbulent flows as fully developed pipe flows and to investigate the boundary layer over a flat plate under a one-dimensional approach. Application of empiricism eliminates the need for a basic turbulence model. Instead, the turbulence effect is implicitly considered at the wall through constitutive laws related to the momentum and energy conservation equations derived under turbulent conditions. In the equation of momentum conservation, the shear stress contribution is studied with the wall shear stress function. In this function, the pressure drop is modeled in terms of the friction loss through the Darcy-Weisbach equation (Eq. 10.64) (Sandá et al., 2019). In the energy equation, the heat flux absorbed by the fluid is estimated by a constitutive model of heat flow rate transmitted by convection from the solid, which depends on the heat transfer coefficient and the wall temperature (Eq. 10.65).

$$\nabla \cdot \boldsymbol{\sigma} \approx \frac{\partial \sigma}{\partial z} = F_w = -\frac{1}{2d_i} f_w \phi V^2 = -\frac{4\sigma_w}{d_i} \tag{10.64}$$

Where d_i is the inner diameter of the conduit, f_w is the Darcy-Weisbach friction factor, F_w is the tube wall friction force per unit volume, and σ_w is the shear stress on the tube wall.

$$\nabla \cdot \mathbf{Q} \approx Q_{c,\,ab-f} = Q_{ab} = h_{c,ab-f}\left(T_{ab} - T_f\right) \tag{10.65}$$

Where $Q_{c,ab-f}$ is the convective heat transfer from the tube wall to the fluid, Q_{ab} is the wall heat flow rate per unit volume, $h_{c,ab-f}$ is the coefficient of convective heat transfer, T_{ab} is the temperature of the tube wall, and T_f is the temperature of the fluid.

10.4.2 Two-phase flows

In solar receivers, the use of some heat transfer fluids (such as water/steam, suspended solid particles) and of porous absorbers involve two phases. The water/steam flow is a liquid-gas flow system. Flows of suspended solids and of fluids through porous media are solid-fluid flow systems, and the involved fluids can be liquids or gases. These flow systems require two-phase flow analysis.

The liquid–gas flows can be analyzed by using one-fluid model or two-fluid model (TFM) approaches. In the one-model technique, the three conservation equations for mass, momentum, and energy conservation are used. This model is also known as the homogeneous equilibrium model (Sandá et al., 2019). It treats the two fluids like a mixture with averaged thermodynamic properties, as a single-phase problem. On the other hand, the two-fluid model describes each phase with three conservation equations, thereby yielding a total of six conservation equations. One advantage of this model is its ability of not being conditioned by any type of temperature or velocity restriction. It permits modeling of two-phase flows with higher accuracy than the three-equation models. Nevertheless, the TFM is more complex because it involves more unknowns and requires supplementary information for closure conditions at the interphase, including the equation of state and constitutive relations. It is difficult to determine the relations at the interphase using physical laws. They are often obtained from empirical correlations (Alobaid et al., 2017), that may not be available for all two-phase problems.

TFM governing equations are derived from the first principles by averaging the local or microscopic formulation for each phase. Three major averaging methods: ensemble averaging, time averaging, and volume or spatial averaging. These averaging techniques could be quite different even though they yield apparently similar final equations (Kleinstreuer, 2003). The conservation equations of mass, momentum, and energy per unit volume for the kth phase presented below are based on the ensemble-averaging techniques derived by Drew and Passman (1998):

Mass conservation

$$\frac{\partial(\alpha_k \phi_k)}{\partial t} + \nabla \cdot (\alpha_k \phi_k \mathbf{V}_k) = \Gamma_k \tag{10.66}$$

Where $\partial(\alpha_k \phi_k)/\partial t$ is the local rate of change of density for the kth phase, $\nabla \cdot (\alpha_k \phi_k \mathbf{V}_k)$ is the convective term of the density for the kth phase, α_k is the fraction of the phase k, and Γ_k is the interfacial mass generation (mass transfer). In case of liquid–gas flow through a tube, α_k is called the void fraction if the kth phase is gaseous. Void fraction varies along the length of the tube because the gas density changes along the tube length. For instance, this type of flow commonly occurs in direct steam generation using solar receivers. Water is fed to the absorber element through the inlet, and as it flows through the absorber, steam is produced progressively until $\alpha_k = 1$. Thus, the outlet fluid is superheated steam that can be exploited in various thermal applications, including electricity generation.

Momentum conservation

$$\begin{aligned}\frac{\partial(\alpha_k \phi_k \mathbf{V}_k)}{\partial t} + \nabla\cdot\left(\alpha_k \phi_k \mathbf{V}_k \mathbf{V}_k^T\right) &= -\alpha_k \nabla P + \nabla\cdot\left[\alpha_k\left(\boldsymbol{\sigma}_k + \boldsymbol{\sigma}_{k,Re}\right)\right] \\ &\quad + \alpha_k \phi_k \mathbf{g} + \Gamma_k \mathbf{V}_{ki} + (P_{ki} - P_k)\nabla\alpha_k \\ &\quad + M_k - \nabla\alpha_k\cdot\boldsymbol{\sigma}_{ki}\end{aligned} \tag{10.67}$$

Where $\partial(\alpha_k \phi_k \mathbf{V}_k)/\partial t$ is the local rate of change of linear momentum for the kth phase per unit volume, $\nabla\cdot\left(\alpha_k \phi_k \mathbf{V}_k \mathbf{V}_k^T\right)$ is the convective term of linear momentum of the kth phase per unit volume, $-\alpha_k \nabla P$ is the contribution of the kth phase pressure to the surface forces acting on kth phase per unit volume, $\nabla\cdot\left[\alpha_k\left(\boldsymbol{\sigma}_k + \boldsymbol{\sigma}_{k,Re}\right)\right]$ is the kth phase viscous and turbulent Reynolds stress $(\boldsymbol{\sigma}_{k,Re})$ contribution to surface forces acting on the kth phase per unit volume, $\alpha_k \phi_k \mathbf{g}$ is the gravitational body force acting on the kth phase per unit volume, $\Gamma_k \mathbf{V}_{ki}$ is the interfacial momentum flux for the momentum change in the kth phase due to mass transfer per unit volume, $(P_{ki} - P_k)\nabla\alpha_k$ is the pressure difference between the area-averaged interfacial pressure and the spatially averaged pressure of the kth phase per unit volume, M_k is commonly known as the interfacial extra momentum source, the generalized interfacial drag, or the averaged interfacial momentum exchange, and this parameter is included in the interfacial drag, the lateral lift force, the wall force, the virtual mass force and the Basset force, and $-\nabla\alpha_k\cdot\boldsymbol{\sigma}_{ki}$ is the interfacial averaged viscous stress of the kth phase per unit volume. The subscript ki stands for interface related to the kth phase.

The terms M_k and $-\nabla\alpha_k\cdot\boldsymbol{\sigma}_{ki}$ in Eq. (10.67) are usually expressed as the interfacial force density or total generalized drag forces $(M_{k,d})$:

$$M_{k,d} = M_k - \nabla\alpha_k\cdot\boldsymbol{\sigma}_{ki} \tag{10.68}$$

Energy conservation

$$\begin{aligned}&\frac{\partial\left[\alpha_k \phi_k\left(e_k + e_{k,Re} + \frac{V^2}{2}\right)\right]}{\partial t} + \nabla\cdot\left[\alpha_k \phi_k\left(e_k + e_{k,Re} + \frac{V^2}{2}\right)\mathbf{V}\right] \\ &= -\nabla\cdot(\alpha_k P_k \mathbf{V}_k) + \nabla\cdot\left[\alpha_k\left(\boldsymbol{\sigma}_k + \boldsymbol{\sigma}_{k,Re}\right)\right] - \nabla.\left[\alpha_k\left(\mathbf{Q}_k + \mathbf{Q}_{k,Re}\right)\right] \\ &\quad + \alpha_k \phi_k \mathbf{g}_k\cdot\mathbf{V}_k + q_k + W_k + \Gamma_k\left(e_{ki} + \frac{V_{ki}^2}{2}\right)\end{aligned} \tag{10.69}$$

Where $\partial\left[\alpha_k \phi_k\left(e_k + e_{k,Re} + \frac{V^2}{2}\right)\right]\Big/\partial t$ is the temporal term or local rate of change of energy for the kth phase per unit volume,

$\nabla\cdot\left[\alpha_k\phi_k\left(e_k+e_{k,Re}+\frac{V^2}{2}\right)\mathbf{V}\right]$ is the rate of convection of energy of the kth phase per unit volume, $-\nabla\cdot(\alpha_k P_k \mathbf{V}_k)$ is the contribution of the kth phase pressure to the work of the surface forces, acting on the kth phase per unit volume, $\nabla\cdot\left[\alpha_k\left(\boldsymbol{\sigma}_k+\boldsymbol{\sigma}_{k,Re}\right)\right]$ is the contribution of the viscous and turbulent Reynolds stress of the kth phase to the work of the surface forces acting on the kth phase per unit volume, $-\nabla\cdot\left[\alpha_k\left(\mathbf{Q}_k+\mathbf{Q}_{k,Re}\right)\right]$ is the contribution of the mean and turbulent heat fluxes by the kth phase, $\alpha_k\phi_k\mathbf{g}_k\cdot\mathbf{V}_k$ is the work of the body forces acting on the kth phase per unit volume, q_k is the interfacial heat flux contribution in the kth phase or heat flow rate per unit volume, W_k is the interfacial work contribution in the kth phase per unit volume, and $\Gamma_k\left(e_{ki}+V_{ki}^2/2\right)$ is the interfacial internal energy and kinetic energy contribution to the energy change in the kth phase due to mass transfer.

A seven-equation model has been developed, as an extension of TFM. In this model, each phase has a different pressure, and an additional equation for the void fraction into the set of the governing equations, thereby ensuring the hyperbolicity of the equation system (Sandá et al., 2019). The additional equation is given by:

$$\frac{\partial\alpha_k}{\partial t}+\mathbf{V}_{ki}\cdot\nabla\alpha_k=\frac{\Gamma_k}{\phi_{ki}} \tag{10.70}$$

10.5 Heat transfer

Most of the absorbed solar thermal energy is evacuated by the heat transfer fluid through hydrodynamic mechanisms in conduits or porous media. However, part of the heat is conducted between walls of the conduit and other components. Thus, the distribution of temperature in receiver structural components can be modeled by a heat conduction model. Heat is also transferred from the outer surfaces of the receiver structures to the environment through convection and radiation. Some surfaces of the receiver elements also exchange radiation. Thus, heat transfer processes can also be modeled using natural forms of equations.

10.5.1 Conduction

Heat conduction is the rate of energy transmission between two points in a medium whereby kinetic energy is transferred between particles or groups of particles. It can take place in gaseous, liquid, and solid phases of media. Heat is conducted from a point at high temperature to a point at low temperature, and the process may be transient or steady.

Heat conduction through a solid wall can be obtained by applying the principle of energy balance. The energy conservation of the wall can be described by means of Fourier's law (Bergman et al., 2011):

$$\phi_w C_{p,w} \frac{\partial T_w}{\partial t} = \nabla \cdot (k_w \nabla T_w) + q_v \tag{10.71}$$

Where q_v is the rate at which energy is generated per unit volume of the solid medium (W m^{-3}), k_w is the thermal conductivity of the wall, T_w is temperature of the wall. When no energy is generated internally, $q_v = 0$.

It is critical to solve an appropriate form of the heat equation to determine the temperature distribution in a medium. Nevertheless, such a solution depends on the physical conditions existing at the boundaries of the medium and, in case of transient processes, on conditions existing in the medium at some initial time. Several common possibilities of boundary conditions exist and can simply be expressed in some mathematical form. Two boundary conditions are required for each coordinate to describe the system because the heat equation is second order in the spatial coordinates. On the other hand, the equation is first order in time, but only one condition, known as the initial condition, must be specified.

Three kinds of boundary conditions are often defined in heat transfer problems (Bergman et al., 2011). These conditions are specified at the surface $x = 0$ for a system in one dimension. Heat transfer is in the positive x-direction with the temperature distribution that may vary with time, represented by T(x, t). The three conditions are: (a) constant surface temperature, (b) constant surface heat flux with: (i) finite heat flux and (ii) adiabatic or insulated surface, and (c) convection surface condition. The first condition represents a situation in which the surface is kept at a fixed temperature, $T(0, t) = T_s$. It is often known as a Dirichlet condition, or a boundary condition of the first kind (Eq. 10.72). This condition is closely approximated, for instance, when the surface is in contact with a medium at its melting or boiling point. In these two processes, the latent heat of fusion or vaporization is transferred while the surface remains at the same temperature of the phase change process. The second condition represents a system in which there is a fixed or constant heat flux at the surface (Eq. 10.73). This heat flux is related to the slope of T(x, t) at the surface. This condition is known as a Neumann condition or a boundary condition of the second kind. It may practically be achieved by fitting a thin film electric heater to the surface. A special case of this condition represents a perfectly insulated or adiabatic surface described by Eq. (10.74). Under steady-state conditions, solar radiation absorbed by the absorber element is approximately constant. The second boundary condition can be applied in such a case. The third boundary condition represents a situation in which there is convection heating (or cooling) at the surface and is obtained from the energy balance of the surface (Eq. 10.74).

$$q_x = -k\frac{\partial T_w}{\partial x}\bigg|_{x=0} = q_s, \text{ with finite heat flux} \tag{10.72}$$

Where q_x is heat flux in the x-direction, and q_s is the heat flux on the surface.

$$q_x = -k\frac{\partial T_w}{\partial x}\bigg|_{x=0} = 0, \text{ with adiabatic or insulated surface} \tag{10.73}$$

$$-k\frac{\partial T_w}{\partial x}\bigg|_{x=0} = h_{c,w-a}[T_a - T(0,t)] \tag{10.74}$$

It is also important to consider the initial and boundary conditions of the heat equation. A possible initial condition is, for instance, to assume that the temperature of the medium is equal to the ambient level.

Consider an evacuated tube shown in Fig. 10.15. The outer surface of the absorber tube absorbs part of the concentrated solar radiation transmitted through the glass tube and converts it to heat, thereby elevating the temperature of the outer surface of the absorber tube to the highest levels compared with other parts of the receiver. Heat ($Q_{cond,ao}$) is conducted from the outer surface to the inner surface of the absorber tube where most of it is removed by the heat transfer fluid through convection ($Q_{c,ao-f}$). The processes of heat and mass transfer through the heat transfer fluid are best modeled by using hydrodynamic principles, as presented and discussed in Chapter 8 and the preceding section. A proportion of $Q_{c,ao-f}$ is delivered as the useful gain from the collector, while the rest is lost to the ambient environment. Only radiative heat transfer ($Q_{r,ao-c}$) occurs between the absorber tube and the inner surface of glass tube. Again, there is heat conduction ($Q_{cond,c}$) from the inner surface to the outer surface of the glass cover exposed to the ambient environment. This is the most significant pathway for heat loss from the collector. Then, the outer surface of the cover loses heat to the environment through convection ($Q_{c,co-a}$) and radiation ($Q_{r,co-sky}$).

For a tubular receiver, it is convenient to express the heat diffusion equation in cylindrical coordinates. In this vein, heat conduction through the walls of the absorber and glass tubes can respectively be represented by Eqs. (10.75) and (10.76).

$$\phi_{ab}C_{p,ab}\frac{\partial T_{ab}}{\partial t} = \frac{1}{r}\frac{\partial}{\partial r}\left(k_{ab}\, r\frac{\partial T_{ab}}{\partial r}\right) + \frac{1}{r^2}\frac{\partial}{\partial \theta}\left(k_{ab}\frac{\partial T_{ab}}{\partial \theta}\right) + \frac{\partial}{\partial z}\left(k_{ab}\frac{\partial T_{ab}}{\partial z}\right) + q_{v,ab} \tag{10.75}$$

$$\phi_c C_{p,c}\frac{\partial T_c}{\partial t} = \frac{1}{r}\frac{\partial}{\partial r}\left(k_c r\frac{\partial T_c}{\partial r}\right) + \frac{1}{r^2}\frac{\partial}{\partial \theta}\left(k_c\frac{\partial T_c}{\partial \theta}\right) + \frac{\partial}{\partial z}\left(k_c\frac{\partial T_c}{\partial z}\right) + q_{v,c} \tag{10.76}$$

Where r is the radial direction, θ is the circumferential direction radial and z is the axial direction.

10.5.2 Convection

Heat convection is the rate of thermal energy transmission between two points in a fluid that involves mixing of the fluid by natural or forced mechanisms. In natural convection, the fluid moves due to density differentials arising from temperature gradients. Forced convection occurs when a moving fluid absorbs heat and transports it away by means of an external pump such as a fan or compressor.

10.5.2.1 Empirical models

Using empirical data, coefficients of heat transfer can be established for calculating the rate of convective heat transfer fluid ($Q_{c,s-f}$) for different surface geometries. In this approach, the convective heat flow from a surface to a fluid is given by:

$$Q_{c,s-f} = A_s h_{c,s-f} \left(T_s - T_f\right) \tag{10.77}$$

Where A_s is the area of the surface, $h_{c,s-f}$ is the local coefficient of convection, T_s is the temperature of the surface, and T_f is the temperature of the fluid.

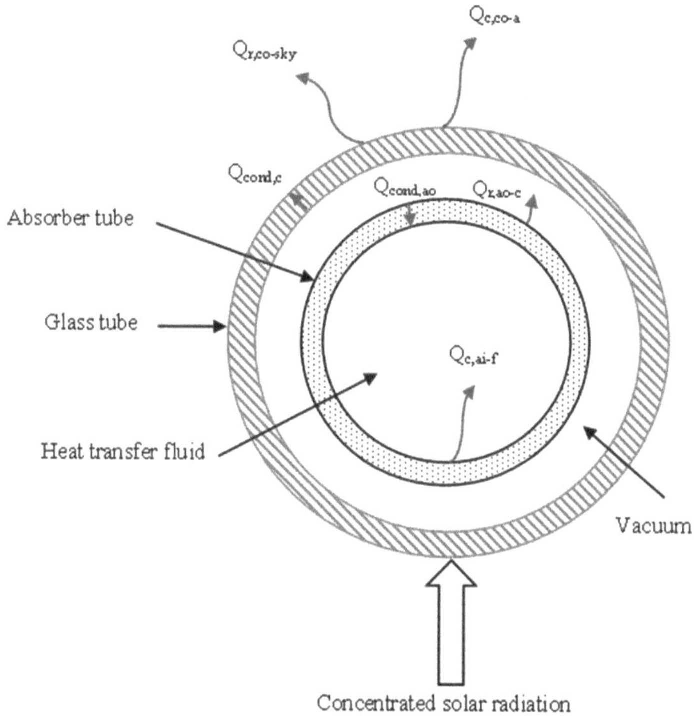

FIGURE 10.15 Schematic representation of heat flow in an evacuated tubular receiver.

The coefficient of convection depends on the Nusselt number, characteristic length, and thermal conductivity of the fluid. For natural convection, Nu can be expressed as a function of the Grashof number (Gr) and Prandtl number (Pr) as shown in Eq. (10.78). Gr is the ratio of buoyancy to viscous force in a fluid, while Pr is the ratio of kinematic viscosity to thermal diffusivity. The general correlation for natural convection is given by (ASHRAE, 2017):

$$\mathrm{Nu} = f(\mathrm{Gr\,Pr}) = f(\mathrm{Ra}) \tag{10.78}$$

Where Ra is the Raleigh number. Specific correlations are found by using empirical data.

$$\mathrm{Pr} = \frac{\nu}{\xi} \tag{10.79}$$

Where ξ is the thermal diffusivity and ν is the kinematic viscosity of the fluid.

For forced convection, the general correlation is given by:

$$\mathrm{Nu} = f(\mathrm{Re},\ \mathrm{Pr}) \tag{10.80}$$

Various correlations for estimating convective heat losses from solar central receivers have been reported by Siebers and Kraabel (1984). Many other authors, including, Bergman et al. (2011) and ASHRAE (2017), have also reported correlations for different geometries and flow regimes.

10.5.2.2 Boundary layer models

A boundary layer is the region close to the wall where the friction of the wall affects the flow of a fluid (ASHRAE, 2017). It occurs in internal and external flows. The boundary layer thickness (often denoted by δ) is thinner than the downstream flow distance. For external flow over a surface, the velocity of the fluid changes from zero at the wall to a maximum level at a distance δ from the wall. In general, boundary layers are laminar near the start of their formation but may become turbulent downstream. Flows of working and heat transfer fluids through receivers are generally internal, and they have been presented in Section 8.4. This section focuses on boundary layers in external flows that are responsible for convective heat transfer from solar receivers to the ambient environment.

At the solid–fluid boundary, heat is transmitted by means of conduction. Velocity, thermal, and concentration boundary layers are created when a fluid flows over a solid surface (Bergman et al., 2011). The velocity boundary layer is created by the difference between the free stream velocity and the zero velocity at the wall. On the other hand, a difference between the free stream and surface temperatures creates the thermal boundary layer. If the fluid is a binary mixture of species A and B, the concentration boundary layer results from the difference between free stream and surface concentrations (Fig. 10.16).

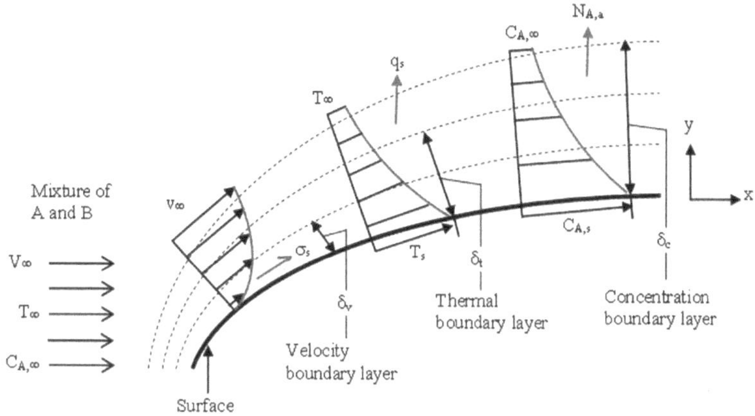

FIGURE 10.16 Development of velocity, thermal, and concentration boundary layers when a fluid flows over an arbitrary surface. The velocity of the fluid increases, while the temperature and species concentration decrease with the distance from the surface and from the edge. *Adapted from Bergman et al. (2011).*

Velocity boundary layer

When fluid particles encounter a surface, their velocity decreases significantly relative to the fluid velocity upstream of the surface, and for most cases, it is reasonable to assume that the particle velocity is zero at the wall. These particles then act to impede the motion of particles in the adjacent fluid layer, which act to slow down the motion of particles in the next layer until at a distance $y = \delta_V$ from the surface where the effect becomes insignificant. This retardation of fluid motion is related to shear stresses (σ_S) acting in planes that are parallel to the velocity of the fluid (see Chapter 8). As the distance y from the surface increases, the x velocity component of the fluid (v) must increase until it is close to the free stream value (v_∞). The subscript ∞ represents conditions in the free stream outside the boundary layer. The parameter δ_V is known as the boundary layer thickness, and it is normally defined as the value of y for which the ratio $v/v_\infty = 0.99$ (Bergman et al., 2011). The profile of the boundary layer velocity refers to the way in which v varies with y through the boundary layer. In this vein, the fluid flow is characterized by two clear regions: (a) a thin fluid layer (boundary layer) with large velocity gradients and shear stresses and (b) a region outside the boundary layer with negligible velocity gradients and shear stresses. The effects of viscosity infiltrate farther into the free stream and the boundary layer grows as the distance from the leading edge increases (δ_V increases with x). This boundary layer may more specifically be called the velocity boundary layer because it relates to the fluid velocity. It is created by fluid flow over a surface and basically critical to problems involving convection transport. Its importance in fluid mechanics comes from its relation to the surface shear stress, and therefore to surface frictional effects. The velocity boundary layer provides the basis for determining the local friction coefficient (C_f) for external flows.

$$C_f = \frac{\sigma_s}{0.5\ \phi\ v_\infty^2} \tag{10.81}$$

Where v_∞ is the free stream velocity, σ_s is the shear stress on the surface, and ϕ is the density of the fluid.

If we assume that the fluid is Newtonian (see Section 8.2.3), then the surface shear stress may be computed from knowledge of the velocity gradient at the surface:

$$\sigma_s = \mu \left.\frac{\partial v}{\partial y}\right|_{y=0} \tag{10.82}$$

Where v is the velocity and μ is the dynamic viscosity of the fluid.

Thermal boundary layer

A thermal boundary layer develops if the free-stream and surface temperatures differ as the fluid flows over a surface. Let us assume that the surface is at a constant temperature. At the leading edge ($x = 0$), the profile is uniform with $T(y) = T_\infty$. Nevertheless, fluid particles that move over the surface achieve thermal equilibrium at the surface temperature. In turn, energy is transferred from these particles to those in the adjacent fluid layer, thereby creating temperature gradients in the fluid. The fluid region in which these temperature gradients exist is referred to as the thermal boundary layer, and its thickness (δ_t) is normally defined as the value of y for which the ratio $[(T_s - T)/(T_s - T_\infty)] = 0.99$ (Alameri & Alkaabi, 2020). The effects of heat transmission infiltrate farther into the free stream, and the thermal boundary layer grows as the distance from the leading edge increases. At any distance x from the leading edge, the local surface heat flux can be determined by applying Fourier's law to the fluid at $y = 0$ (Bergman et al., 2011):

$$q_s = -k_f \left.\frac{\partial T_w}{\partial y}\right|_{y=0} \tag{10.83}$$

Where k_f is the thermal conductivity of the fluid and T_w is the temperature of the wall.

At the surface, Eq. (10.83) is appropriate because the fluid particles are stationary and energy transfer takes place only by conduction. Applying Newton's law of cooling, we observe that:

$$q_s = h_c(T_s - T_\infty) \tag{10.84}$$

Where h_c is the coefficient of convective heat transfer.

Combining Eqs. (10.83) and (10.84) yields:

$$h_c = \frac{-k_f \left.\frac{\partial T_w}{\partial y}\right|_{y=0}}{T_s - T_\infty} \tag{10.85}$$

Eq. (10.85) shows that conditions in the thermal boundary layer that strongly affect the wall temperature gradient at $y = 0$, are determinants of the rate of heat transfer across the boundary layer. Temperature gradients in the boundary layer decreases with increasing x because $(T_s - T_\infty)$ is constant and independent of x. Thus, the magnitude of the gradient at $y = 0$ decreases with increasing x, and h_c decreases with increasing x. In contrast, δ_t increases with x.

Concentration boundary layer

We now consider a binary mixture of chemical species A and B that flows over a surface. The molar concentration (kmol m^{-3}) of species A at the surface is $C_{A,s}$, and in the free stream, it is $C_{A,\infty}$. Transfer of species A by convection occurs when $C_{A,s}$ is different from $C_{A,\infty}$. For instance, a wet air stream contains vapor. As the air flows over a water surface, vapor is transferred from the surface because of evaporation at a liquid surface or due to sublimation at a solid surface. In this case, a concentration boundary layer is created. It is the region of the fluid in which concentration gradients exist, and its thickness δ_c is normally defined as the value of y for which $[(C_{A,s} - C_A)/(C_{A,s} - C_{A,\infty})] = 0.99$ (Bergman et al., 2011). The effects of species transfer penetrate farther into the free stream and the concentration boundary layer grows as the distance from the leading edge increases. The convective mass transfer between the surface and the free stream fluid is determined by conditions in the boundary layer. Of interest is the molar flux of species A (kmol $s^{-1}m^{-2}$). The molar flux associated with species transfer by diffusion is described by an expression that is similar to Fourier's law. This expression is known as Fick's law, given by (ASHRAE, 2017):

$$N_A = -D_{AB}\frac{\partial C_A}{\partial y} \tag{10.86}$$

Where N_A is the mass flux of species A, C_A is the concentration of species A, and D_{AB} is the binary diffusion coefficient.

The transmission of species is due to both bulk fluid motion (advection) and diffusion at any point corresponding to $y > 0$ in the concentration boundary layer. Nevertheless, fluid motion at the surface can be neglected because of the absence of nano- or microscale effects and the influence of species diffusion on the velocity normal to the surface. Consequently, the molar flux at $y = 0$ can be expressed as:

$$N_{A,s} = -D_{AB}\left.\frac{\partial C_A}{\partial y}\right|_{y=0} \tag{10.87}$$

Similar to Newton's law of cooling, an expression that relates the molar flux to the concentration difference across the boundary layer can be written as:

$$N_{A,s} = h_m \left(C_{A,s} - C_{A,\infty} \right) \quad (10.88)$$

Where h_m is the coefficient of convective mass transfer.

Boundary layer equations

External flows are associated with coexisting velocity, temperature, and concentration gradients. Wind is an example of an external flow that involves these three kinds of boundary layer variables. Conservation of mass, momentum, energy, and chemical species must be satisfied at any given point in the fluid. Like the hydrodynamic models of internal flows, governing equations of mass, momentum, and energy can be derived by applying the laws to a differential control volume located in the flow.

Let us consider the flow of a viscous incompressible fluid with simultaneous heat and mass transfer. The fluid has constant properties in the x- and y-directions of a Cartesian coordinate system. Assuming that the flow is steady and two-dimensional, then partial differential equations may be used to predict velocity and temperature distributions within the flow field. The governing boundary equations for laminar flow are (Bergman et al., 2011):

Mass conservation

$$\frac{\partial u}{\partial x} + \frac{\partial v}{\partial y} = 0 \quad (10.89)$$

Where u and v are respectively the x- and y-components of the mass average velocity.

Eq. (10.89) is a general expression of the overall mass conservation requirement, and it must be fulfilled at every point in the fluid. This equation is applicable to a single species fluid, and mixtures in which species diffusion and chemical reactions may be taking place, on condition that the fluid can be approximated as incompressible.

Momentum conservation

$$\phi \left(u \frac{\partial u}{\partial x} + v \frac{\partial u}{\partial y} \right) = -\frac{\partial P}{\partial x} + \mu \left(\frac{\partial^2 u}{\partial x^2} + \frac{\partial^2 u}{\partial y^2} \right) + f_x \quad (10.90)$$

Where f_x is the body force in the x-direction, P is the pressure and μ is the coefficient of dynamic viscosity.

$$\phi \left(u \frac{\partial v}{\partial x} + v \frac{\partial v}{\partial y} \right) = -\frac{\partial P}{\partial y} + \mu \left(\frac{\partial^2 v}{\partial x^2} + \frac{\partial^2 v}{\partial y^2} \right) + f_y \quad (10.91)$$

Where f_y is the body force in the y-direction.

In Eqs. (10.90) and (10.91), the two terms on the left-hand side of each equation represent the net rate of momentum flow from a control volume (see Section 8.3.1). Taken in order, the terms on the right-hand side of the two equations account for the net pressure force, the net viscous forces, and the body force. Eqs. (10.90) and (10.91) can be solved for the velocity field.

Energy conservation

$$\phi C_p\left(u\frac{\partial T}{\partial x}+v\frac{\partial T}{\partial y}\right)=k\left(\frac{\partial^2 T}{\partial x^2}+\frac{\partial^2 T}{\partial y^2}\right)+\mu\Phi+q_v \quad (10.92)$$

Where C_p is the specific heat capacity, k is thermal conductivity of the fluid, q_v is the volumetric rate of thermal energy generation, and $\mu\Phi$ is the viscous dissipation.

The viscous dissipation term is given by:

$$\mu\Phi=\mu\left\{\left(\frac{\partial u}{\partial y}+\frac{\partial v}{\partial x}\right)^2+2\left[\left(\frac{\partial u}{\partial x}\right)^2+\left(\frac{\partial v}{\partial y}\right)^2\right]\right\} \quad (10.93)$$

Eq. (10.92) also applies to an ideal gas with negligible changes in pressure. The terms on the left-hand side of this equation account the net rate of thermal energy that leaves the control volume due to the bulk fluid motion (advection). On the other hand, when taken in order, the terms on the right-hand side account for the net inflow of energy due to conduction, viscous dissipation, and generation. The viscous dissipation term accounts for the net rate at which mechanical work is irreversibly converted to thermal energy because of viscous effects in the fluid. The generation term typifies the conversion of other forms of energy to heat. For instance, chemical, electrical, electromagnetic, or nuclear energy may be converted to heat.

There are also governing boundary layer equations for turbulent flows. For engineers, the main concern is about the time-mean properties. So, the difficulty of solving the time-dependent governing equations is usually eliminated by averaging the equations over time by using well-established time-averaging techniques (Kays et al., 2005). For steady (in the mean values), incompressible, constant property, boundary layer flow with negligible viscous dissipation, the continuity, x-momentum, and energy can be expressed according to Bergman et al. (2011):

Mass conservation

$$\frac{\partial \bar{u}}{\partial x}+\frac{\partial \bar{v}}{\partial y}=0 \quad (10.94)$$

Where $\bar{u}$ is the average velocity in the x-direction and $\bar{v}$ is the average velocity in the y-direction.

Momentum conservation

$$\phi\left(\bar{u}\frac{\partial \bar{u}}{\partial x}+\bar{v}\frac{\partial \bar{u}}{\partial y}\right)=-\frac{dP_\infty}{dx}+\frac{\partial}{\partial y}\left(\mu\frac{\partial \bar{u}}{\partial y}-\phi\,\overline{u'v'}\right) \quad (10.95)$$

Energy conservation

$$\phi C_p\left(\bar{u}\frac{\partial T}{\partial x}+\bar{v}\frac{\partial T}{\partial y}\right)=\frac{\partial}{\partial y}\left(k\frac{\partial T}{\partial y}-\phi\ C_p\overline{v'T'}\right) \tag{10.96}$$

It should be noted that equations for the turbulent flow are similar to those for the laminar boundary layer (after ignoring the viscous dissipation), except for the presence of additional terms of the form (a′b′) that account for the effect of the turbulent fluctuations on momentum and energy. Based on Eqs. (10.96) and (10.97), it is usual to express the total shear stress (σ_{to}) and total heat (q_{to}) as:

$$\sigma_{to}=\frac{\partial}{\partial y}\left(\mu\frac{\partial\bar{u}}{\partial y}-\phi\ \overline{u'v'}\right) \tag{10.97}$$

$$q_{to}=-\left(k\frac{\partial T}{\partial y}-\phi\ C_p\overline{v'T'}\right) \tag{10.98}$$

10.5.3 Radiation

Thermal radiation is energy emitted by matter at a nonzero absolute temperature. Emission occurs from solids, liquids, and gases, and it may be ascribed to changes in the electron configurations of the constituent atoms or molecules. The energy of the radiation field is transferred by electromagnetic waves (ASHRAE, 2017). Unlike heat conduction and convection, radiative heat transfer does not require a medium for propagation to occur. Consequently, the use of a vacuum to reduce heat loss only eliminates convective and conductive modes of heat transmission. The presence of a medium between a radiator and receiver provides resistance to radiative heat transfer. The amount of energy emitted by a radiator is influenced by the nature of the material, microscopic structure, and temperature of the radiator and its surroundings. A black body is an ideal emitter and absorber, and it serves as a standard against which the radiative properties of real surfaces may be compared. So, a blackbody emits the upper limit of thermal power (E_b) that can be described by Stefan–Boltzmann law (Crepeau, 2007):

$$E_b=\sigma_{sb}\ T_s^4 \tag{10.99}$$

Where T_s is the temperature of the surface and σ_{sb} is the Stefan–Boltzmann constant.

The thermal radiation emitted by a real surface (E) is lower than that of a blackbody at the same temperature and is given by:

$$E = \varepsilon\ \sigma_{sb} T_s^4 \tag{10.100}$$

Where ε is the emittance of the surface.

Surfaces can exchange radiation, depending on their emittance and geometric view factor. For instance, consider an evacuated tubular receiver shown in Fig. 10.14. Due to the presence the vacuum space, heat transfer between the absorber and glass tubes is only through radiation. The radiative heat is transferred from the outer surface of the absorber tube to the inner surface of the glass tube ($Q_{r,\ ao-ci}$), given by:

$$Q_{r,ab-ci} = A_{ao}\ h_{r,ao-ci}(T_{ao} - T_{ci}) \tag{10.101}$$

$$h_{r,ab-ci} = \frac{\sigma_{sb}\left(T_{ao}^2 + T_{ci}^2\right)(T_{ao} + T_{ci})}{\dfrac{1-\varepsilon_{ao}}{\varepsilon_{ao}} + \dfrac{A_{ao}(1-\varepsilon_{ci})}{A_{ci}\varepsilon_{ci}} + \dfrac{1}{F_{ao-ci}}} \tag{10.102}$$

Where A_{ao} is the area of the outer surface of the absorber tube, T_{ci} is the temperature of the inner surface of the glass cover, T_{ao} is the temperature of the outer surface of the absorber, F_{ao-ci} is the view factor of the absorber relative to the glass cover (F_{i-j} is the fraction of all radiant energy leaving a surface i that is directly incident on surface j), ε_{ao} is the emittance of the outer surface of the absorber, and ε_{ci} is the emittance of the inner surface of glass cover.

Values of F_{i-j} depend on the geometry of the surfaces, and techniques for calculating this parameter are reported by many authors including ASHRAE (2017), Duffie and Beckman (2013) and Bergman et al. (2011). For the design in Fig. 10.14, $F_{ao-ci} = 1$, while $F_{ci-ao} = A_{ao}/A_{ci}$. Therefore, Eq. (10.102) can be simplified to:

$$h_{r,ab-ci} = \frac{\sigma_{sb}\left(T_{ao}^2 + T_{ci}^2\right)(T_{ao} + T_{ci})}{\dfrac{1}{\varepsilon_{ao}} + \dfrac{A_{ao}(1-\varepsilon_{ci})}{A_{ci}\varepsilon_{ci}}} \tag{10.103}$$

Similarly, radiative heat transfer also occurs between the outer surface of the glass tube and the sky. In this case, these two surfaces exchange radiation defined by:

$$Q_{r,ci-sky} = A_{co}\ h_{r,co-sky}\left(T_{co} - T_{sky}\right) \tag{10.104}$$

$$h_{r,co-sky} = \frac{\sigma_{sb}\left(T_{co}^2 + T_{sky}^2\right)\left(T_{co} + T_{sky}\right)}{\dfrac{1-\varepsilon_{co}}{\varepsilon_{co}} + \dfrac{1}{F_{co-sky}}} \tag{10.105}$$

10.6 Thermomechanical performance

The thermal efficiency of a receiver and overall performance of a CSP system increase with the level of operating temperature. From a thermodynamic perspective, some types of solar receivers can achieve temperatures of up to 1773 K. However, most of the CSP plants in the world are operating at temperatures below 900 K due to material limitation. Consequently, it is important to model the thermomechanical performance of receivers.

Generally, application of heat to a structure causes temperature differentials that create thermal strains and stresses. Thus, knowledge about the distribution of temperature in a solid is required. In this vein, the heat diffusion equation is suitable for mapping stresses in the various elements of a receiver. Relationships among temperature, strain, and stresses have been reported in Chapter 9.

For instance, consider an external tubular receiver in which molten salt is used as a heat transfer medium. The effective thermomechanical stress (σ_{eff}) in the absorber tubes of the receiver can be computed from Rodríguez-Sánchez et al. (2018):

$$\sigma_{eff} \approx \gamma\, E\left[\left(\frac{T_w + T_{fi}}{2}\right) - T_m\right] + \frac{T_w - T_{fi}}{2(1-\nu_{pr})} \tag{10.106}$$

$$T_m = T_f + \frac{1}{\pi}\left(\frac{T_{wa} + T_{fi}}{2} - T_f\right) \tag{10.107}$$

Where T_{fi} is the film temperature (on the inner surface of tube), T_w is the temperature of the tube wall, E is the Young's modulus, T_f is the temperature of the heat transfer fluid (molten salt), ν_{pr} is the Poisson's ratio, and γ is the volumetric coefficient of thermal expansion of the tube material. All the properties of the tube (E, ν, and γ) depend on the tube wall temperature.

There is thermal coupling among heat diffusion in the tube wall, hydrodynamics of the heat transfer fluid, and the thermal stress in the tube wall. Solving the heat diffusion equation yields the distribution of the temperature (T_w), at any given point inside the wall. The fluid evacuates heat from the inner surface of the wall, thereby affecting T_w. The distribution of temperature (T_f) within the heat transfer fluid is best modeled by using the hydrodynamic equations. For accurate results, heavily coupled systems of equations must be solved simultaneously.

10.7 Discretization of differential equation systems

Thermodynamic, hydrodynamic, heat transfer, and thermomechanical models involve partial differential equations (PDEs) that can be solved using analytical or numerical methods. Analytical solutions to differential equations for transient systems are confined to simple geometries and boundary conditions

(Incropera et al., 2007). Consequently, most of the systems of differential equations can only be solved by numerical methods that yield approximate solutions.

A system of PDEs cannot be directly solved by modeling and simulation (M&S) tools. Consequently, it requires discretization to convert the PDEs into a system of differential algebraic equations (DAEs). Computers use computational fluid dynamics (CFD) set of methods to solve the fluid motion equations. There are well-established methods for solving PDEs numerically, and they can be classified into three different categories (Sandá et al., 2019): finite difference method (FDM), finite element method (FEM), and finite volume method (FVM). These three methods are classical, and they have been extensively applied in the development of solar thermal technology. In addition to the classical methods (also called distributed parameters methods), the lumped parameter method (LPM), also known as the lumped capacitance method, is usually applied in thermohydraulic modeling of systems with negligible temperature gradients (distribution of temperature is approximately uniform). Other methods also exist for specific applications such as spectral methods, cellular automaton, moving boundary method (MBM), and the method of lines (MOL). In this book, only classical methods are presented and discussed.

10.7.1 Finite difference

Among the three classical methods, the finite difference method is the simplest and oldest one. It was known by Euler (1707–83), but the first work that established it as a numerical technique was published in 1911 by Richardson (1911). However, it only became popular in the early 1950s when computers were introduced into the computational field. The FDM is based on finite approximations of PDEs at each node in the Taylor series expansion form. The DAE obtained by FDM yields a variable unknown in the central node and in some of its neighbors, depending on the procedure of discretization (forward, backward, or central).

Some disadvantages of the FDM are: (a) the coefficients of the DAE system obtained from its application for complex geometries, multidimensional problems, and nonhomogenous media can present discontinuities, and (b) requirement of more effort to attain a high degree of accuracy.

10.7.1.1 Nodal network

A numerical solution enables determination of the variable at only discrete points. Consequently, the first step in any numerical analysis must be selection of these points. In this vein, the medium of interest is divided into several regions, and a reference point is assigned to the center of each region. Usually, the reference point is known as a node, and the collection of points is called a

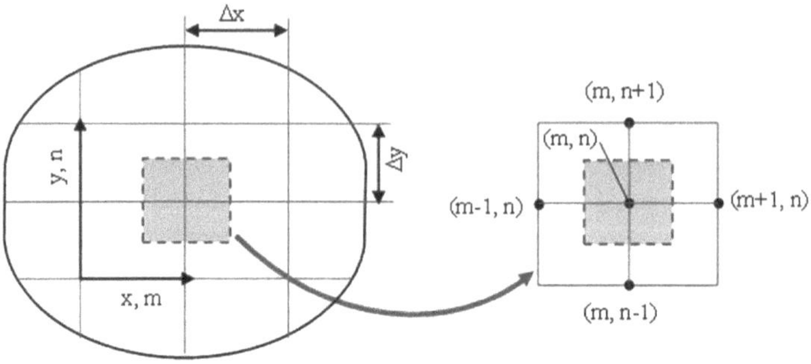

FIGURE 10.17 Schematic representation of the nodal network for a two-dimensional heat conduction problem. *Adapted from Bergman et al. (2011).*

nodal network, grid, or mesh. For instance, consider the two-dimensional heat diffusion equation. The x and y locations of a node can respectively be represented by m and n indices (Fig. 10.17). The temperature of a node is a measure of the average temperature of the region. For instance, the temperature of the node (m, n) may be considered as the average temperature of the surrounding shaded area. The selection of nodal points is often influenced by factors such as geometric convenience and the desired accuracy. Generally, the accuracy of the solution increases with number of nodes.

For the first derivative, a finite difference can be expressed in forward, backward, or centered forms:

$$\frac{\partial T}{\partial x} \approx \begin{cases} \dfrac{T_{m+1} - T_m}{\Delta x}, & \text{forward difference} \\ \dfrac{T_m - T_{m-1}}{\Delta x}, & \text{backward difference} \\ \dfrac{T_{m+1} - T_{m-1}}{2\Delta x}, & \text{centred difference} \end{cases} \tag{10.108}$$

These finite differences have associated truncation errors.

10.7.1.2 Finite difference form of the heat equation

An appropriate conservation equation must be written for each nodal point of unknown temperature to numerically determine the temperature distribution in the medium. The resulting set of equations can then be solved simultaneously for each nodal temperature. For steady heat conduction with uniform thermal conductivity and no heat generation, the exact form of the energy conservation requirement is given by (Bergman et al., 2011):

$$\frac{\partial^2 T}{\partial x^2}+\frac{\partial^2 T}{\partial y^2}=0 \tag{10.109}$$

Approximate solutions can be found using finite difference. We consider derivatives at node (m, n):

$$\left.\frac{\partial^2 T}{\partial x^2}\right|_{(m,n)} \approx \frac{T_{m+1,n}+T_{m-1,n}-2T_{m,n}}{(\Delta x)^2} \tag{10.110}$$

$$\left.\frac{\partial^2 T}{\partial y^2}\right|_{(m,n)} \approx \frac{T_{m,n+1}+T_{m,n-1}-2T_{m,n}}{(\Delta y)^2} \tag{10.111}$$

Using Eqs. (10.110) and (10.111), the derivative at node (m,n) can be expressed as:

$$\frac{\partial^2 T}{\partial x^2}+\frac{\partial^2 T}{\partial y^2} \approx \frac{T_{m+1,n}+T_{m-1,n}-2T_{m,n}}{(\Delta x)^2}+\frac{T_{m,n+1}+T_{m,n-1}-2T_{m,n}}{(\Delta y)^2}=0 \tag{10.112}$$

If $\Delta x=\Delta y$, then Eq. (10.112) reduces to:

$$T_{m+1,n}+T_{m-1,n}+T_{m,n+1}+T_{m,n-1}-4T_{m,n}=0 \tag{10.113}$$

10.7.1.3 Explicit and implicit methods of discretization

A finite difference equation can be expressed in an explicit or implicit form. One drawback of the explicit method is its conditional stability. As the time step increases, the solution may significantly swing from the steady-state conditions resulting in large errors. For instance, the two-dimensional heat equation (with uniform thermal conductivity and no generation) in Cartesian coordinates can be expressed as:

$$\frac{1}{\xi}\frac{\partial T}{\partial t}=\frac{\partial^2 T}{\partial x^2}+\frac{\partial^2 T}{\partial y^2} \tag{10.114}$$

Where ξ is the thermal diffusivity.

The time derivative on the left hand side of Eq. (10.114) can be approximated by a forward finite difference:

$$\frac{1}{\xi}\frac{\partial T}{\partial t} \approx \frac{T^{j+1}_{m,n}-T^{j}_{m,n}}{\xi\,\Delta t} \tag{10.115}$$

Where Δt is the time step, and time $t=j\Delta t$.

Nodal temperatures at the jth time step can be determined from Eq. (10.112). This implies that the heat equation can be expressed as:

$$\frac{T_{m,n}^{j+1} - T_{m,n}^{j}}{\xi \, \Delta t} = \frac{T_{m+1,n}^{j} + T_{m-1,n}^{j} - 2T_{m,n}^{j}}{(\Delta x)^2} + \frac{T_{m,n+1}^{j} + T_{m,n-1}^{j} - 2\,T_{m,n}^{j}}{(\Delta y)^2} \tag{10.116}$$

Assuming $\Delta x = \Delta y$, the nodal temperatures at the new $(j + 1)$ time can be expressed as:

$$T_{m,n}^{j+1} = F_o\left(T_{m+1,n}^{j} + T_{m-1,n}^{j} + T_{m,n+1}^{j} + T_{m,n-1}^{j}\right) + (1 - 4F_o)T_{m,n}^{j} \tag{10.117}$$

$$F_o = \frac{\xi \, \Delta t}{(\Delta x)^2} \tag{10.118}$$

Where F_o is the Fourier number.

Using a similar approach, explicit forms of the finite difference equations in one or three dimensions can be obtained. For instance, the one-dimensional explicit form of the finite difference equation for the mth interior node is:

$$T_{m,n}^{j+1} = F_o\left(T_{m+1}^{j} + T_{m-1}^{j}\right) + (1 - 2F_o)T_{m,n}^{j} \tag{10.119}$$

Eqs. (10.117) and (10.119) are in the explicit form of finite difference because unknown nodal temperatures for the new time are computed exclusively by using known nodal temperatures at the previous time. Consequently, computation of the unknown temperatures is simple. The calculations commence at $t = 0$ $(j = 0)$ because the temperature of each interior node is known at $t = 0$ $(j = 0)$ from specified initial conditions. Nevertheless, solutions obtained by the explicit approach may be characterized by numerically induced oscillations that are practically impossible (Bergman et al., 2011). These oscillations may become unstable, thereby causing the solution to diverge from the real steady-state conditions. To avert such errors, the set value of Δt must be maintained below a certain limit that depends on Δx and other parameters of the system. This relationship is known as a stability criterion, which may be derived mathematically or demonstrated by using a thermodynamic argument. For the problems of interest in this chapter, the criterion is determined by stipulating that the coefficient (F_o) associated with the node of interest at the previous time should be greater than or equal to zero. In general, this is done by collecting all terms involving $T_{m,n}^{j}$ to find the form of the coefficient. This outcome is then used to get a limiting relation involving F_o, from which the maximum permissible value of Δt may be computed. For instance, Eqs. (10.117) and (10.119) are already expressed in the desired form for deriving the stability criteria. Thus, the choice of mesh size is governed by some criteria to ensure the solutions are stable. For the one-dimensional finite-difference equation, the stability criterion is $(1-2F_o \geq 0)$. Therefore,

$$F_o \leq \frac{1}{2} \tag{10.120}$$

Similarly, the stability criterion for the two-dimensional finite-difference equation is:

$$F_o \leq \frac{1}{4} \tag{10.121}$$

Even though the explicit finite-difference scheme offers computational convenience, it limits the selection of Δt. The time interval must satisfy stability requirements for a given space increment. Usually, this entails the use of extremely small values of Δt, which may require a very large number of time intervals to obtain a solution. A decrease in the amount of computation time may usually be obtained by employing an implicit finite-difference scheme. The implicit form of a finite-difference equation can be derived by using Eq. (10.115) to approximate the time derivative, while all other temperatures are evaluated at the new $(j+1)$ time, instead of the previous (j) time. Thus, this equation is considered to give a forward-difference approximation to the time derivative, and the implicit form of the finite-difference equation for the interior node of a two-dimensional system can be expressed as:

$$\frac{T^{j+1}_{m,n} - T^{j}_{m,n}}{\xi\,\Delta t} = \frac{T^{j+1}_{m+1,n} + T^{j+1}_{m-1,n} - 2T^{j+1}_{m,n}}{(\Delta x)^2} + \frac{T^{j+1}_{m,n+1} + T^{j+1}_{m,n-1} - 2\,T^{j+1}_{m,n}}{(\Delta y)^2} \tag{10.122}$$

If we assume that $\Delta x = \Delta y$ and rearrange Eq. (10.122), we obtain:

$$(1+4F_o)T^{j+1}_{m,n} + F_o\left(T^{j+1}_{m+1,n} + T^{j+1}_{m-1,n} + T^{j+1}_{m,n+1} + T^{j+1}_{m,n-1}\right) = T^{j}_{m,n} \tag{10.123}$$

Eq. (10.123) shows that the new temperature of the (m,n) node is a function of the new temperatures of its adjoining nodes that are generally unknown. Consequently, all the corresponding nodal equations must be solved simultaneously to determine the unknown nodal temperatures at $t+\Delta t$. These equations can be solved directly or iteratively. Direct methods include backward substitution, Gaussian elimination, and matrix inversion, and they are suitable for a small number of equations. Each interior node must have a corresponding equation. In addition, direct methods require a large computer memory and time. Consequently, iterative methods may be more suitable and efficient than the direct ones. Iterative schemes are required to solve a nonlinear system of differential equations (Burden & Faires, 1985). These schemes include the Jacobi and Gauss-Seidel methods, and generally, the latter iterative method is more superior than the former. The marching solution involves simultaneously solving the nodal equations at each time until the specified final time is attained.

One advantage of the implicit scheme is its unconditional stability. The solution remains stable for all space and time intervals, and so, there are no restrictions on the values of Δx and Δt. In view of this, many authors have used the finite-difference scheme to discretize systems of differential equations in models for solar receivers. For instance, Tan and Chen (2010) used time-averaged governing equations for the conservation of gas phase mass, momentum, and energy of a solid particle receiver with air jet flow. These authors applied a realizable k−ε model (turbulent k−ε model) to close the turbulent Navier−Stokes system of equations. The governing differential equations were discretized in the Cartesian coordinate system by using a control-volume finite difference method. Discrete and nonlinear equations were linearized using an implicit technique with respect to the set of dependent variables. An additive correction multigrid method with a Gauss-Seidel relaxation procedure was applied to solve the algebraic equations iteratively. Based on the results, they concluded that a smaller size of the solid particles leads to a higher cavity with the same operating conditions. Nevertheless, the selection of solid particle size should also consider the manufacturing issues. Too small a size of the particle may cause the leakage of the particles from the aperture. From the simulation results, the effect of the wind on the performance of the receiver is very complex. Basically, formation of an aerowindow in front of the aperture of the particle solar receiver is useful in enhancing the cavity efficiency and the particle exiting temperature. However, the combined effect of the wind and the air jet flow may cause the efficiency to decline if the wind speed is larger than $1\ m\ s^{-1}$. The increase of the air jet temperature marginally affects the cavity efficiency. Other authors, including, Goebel (1999), Serrano-Aguilera et al. (2014) and Gil et al. (2015) have also used the finite-difference method in their studies.

10.7.2 Finite element

The finite element method is based on the subdivision into elements (finite elements) of the domain of the integral form of the partial differential equation (PDE). It was developed in the 1950s by Jonathan Turner and his colleagues at Boeing in the aerospace industry. During the 1950 and 1960s, three academics (J.H. Argyris, R.W. Clough, and O. C. Zienkiewic) promoted the method brought from the aerospace industry to energy applications (Clough, 2004). The FEM is more commonly used in structural analysis although it is also applied in fluid dynamics. Solar receivers have structural elements, and so this method is relevant to their analysis.

Solutions can be approximated by piece-wise functions (linear combinations of approximation functions and unknown parameters) that satisfy the PDE. Piece-wise functions are substituted into the equations, and residuals are defined to measure the errors, that are minimized by multiplying them by a set of weighting functions and then integrating them accordingly. The sum of these integrals is zero. A set of DAE for the unknown coefficients of the piecewise functions is obtained and solved. The advantage of the finite-element method is its application to more complex geometries. Although this method presents some difficulties in fluids due to the coupling of conservations laws, it has found application even in the field of computational fluid dynamics.

For instance, consider the one-dimensional transport equation:

$$\frac{\partial T}{\partial t}+u\frac{\partial T}{\partial x}=D\frac{\partial^2 T}{\partial x^2},\ 0\leq x\leq L \tag{10.124}$$

Where u is a convective velocity and D is a diffusion coefficient.

Eq. (10.124) is the classic convection-diffusion problem for T (x, t). It can be simplified by assuming that u and D are constant, with the following boundary conditions (Hu, 2007, pp. 411–465):

$$T(0,t)=g \text{ and } \frac{\partial T\ (L,t)}{\partial t}=q \tag{10.125}$$

Where g and q are constants.

The initial conditions are:

$$T(x,0)=T_o(x) \text{ for } 0\leq x\leq L \tag{10.126}$$

10.7.2.1 Weak or variational form of partial differential equations

Let us first define a collection of trial solutions that consist of all function square-integrable first derivatives (H^1 functions, that is: $\int_0^L (T,x)^2dx<\infty,\ T\in H^1$) and satisfy the Dirichlet type of boundary condition (where the value of the variable is specified) at $x=0$. This is expressed as a trial space (S):

$$S=\{T|T\in H^1,\ T(0)=g\} \tag{10.127}$$

The variational space (V) of the trial solution is defined as:

$$V=\{w|\ w\in H^1,\ w(0)=0\} \tag{10.128}$$

Where w is a function in the variational space.

Eq. (10.128) requires a homogeneous boundary condition. Multiplying Eq. (10.124) by w and integrating the product over the domain where the problem is defined yields:

$$\int_0^L\left(\frac{\partial T}{\partial t}w\right)dx+u\int_0^L\left(\frac{\partial T}{\partial x}w\right)dx=D\int_0^L\left(\frac{\partial^2 T}{\partial x^2}w\right)dx \tag{10.129}$$

When we integrate the right-hand side of Eq. (10.129) by parts, we obtain:

$$\int_0^L \left(\frac{\partial T}{\partial t} w\right) dx + u \int_0^L \left(\frac{\partial T}{\partial x} w\right) dx + D \int_0^L \left(\frac{\partial T}{\partial x}\frac{\partial w}{\partial x}\right) dx = D\left[\frac{\partial T}{\partial x} w\right]_0^L \quad (10.130)$$

Applying the boundary conditions $\partial T/\partial x = q$ and $w(0) = 0$. Eq. (10.130) reduces to:

$$\int_0^L \left(\frac{\partial T}{\partial t} w\right) dx + u \int_0^L \left(\frac{\partial T}{\partial x} w\right) dx + D \int_0^L \left(\frac{\partial T}{\partial x}\frac{\partial w}{\partial x}\right) dx = D\, q\, w(L) \quad (10.131)$$

Eq. (10.131) is known as the weak form of the two-dimensional transport problem.

10.7.2.2 Galerkin's approximation and finite element interpolations

The strong and weak forms of the problem are equivalent. Finite element methods commence with the weak form of the problem. Finite-dimensional approximations of S and V, respectively denoted by S^h and V^h can be constructed. The superscript represents a discretization with a characteristic grid size h. Eq. (10.131) can be rewritten using these new spaces, as: Find $T^h \in S^h$ such that for all $w^h \in V^h$,

$$\int_0^L \left(\frac{\partial T^h}{\partial t} w^h\right) dx + u \int_0^L \left(\frac{\partial T^h}{\partial x} w^h\right) dx + D \int_0^L \left(\frac{\partial T^h}{\partial x}\frac{\partial w^h}{\partial x}\right) dx = D\, q\, w^h(L) \quad (10.132)$$

Usually, S^h and V^h are respectively subsets of S and V. This implies that if a function $\phi \in S^h$ then $\phi \in S$, and if another function $\psi \in V^h$ then $\psi \in V$. Therefore, Eq. (10.132) defines an approximate solution of T^h to the exact weak form of the problem. Up to the boundary condition $T(0) = g$, the function spaces S^h and V^h comprise identical collections of functions. This boundary condition can be removed by defining a new function:

$$v^h(x,t) = T^h(x,t) - g^h(x) \quad (10.133)$$

Where g^h is a specific function that satisfies the boundary condition $g^h(0) = g$. Consequently, v^h and w^h belong to the same space V^h.

Eq. (10.132) can be rewritten in terms of the new function v^h. Find $T^h = v^h + g^h$, where $v^h \in V^h$, such that for all $w^h \in V^h$,

$$\int_0^L \left(\frac{\partial v^h}{\partial t} w^h\right) dx + a\left(v^h, w^h\right) = Dqw^h(L) - a\left(v^h, w^h\right) \quad (10.134)$$

The operator a(.,.) is defined as:

$$a\left(v^h, w^h\right) = \int_0^L \left(\frac{\partial v}{\partial t} w\right) dx + D \int_0^L \left(\frac{\partial v}{\partial x} \frac{\partial w}{\partial x}\right) dx \tag{10.135}$$

Eq. (10.134) is called a Galerkin formulation because the solution and the variational functions belong to the same space. It should be mentioned that the Galerkin formulation of the problem is an approximation to the weak formulation. The solution function may be contained in a collection of functions other than V^h in other classes of approximation methods known as Petrov-Galerkin methods.

The next step is to explicitly construct the finite-dimensional variational space V^h. We assume that the dimension of the space is n and that the basis (shape or interpolation) functions for the space are:

$$N_m(x), \ m = 1, \ 2, \ 3, \ldots n \tag{10.136}$$

Each shape function must satisfy the boundary condition at $x = 0$, that is required by the space V^h:

$$N_m(0) = 0, \ m = 1, \ 2, \ 3, \ldots n \tag{10.137}$$

Any function $w^h \in V^h$ can be expressed as a linear combination of these shape functions:

$$w^h = \sum_{m=1}^{n} C_m N_m(x) \tag{10.138}$$

Where C_1, C_2, C_3,… C_n are coefficients that are independent of x and they uniquely define this function.

Let us introduce one additional function N_0 to specify the function g^h in Eq. (10.133) related to the critical boundary condition. This shape function has the following property:

$$N_0(0) = 1 \tag{10.139}$$

In view of this, the function g^h can be expressed as:

$$g^h = gN_0(x) \text{ and } g^h(0) = g \tag{10.140}$$

Using these definitions, we can express the approximate solution as:

$$v^h(x, t) = \sum_{m=1}^{n} d_m(t) N_m(x) \tag{10.141}$$

$$T^h(x,t) = \sum_{m=1}^{n} d_m(t)\, N_m(x) + gN_0(x) \qquad (10.142)$$

Where d_1, d_2, d_3, $\ldots d_m$ are functions of time only for time dependent problems.

10.7.2.3 Comparison with finite difference method

The Galerkin formulation (Eq. 10.134) leads to a coupled system of ordinary differential equations. Substitution of the expressions for the variational function (Eq. 10.138) and for the approximate solution formulation (Eq. 10.141) into the Galerkin formulation yields:

$$\begin{aligned}\int_0^L \left(\sum_{r=1}^{n} \dot{d}_r N_r \sum_{m=1}^{n} C_m N_m\right) dx + a\left(\sum_{m=1}^{n} C_m N_m, \sum_{r=1}^{n} d_r N_r\right) \\ = Dq \sum_{m=1}^{n} C_m N_m(L) - a\left(\sum_{m=1}^{n} C_m N_m, gN_0\right)\end{aligned} \qquad (10.143)$$

Where $\dot{d}_r = d(d_r)/dt$

Upon rearrangement of terms, Eq. (10.143) can be simplified to:

$$\sum_{m=1}^{n} C_m G_m = 0 \qquad (10.144)$$

$$G_m = \sum_{r}^{n} \dot{d}_r \int_0^L (N_m N_r) dx + \sum_{r=1}^{n} d_r a(N_m, N_r) - Dq\, N_m(L) + ga(N_m, N_0) \qquad (10.145)$$

The coefficients, $C_1, C_2, C_3, \ldots, C_n$ should be arbitrary because the Galerkin formulation must hold for all possible functions of $w^h \in V^h$. So, the necessary condition for Eq. (10.145) to hold is that each G_m must be equal to zero:

$$\sum_{r=1}^{n} \dot{d}_r \int_0^L (N_m N_r) dx + \sum_{r=1}^{n} d_r a(N_m, N_r) = Dq\, N_m(L) + ga(N_m, N_0) \qquad (10.146)$$

Where $m = 1, 2, 3, \ldots, n$.

Eq. (10.146) constitute a system of n first-order ordinary differential equations (ODEs) for $d_1, d_2, \ldots, d_n$.

We can use a finite difference approximation for the time derivative term in the equation at the start of the process, say, by replacing $\partial T/\partial t$ with $(T^{j+1} - T^j)/\Delta t$, and apply the finite element scheme to discretize the resulting equation. Let us consider the real construction of the shape functions for the finite dimensional variational space. The easiest example is to use piecewise-linear finite element space. First, we partition the domain [0, L] into n distinct subintervals (elements). A typical space interval is denoted by $[x_m, x_{m+1}]$. The shape functions associated with the interior nodes, $m = 1, 2, \ldots, n - 1$, are defined as:

$$N_m(x) = \begin{cases} \dfrac{x - x_{m-1}}{x_m - x_{m-1}}, & \text{for } x_{m-1} \le x \le x_m \\ \dfrac{x_{m+1} - x}{x_{m+1} - x_m}, & \text{for } x_m \le x \le x_{m+1} \\ 0, & \text{elsewhere} \end{cases} \quad (10.147)$$

In addition, for the boundary nodes, the shape functions are defined as:

$$N_n(x) = \frac{x - x_{n-1}}{x_n - x_{n-1}}, \quad x_{n-1} \le x \le x_n \quad (10.148)$$

$$N_0(x) = \frac{x_1 - x}{x_1 - x_0}, \quad x_0 \le x \le x_1 \quad (10.149)$$

It should be noted that these shape functions have very compact (local) support and they satisfy $N_m(x_r) = \delta_{mr}$, where δ_{mr} is the Kronecker delta (that is, $\delta_{mr} = 1$ if $m = r$, whereas $\delta_{mr} = 0$ if $m \neq r$). The coefficients, $d_1, d_2, \ldots, d_n$, in the expression for the approximate solution (Eq. 10.142) represent the values of T^h at the nodes

$$x = x_m(i = 1, 2, \ldots, n), \text{ or}: \ d_m = T^h(x_m) = T_m \quad (10.150)$$

We substitute Eq. (10.147) into Eq. (10.146) and evaluate the integrals to compare the discretized equations generated from the finite element and difference methods. For an interior node x_m ($m = 1, 2, \ldots, n - 1$), we obtain:

$$\frac{d}{dt}\left(\frac{T_{m-1}}{6} + \frac{2T_m}{3} + \frac{T_{m+1}}{6}\right) + \frac{u}{2h}(T_{m+1} - T_{m-1}) - \frac{D}{h^2}(T_{m-1} - 2T_m + T_{m+1}) = 0 \quad (10.151)$$

Where h is the uniform mesh size.

Using the finite difference method, the transport equation can be discretized as:

$$\frac{T_m^j - T_m^{j-1}}{\Delta t} + u\frac{T_{m+1}^j - T_{m-1}^j}{2\Delta x} = D\frac{T_{m+1}^j - 2T_m^j + T_{m-1}^j}{(\Delta x)^2} \quad (10.152)$$

The convective and diffusive terms in Eq. (10.151) have the same forms as those discretized using the standard second-order finite difference method (Eq. 10.152). Nevertheless, the time derivative term is presented with a three-point spatial average of the variable T in the finite-element method that differs from the finite-difference method. The Galerkin finite-element formulation is generally equivalent to the finite difference method.

Many authors have applied the finite-element method in the development of solar receivers. For instance, Teng and Xuan (2018) proposed a novel volumetric receiver with mirrors to reflect radiative heat from the porous medium of the receiver. The use of mirrors was aimed at reducing radiative heat loss from the receiver. The design of this receiver was relatively complex, and these authors used a finite element-based software to investigate the thermal and hydrodynamic performance of the receiver. Theoretically, the receiver had a high thermal efficiency of 88.6% at a high outlet temperature of 1352 K. This good performance was ascribed to the recycle of radiative heat.

10.7.3 Finite volume

The finite volume method (FVM) started as an extension of the finite difference method from the integral formulation of the conservation equations. This work was mostly led by the group of Brian Spalding at the Imperial College from mid-1960s (two-dimensional flows) to mid-1970s (three dimensional flows) and ended in 1972 with the formulation of the Semi-Implicit method for pressure linked equations (SIMPLE) algorithm that revolutionized the computational fluid dynamics (CFD) practice (Sandá et al., 2019). CFD is a branch of science that, aided by digital computers, yields quantitative predictions of fluid-flow phenomena based on conservation laws (of mass, momentum, and energy) governing fluid motion.

In the FVM, the domain is divided into finite volumes, where the partial differential equation is integrated, and transformed into a system of differential algebraic equations. The variables are located at the control volume center. Some kind of interpolation is applied to express the values of those on the surfaces. The discontinuities in the variables do not pose challenges if the mesh is defined in such a way that they take place in the borders of the control volume.

The finite volume scheme is probably the most widespread method in CFD because of its simplicity as all the terms that must be approximated have a physical meaning. For instance, Becker et al. (2006) studied flow instabilities in a volumetric receiver with porous medium. They used a finite volume method in combination with a heterogeneous heat transfer model for the porous body. Their findings show that heat conductivity and permeability properties of the porous materials have significant influence on the probability of the occurrence of flow instabilities.

10.8 Economic performance

Sustainable development involves interrelations among society, economy, and environment. In view of this, the United Nations established the Sustainable Development Goal 7 (SDG 7) to "ensure access to affordable, reliable, sustainable and modern energy for all" by 2030 (Munro et al., 2017). The goal of increasing access to electricity is also connected to climate change policies that aim at achieving a balance between anthropogenic emissions by sources and removals by sinks of greenhouse gases by the second half of the 21st century at global level (UNFCCC, 2015). Worldwide, coal is the largest carbon emitter in the power sector. Consequently, replacement of this fossil fuel is necessary to achieve deep decarbonization. Renewable energy can play a crucial role in mitigating climate change and raising energy security. Nevertheless, the main challenge is how to meet these targets in a cost-effective manner. Carbon-neutral electricity can compete against coal power if its generation costs are lower than approximately 0.08 USD/kWh (IRENA, 2013). Solar receivers contribute to the cost of electricity produced by a CSP system. For instance, Schwarzbözl et al. (2006) reported costs (per unit area of receiver aperture) of 16,000 EURO/m^2, 33,000 EURO/m^2 and 37,500 EURO/m^2 for low-, medium-, and high-temperature receivers. Therefore, the economic performance of a receiver is usually analyzed as an integral part of the whole CSP system and decreasing the cost of the receiver results in the reduction of the overall system cost.

There are different methods for determining indicators of economic performance. They include the net present value (NPV), discounted pay-back (DPB), internal rate of return (IRR), and levelized cost of energy (LCOE). Nevertheless, many researchers have used the LCOE method to evaluate the economic competitiveness of energy technologies. It is the ratio of the sum of all costs incurred to the sum of all the energy produced during the lifetime of the project. This parameter yields the average unit cost of energy generated by a power plant over its lifetime and enables the comparison of different energy technologies of unequal lifespans and capacities. The LCOE can be given by Zhao et al. (2017):

$$\mathrm{LCOE} = \left[\sum_{t=0}^{t=N} C_t(1+i)^{-t}\right]\left[\sum_{t=0}^{t=N} E_t(1+i)^{-t}\right]^{-1} \tag{10.153}$$

Where C_t is the total incurred cost (capital, operational and maintenance) and E_t is the energy produced during the period t.

For a CSP system, E_t is equal to the product of the effective solar radiation measured on the aperture plane (G_{ap}), optical efficiency (η_{opt}), receiver thermal efficiency (η_{rec}), and power block efficiency (η_{pb}) as shown in Eq. (1.31). Thus, E_t increases with the thermal efficiency of the receiver, thereby reducing the LCOE. It can be predicted from the design, operational, and environmental parameters.

Some advantages of the LCOE over NPV are: (a) absence of restrictions on project scale, (b) LCOE is independent of power generation technologies, and (c) LCOE is applicable even when energy technologies are different. A range of LCOE values have been reported, from 0.06 US$/kWh for a combined cycle gas turbine driven by a solar tower (Kribus et al., 1998) to 0.40 US$/kWh for solar hybrid steam injected gas turbine driven by a parabolic trough concentrator (Selwynraj et al., 2015a, 2015b).

10.9 System optimization

The aim of optimization is to maximize or minimize a given function relative to a set of available choices under a specified situation. The function being optimized is known as the objective function. Choices may include several variables with their possible ranges (say, area of collector aperture). So, it is important to answer pertinent questions before formulating an optimization problem, and some of these questions may relate to processes and components (Bejan et al., 1996). In this vein, the first step in the optimization process is to consider the elements of a receiver and the processes that occur in it. Optimization problems fall into single-objective and multi-objective categories (Cui et al., 2017). In a single-objective optimization problem, only one objective function is defined. Two or more objective functions are defined in multiobjective optimization problems, and these functions are simultaneously solved. In addition, an objective function may be linear or nonlinear. Minimization and maximization functions are transferrable (Gong et al., 2009) as shown in Eq. (10.154). The transferability of a function is critical in multiobjective optimization problems with mixed extrema (where some objective functions require minimization while others simultaneously require maximization).

$$\max \{f(\mathbf{x})\} \Leftrightarrow \min \{-f(\mathbf{x})\} \tag{10.154}$$

Where $\mathbf{x}$ is a matrix.

Boundaries of the system are defined in the second step of the optimization process. It is also necessary to select optimization criteria. Thirdly, independent variables are selected, while a mathematical model is formulated based on the objective and independent variables in the fourth step. Finally, it is necessary to find solutions to the optimization problem.

10.9.1 Elements and processes

As shown in Section 10.1.1, the central element of a solar receiver is the absorber. A heat transfer fluid flows through the receiver to extract useful heat. In addition, some receiver designs have transparent and insulation elements. Major technical processes of this device can broadly be classified into optical and thermal processes.

The optical characteristics of a receiver significantly influence the thermal efficiency of the receiver and overall performance of the CSP system. A high level of solar absorption is desirable. Details of parameters that affect the optical efficiency of a solar receiver are discussed in Section 10.2. Useful heat output is an important indicator of the receiver performance. In this context, Chapter 7 and Section 10.3 focus on thermodynamic parameters. A heat transfer fluid is passed through the receiver for heat evacuation, and consequently, it is important to consider thermal hydraulic parameters when optimizing the performance of receivers. Governing equations of hydrodynamics are presented and discussed in Chapter 8 and Section 10.4. Heat transfer processes are vital in the distribution of the useful heat from the receiver to other components of the CSP system. Nevertheless, heat losses are also incurred through mechanisms of heat transfer. Models for computation of heat transfer are included in Section 10.5. Solar receivers operate at elevated temperatures that induce thermal stresses. In this context, mathematical models of thermo-mechanical performance have been presented in Chapter 9 and Section 10.5.

10.9.2 System boundaries

A solar receiver comprises several elements that interact with one another. So, it is important to define the boundaries of the system that requires optimization. All the elements (such as the absorber, heat transfer fluid, insulation and transpaent cover) that have significant impact on the performance of the system must be included in the optimization exercise to achieve a high level of system performance. The number of elements in a receiver varies with the design. For example, some receiver designs do not have transparent covers.

Due to the complexity of the receiver, it may be necessary to decompose the system into smaller subsystems that are amenable to sub-optimization, a process of optimizing one aspect of the problem while neglecting other factors that also influence the performance of the system. Nevertheless, sub-optimization does not necessarily guarantee optimization of the entire system. A possible solution to this constraint is the application of an iterative approach (Bejan et al., 1996).

10.9.3 Optimization criteria

Optimization criteria are properties based on which a system design can be evaluated and optimized. These criteria may be classified into social, economic, technological, and environmental categories. Social criteria may include technology acceptance and other factors. The economic dimension is important in promoting the diffusion of the technology. Technological factors of a receiver influence its economic and other aspects of performance, and they include optical and thermal factors. Protection of the environment is also an important consideration in receiver development. For instance, the effect of receiver efficiency on environmental protection (such as avoidance cost of carbon dioxide).

An optimized receiver design is expected to exhibit minimum values of variables such as total capital investment, levelized cost of energy, production time, and rate of pollution. On the other hand, optimal system design should give maximum values of variables such as levelized profit, power output, and thermodynamic efficiency. Some of these criteria conflict with each other. For instance, attempts to increase the thermodynamic efficiency of a receiver tends to raise the capital investment of the system. Consequently, optimization with one objective is not adequate to handle a real system with a high degree of complexity (Nguyen et al., 2017). Multicriteria optimization is suitable to ensure that the influences of all important criteria are simultaneously considered.

10.9.4 Optimization models

Solar receivers have elements and processes that interact in complex ways, and so, they require suitable mathematical models for design and performance appraisal. Considerations for selecting independent variables are (Bejan et al., 1996): (a) inclusion of all the variables that significantly affect the performance and cost of the system, (b) exclusion of fine detail or variables with insignificant effect, and (c) distinguishing decision variables from parameters among independent variables. The values of decision variables can be changed while those of parameters remain fixed in each model statement. It is necessary to select independent variables that fully characterize the potential options of the system design. Basic principles are usually applied in formulating a suitable optimization model that is made up of: (a) an objective function that needs maximization or minimization and (b) constraints that may comprise equations and inequalities.

An objective function can be defined as (Cui et al., 2017):

$$\min \mathbf{y} = f(\mathbf{x}) = [f_1(\mathbf{x}), f_2(\mathbf{x}), \ldots, f_m(\mathbf{x})]^T \text{ objective functions} \tag{10.155}$$

Subject to (s.t.):

$$\begin{cases} g_j(\mathbf{x}) \le 0, \ j = 1,\ 2,\ \ldots,\ p \\ h_k(\mathbf{x}) = 0, \ k = 1,\ 2,\ \ldots,\ q \\ x_i^{min} \le x_i \le x_i^{max}, \ i = 1,\ 2,\ \ldots,\ n \text{ constraint functions} \\ x = [x_1,\ x_2, \ldots, x_n]^T \in \Theta \\ y = [y_1,\ y_2, \ldots, y_m]^T \in \Lambda \end{cases} \tag{10.156}$$

Where m is the number of objective functions, the superscript T represents the transpose of the row matrix, $\mathbf{\Theta}$ is an n-dimensional search space which is determined by the upper limit of $\mathbf{x}^{max} = [x_1^{max},\ x_2^{max}, \ldots, x_n^{max}]^T$ and the lower limit $\mathbf{x}^{min} = [x_1^{min},\ x_2^{min}, \ldots, x_n^{min}]^T$ of the decision variables x_i, $\mathbf{\Lambda}$ is an m-dimensional vector space of objective functions that is determined by $\mathbf{\Theta}$ and the objective function f($\mathbf{x}$). There are p inequality and q equation constraints. If $p = q = 0$, then the optimization problem is unconstrained. If all the functions are linear, then the problem is known a linear optimization problem. If some of the functions are nonlinear then the problem is said to be nonlinear. For a single objective optimization problem, $m = 1$.

In a solar receiver, the inequality constraints $g_j(\mathbf{x}) = 0$ may arise from consideration of material, equipment, safety, and other limits. For example, the temperature of a structure such as an absorber element must be less than a certain level to prevent failure of the absorber structure and other associated problems. Similarly, the working pressure in a receiver is confined to a certain limit. The equality constraints $h_k(\mathbf{x}) = 0$ may result from variables such as mass, energy and momentum balances. These equations can be in algebraic and/or differential forms.

Example

Consider a CSP plant driven by a solar tower. It is desirable to optimise the output power (Q) and cost of energy (C) subject to changes in the aperture (A_a) of concentrator, mass flow rate of the heat transfer fluid ($\dot{m}$), and thermal efficiency (η_{rec}) of the receiver.

Solution:

The output power and cost of energy are conflicting objectives. Generally, attempts to increase output power tend to raise the cost of energy. Therefore, output power must be maximized while the cost of energy must be minimized. It is noted that maximizing Q is equivalent to minimizing $(1/Q)$. So, this optimization problem can be formulated as a multiobjective minimization problem:

$$\min \mathbf{y} = \begin{pmatrix} 1/Q \\ C \end{pmatrix} \text{ objective function} \tag{10.157}$$

$$\mathbf{x} = \begin{pmatrix} A_a \\ \dot{m} \\ \eta_{rec} \end{pmatrix} \text{ decision variables} \tag{10.158}$$

10.9.5 Selected studies

Mlatho and McPherson (2012) modeled the performance of two designs of volumetric receivers with a wire mesh absorber by using CFD. The first receiver was designed based on the geometry of a round-bottomed flask (with parabolic surface), while the second one was based on the geometry of a cylindrical cavity. Oil was used as a heat transfer medium in both designs. Three-dimensional models of the receivers were created and meshed by using Gambit software and then exported to the Fluent software for solving the governing equations of mass, momentum, and energy. The model was validated by using experimental data. In this study, an optimal mesh was found to have a porosity of about 0.95%, with good agreement between simulated and experimental results.

Capuano et al. (2017) studied the performance of a volumetric solar receiver with a square channel made of siliconized silicon carbide (SiSiC). They observed that such geometry was the best compromise for large-scale application because of the low production costs, easy manufacturability through extrusion procedure, and overall acceptable performance. Nevertheless, this geometry is not able to suppress the radiative heat losses, mainly from the front surface. So, these authors aimed at optimizing the geometry to reduce heat losses and improve the efficiency of the system. In this vein, they formulated a numerical model based on the Navier–Stokes equations for the fluid phase and Fourier law for the solid phase in three dimensions. They achieved an optimized absorber geometry that could raise the final thermal efficiency by more than 12% compared to the current state-of-the-art. The outlet fluid temperature of the optimized receiver was higher than the solid inlet temperature, which indicates achievement of the volumetric effect. A laboratory scale-prototype type was developed and tested, with good agreement (deviation of 3%) between experimental and numerical results.

Zhang et al. (2020) used a combined optics and heat transfer model to simulate the influence of the conical angle, loop number, and insulation thickness on the performance of a conical cavity receiver. The objective of this study was to optimize the performance of the receiver. Their model was validated by comparing the simulation results with published experiment data, and there was good agreement between the two data set. Findings also indicate that the optimal conical angle is 5 degrees with an overall efficiency of 63.6%. Both thermal efficiency and overall efficiency increase as the loop number rises from 4 to 15. Nevertheless, the optical efficiency decreases from 87.5% to 87.1% as the number of loops increase from 4 to 15. From this study, the optimal value of insulation thickness is 175 mm, which achieves an overall efficiency of 70.4%.

10.10 Simulation programs

Systems of equations generated from discretization and optimization processes are often so complex that solutions can only be obtained by using computer programs. For instance, heavily coupled equations of thermodynamic, hydrodynamic, heat transfer, and thermomechanical variables must be solved simultaneously. This computation cannot be done by hand, and so, many simulation programs have been developed over the time.

Basically, simulation programs fall into two main categories: (a) special purpose and (b) general purpose (Nafey, 2005). A special purpose program is developed to simulate a specific process, with no flexibility. Any change to the process may demand extensive modifications to the program. Nonetheless, the advantage of a special purpose program is the simplicity of developing a mathematical model that adequately describes a real system. These programs or codes are commonly written by using programming languages such as

BASIC, FORTRAN, Pascal, and MATLAB (Burden & Faires, 1985; Vestlund et al., 2009). Tailored details can be included in the code for a specific purpose.

A general-purpose program is developed for simulation of different system configurations and operating conditions. The model, in this case, comprises a set of equations for a given unit of the system. Each set of equations constitutes a module that may be stand-alone and used in any process where it is required. For example, a set of equations that cover the thermodynamics of a system. Hence, a general purpose program comprises different modules, and it may be irrelevant to other computational problems due to its generality. Examples of well-known general purpose programs for simulating energy systems include ANSYS CFX, EMPG2, EnergyPlus, and TRNSYS.

ANSYS CFX is a CFD software that can solve a complex system of equations (including Navier–Stokes equations) with a high degree of accuracy. This software finds application in single-phase nonreacting flows, heat transfer, shape optimization, multiphase surface flows, reacting flows and combustion, dispersed multiphase flows, turbomachinery, aircraft icing, particle flows, high performance computing (HPC) for fluids, fluid-structure interaction (FSI), high rheology material, and thermal FSI (ANSYS, 2017). Consequently, the software has been used for studying systems with coupled aero-thermal-structure characteristics (Xiaodong & Xiuli, 2015). EMPG2 has features and capabilities for simulating a range of thermal systems, system control strategies and load types (Nafey, 2005), while TRNSYS is a modular program for simulating solar thermal systems (Duffie & Beckman, 2013; Klein et al., 1975). EnergyPlus is mainly suitable for building simulation (Loutzenhiser et al., 2007). Nevertheless, the choice to use a specific program depends on factors such as cost and computational capabilities of the program.

Nomenclature

A area (m^2)
C_p specific heat capacity ($J\ kg^{-1}K^{-1}$)
E energy flow (Wm^{-2})
f focal length (m)
F_{i-j} view factor of the ith surface relative to the jth surface
F_R flow factor (dimensionless)
F_s Fraction of absorbed solar radiation (dimensionless)
G flux (Wm^{-2})
g acceleration due to gravity (ms^{-2})
Gr Grashof number (dimensionless)
H depth of nanofluid (m)
h heat transfer coefficient ($Wm^{-2}K^{-1}$)
h_m coefficient of convective mass transfer (ms^{-1})
I intensity (Wm^{-2})
k thermal conductivity ($Wm^{-1}K^{-1}$)
K coefficient of absorption (m^{-1})
L length (m)

Nu Nusselt number (dimensionless)
Pr Prandtl number (dimensionless)
Q power (W)
q flux (Wm^{-2})
$\mathbf{q_v}$ rate at which energy is generated per unit volume of the solid medium ($W\ m^{-3}$)
q′ thermal power per unit length of collector (Wm^{-1})
S absorbed solar radiation (Wm^{-2})
t time (s)
T temperature (K)
U coefficient of heat loss ($Wm^{-2}K^{-1}$)
V velocity ($m\ s^{-1}$)
$\hat{\mathbf{E}}$ exergy flow (W)
$\dot{\mathbf{m}}$ mass flowrate ($kg\ s^{-1}$)
α absorptance (dimensionless)
β coefficient of thermal expansion (K^{-1})
γ volumetric coefficient of thermal expansion (K^{-1})
ε emittance (dimensionless)
η efficiency (dimensionless)
ϕ density ($kg\ m^{-3}$)
λ wavelength (m)
μ dynamic viscosity (Pa s)
ρ reflectance (dimensionless)
τ transmittance (dimensionless)
ν kinematic viscosity ($m^2\ s^{-1}$)
ξ thermal diffusivity ($m^2\ s^{-1}$)

Subscripts

a air/ambient
ab absorbed/absorber
ai inner surface of absorber tube
ao outer surface of absorber tube
ap aperture
c convective/cover
ci inner surface of transparent cover
co outer surface of transparent cover
eff effective
f fluid
r radiative
s surface
w wall

References

Alameri, S. A., & Alkaabi, A. K. (2020). Fundamentals of nuclear reactors. In S. U.-D. Khan, & A. Nakhabov (Eds.), *Nuclear Reactor Technology Development and Utilization* (pp. 27–60). Elsevier Ltd.

Alobaid, F., Mertens, N., Starkloff, R., Lanz, T., Heinze, C., & Epple, B. (2017). Progress in dynamic simulation of thermal power plants. *Progress in Energy Combustion Science, 59*, 79–162. https://doi.org/10.1016/j.pecs.2016.11.001

ANSYS. (2017). *Fluids: Computational fluid dynamics*. http://www.ansys.com/products/fluids.

ASHRAE. (2017). *ASHRAE handbook-fundamentals*. American Society of Heating, Refrigerating and Air-Conditioning Engineers (ASHRAE).

Becker, M., Fend, T., Hoffschmidt, B., Pitz-Paal, R., Reutter, O., Stamatov, V., Steven, M., & Trimis, D. (2006). Theoretical and numerical investigation of flow stability in porous materials applied as volumetric solar receivers. *Solar Energy, 80*, 1241–1248. https://doi.org/10.1016/j.solener.2005.11.006

Behar, O., Khellaf, A., & Mohammedi, K. (2013). A review of studies on central receiver solar thermal power plants. *Renewable and Sustainable Energy Reviews, 23*, 12–39. https://doi.org/10.1016/j.rser.2013.02.017

Bejan, A., Tsatsaronis, G., & Moran, M. (1996). *Thermal design and optimization*. New York: John Wiley & Sons, Inc.

Bergman, T. L., Lavine, A. S., Incropera, F. P., & DeWitt, D. P. (2011). *Fundamentals of heat and mass transfer* (7th ed.). New Jersey: John Wiley & Sons.

Bohren, C. F., & Huffman, D. R. (1998). *Absorption and scattering of light by small particles*. New York: Wiley-VCH.

Bonilla, J., Yebra, L. J., Dormido, S., & Zarza, E. (2013). *Modeling and simulation of two-phase flow evaporators for parabolic-trough solar thermal power plants*. Madrid: CIEMAT.

Burden, R. L., & Faires, J. D. (1985). *Numerical analysis* (3rd ed.). Boston: PWS Publishers.

Capuano, R., Fend, T., Schwarzbözl, P., Smirnova, O., Stadler, H., Hoffschmidt, B., & Pitz-Paal, R. (2016). Numerical models of advanced ceramic absorbers for volumetric solar receivers. *Renewable and Sustainable Energy Reviews, 58*, 656–665. https://doi.org/10.1016/j.rser.2015.12.068

Capuano, R., Fend, T., Stadler, H., Hoffschmidt, B., & Pitz-Paal, R. (2017). Optimized volumetric solar receiver: Thermal performance prediction and experimental validation. *Renewable Energy, 114*, 556–566. https://doi.org/10.1016/j.renene.2017.07.071

Clough, R. W. (2004). Early history of the finite element method from the view point of a pioneer. *International Journal of Numerical Methods in Engineering, 60*, 283–287. https://doi.org/10.1002/nme.962

Collado, F. J., & Guallar, J. (2012). Campo: Generation of regular heliostat fields. *Renewable Energy, 46*, 49–59.

Crepeau, J. (2007). Joseph Stefan: His life and legacy in the thermal sciences. *Experimental Thermal Fluid Science, 31*, 795–803.

Cui, Y., Geng, Z., Zhu, Q., & Han, Y. (2017). Review: Multi-objective optimization methods and application in energy saving. *Energy, 125*, 681–704.

Drew, D. A., & Passman, S. L. (1998). *Theory of multicomponents fluids*. New York: Springer.

Duffie, J. A., & Beckman, W. A. (2013). *Solar engineering of thermal processes* (4th ed.). New Jersey: John Wiley & Sons Inc.

Evans, D. L. (1977). On the performance of cylindrical parabolic solar concentrators with flat absorbers. *Solar Energy, 19*, 379.

Fend, T., Hoffschmidt, B., Pitz-Paal, R., & Reutter, O. (2005). Cellular ceramics use in solar radiation conversion. In M. Scheffer, P. Colombo, & edts) (Eds.), *Cellular ceramics: Structure, manufacturing and applications*. Weinheim: Willey-VCHGmbH & Co. KgA.

Gil, R., Monné, C., Bernal, N., Muñoz, M., & Moreno, F. (2015). Thermal model of a dish stirling cavity-receiver. *Energies, 8*, 1042–1057. https://doi:10.3390/en8021042

Goebel, O. (1999). Thermohydraulics of direct steam generation. *Le Journal de Physique IV, 9*(Pr3), 481–487. https://doi.org/10.1051/jp4:1999376

Gong, M., Jiao, L., Yang, D., & Ma, W. (2009). Research on evolutionary multi-objective optimization algorithms. *Journal of Software, 20*(2), 271–289.

Houbak, N. (1995). Simulation of energy systems—guidelines and pitfalls. *Mathematics and Computers in Simulation, 39*, 559–564.

Hu, H. H. (2007). *Computational fluid dynamics. Fluid mechanics* (pp. 411–465). Elsevier Science & Technology.

Incropera, F. P., DeWitt, D. P., Bergman, T. L., & Lavine, A. S. (2007). *Fundamentals of heat and mass transfer* (6th ed.). New Jersey: John Wiley & Sons.

IRENA. (2013). *Southern African power pool: Planning and prospects for renewable energy*. Abu Dhabi: International Renewable Energy Agency (IRENA).

Kays, W. M., Crawford, M. E., & Weigand, B. (2005). *Convective heat and mass transfer* (4th ed.). Boston: McGraw-Hill Higher Education.

Klein, S. A. (1975). Calculation of flat-plate collector loss coefficients. *Solar Energy, 17*, 79–80.

Kleinstreuer, C. (2003). *Two-phase flow: Theory and applications* (1st ed.). CRC Press.

Kribus, A., Zaibel, R., Carey, D., Segal, A., & Karni, J. (1998). A solar-driven combined cycle power plant. *Solar Energy, 62*, 121–129.

Lenert, A., & Wang, E. N. (2012). Optimization of nanofluid volumetric receivers for solar thermal energy conversion. *Solar Energy, 86*(1), 253–265. https://doi.org/10.1590/1980-5373-MR-2017-0556

Loutzenhiser, P. G., Manz, H., Felsmann, C., Strachan, P. A., Frank, T., & Maxwell, G. M. (2007). Empirical validation of models to compute solar irradiance on inclined surfaces for building energy simulation. *Solar Energy, 81*, 254–267. https://doi.org/10.1016/j.solener.2006.03.009

Mlatho, J. S. P., & McPherson, M. (2012). Performance optimisation of solar receivers that use oil as a heat transfer fluid. *International Journal of Sustainable Energy, 2012*, 1–19. https://doi.org/10.1080/14786451.2012.716060

Munro, P., Van der Horst, G., & Healy, S. (2017). Energy justice for all? Rethinking sustainable development goal 7 through struggles over traditional energy practices in Sierra Leone. *Energy Policy, 105*, 635–641. https://doi.org/10.1016/j.enpol.2017.01.038

Nafey, A. S. (2005). Simulation of solar heating systems—an overview. *Renewable and Sustainable Energy Reviews, 9*, 576–591.

Nguyen, H. A., Iperen, Z. V., Raghunath, S., Abramson, D., Kipouros, T., & Somasekharan, S. (2017). Multi-objective optimisation in scientific workflow. *Procedia of Computer Science, 108*, 1443–1452.

Querol, E., Gonzalez-Regueral, B., & Perez-Benedito, J. L. (2013). *A practical to exergy and Thermoeconomic analyses of industrial processes*. London: Springer.

Richardson, L. F. (1911). The approximate arithmetical solution by finite differences of physical problems involving differential equations, with an application to the stresses in a masonry dam. *Philosophical Transactions of the Royal Society of London. Series A, Containing Papers of a Mathematical or Physical Character, 210*, 307–357. https://doi.org/10.1098/rsta.1911.0009

Rodríguez-Sánchez, M. R., Sánchez-González, A., & Santana, D. (2018). Feasibility study of a new concept of solar external receiver: Variable velocity receiver. *Applied Thermal Engineering, 128*, 335–344. https://doi.org/10.1016/j.applthermaleng.2017.08.173

Sánchez, M., & Romero, M. (2006). Methodology for generation of heliostat field layout in central receiver systems based on yearly normalized energy surfaces. *Solar Energy, 80*, 861–874.

Sandá, A., Moya, S. L., & Valenzuela, L. (2019). Modelling and simulation tools for direct steam generation in parabolic trough solar collectors: A review. *Renewable and Sustainable Energy Reviews, 113*, 109226. https://doi.org/10.1016/j.rser.2019.06.033

Schwarzbözl, P., Buck, R., Sugarmen, C., Ring, A., Crespo, M. J., Altwegg, P., & Enrile, J. (2006). Solar gas turbine systems: Design, cost and perspectives. *Solar Energy, 80*, 1231–1240.

Selwynraj, A. I., Iniyan, S., Polonsky, G., Suganthi, S., & Kribus, A. (2015). Exergy analysis and annual exergetic performance evaluation of solar hybrid STIG (steam injected gas turbine) cycle for Indian conditions. *Energy, 80*, 414–427. https://doi.org/10.1016/j.energy.2014.12.001

Selwynraj, A. I., Iniyan, S., Polonsky, G., Suganthi, L., & Kribus, A. (2015). An economic analysis of solar hybrid steam injected gas turbine (STIG) plant for Indian conditions. *Applied Thermal Engineering, 75*, 1055–1064.

Serrano-Aguilera, J. J., Valenzuela, L., & Parras, L. (2014). Thermal 3D model for direct solar steam generation under superheated conditions. *Applied Energy, 132*, 370–382. https://doi.org/10.1016/j.apenergy.2014.07.035

Siebers, D. L., & Kraabel, J. S. (1984). *Estimating convective energy losses from solar central receivers. Sandia Report SAND84-8717*. Albuquerque: Sandia National Laboratories.

Stuetzle, T., Blair, N. J., Mitchell, J. W., & Beckman, W. A. (2004). Automatic control of a 30 MWe SEGS VI parabolic trough plant. *Solar Energy, 76*, 187.

Tan, T., & Chen, Y. (2010). Review of study on solid particle solar receivers. *Renewable and Sustainable Energy Reviews, 14*, 265–276. https://doi.org/10.1016/j.rser.2009.05.012

Teng, L., & Xuan, Y. (2018). Thermal and hydrodynamic performance of a novel volumetric solar receiver. *Solar Energy, 163*, 177–188. https://doi.org/10.1016/j.solener.2018.01.087

Too, Y. C. S., García, J., Padilla, R. V., Kim, J.-S., & Sanjuan, M. (2019). A transient optical-thermal model with dynamic matrix controller for solar central receivers. *Applied Thermal Engineering, 154*, 686–698. https://doi.org/10.1016/j.applthermaleng.2019.03.086

UNFCCC. (2015). *Paris agreement*. Paris: United Nations Framework Convention on Climate Change (UNFCCC).

Versteeg, H., & Malalasekera, W. (2007). *An introduction to computational fluid dynamics* (2nd ed.). Harlow: Pearson Education.

Vestlund, J., Rönnelid, M., & Dalenbäck, J. (2009). Thermal performance of gas-filled flat plate solar collectors. *Solar Energy, 83*, 896–904.

Wu, Z., Gong, B., Wang, Z., Li, Z., & Zang, C. (2010). An experimental and numerical study of the gap effect on wind load on heliostat. *Renewable Energy, 35*, 797–806. https://doi.org/10.1016/j.renene.2009.09.009

Wei, X., Lu, Z., Wang, Z., Yu, W., Zhang, H., & Yao, Z. (2010). A new method for the design of the heliostat field layout for solar tower power plant. *Renewable Energy, 35*, 1970–1975.

Xiaodong, Q., & Xiuli, S. (2015). Multidisciplinary design optimization of turbine disks based on ANSYS workbench platform. *Procedia Engineering, 99*, 1275–1283.

Yılmaz, İ. H., & Mwesigye, A. (2018). Modelling, simulation and performance analysis of parabolic trough solar collectors: A comprehensive review. *Applied Energy, 225*, 135–174. https://doi.org/10.1016/j.apenergy.2018.05.014

Zhang, Y., Xiao, H., Zou, C., Falcoz, Q., & Neveu, P. (2020). Combined optics and heat transfer numerical model of a solar conical receiver with built-in helical pipe. *Energy, 193*, 116775. https://doi.org/10.1016/j.energy.2019.116775

Zhao, Z., Chen, Y., & Thomson, J. (2017). Levelized cost of energy modeling for concentrated solar power projects: a China study. *Energy, 120*, 117–127. https://doi.org/10.1016/j.energy.2016.12.122

Chapter 11

Testing of solar receivers

11.1 Introduction

A concentrating solar collector is made up of concentrator and receiver components. The concentrator focuses solar radiation onto the receiver where it is absorbed and converted to heat. Evacuation of the heat is achieved by passing a heat transfer fluid through the receiver to other components of a concentrating solar power (CSP) technology. Thus, a solar receiver is central to CSP systems. The advancement of receivers involves designing and testing stages. A new product or process is often designed, tested, modified, and retested until a satisfactory final product is achieved (Grove & Davis, 1992). Testing is helpful in evaluating the effect of changes in design, operating, and environmental variables on the performance of receivers. In this vein, thermal efficiency is often used as a critical indicator of the receiver performance.

The useful thermal output power from a solar receiver can be calculated from the measurement of mass flow rate ($\dot{m}_{htf}$) and inlet ($T_{htf,in}$) and outlet ($T_{htf,out}$) temperatures of the heat transfer fluid. To compute the thermal efficiency of the receiver, the intensity of solar radiation intercepted by the receiver is required (Ebert et al., 2016). Empirical data on the magnitude of direct normal irradiance on the aperture of the concentrator can also be used to compute the level of solar irradiance on the receiver. This set of data can be used to compute the thermal efficiency of the receiver as follows:

$$\eta_{rec} = \frac{\dot{m}_{htf}\, C_{p,htf}\left(T_{htf,out} - T_{htf,in}\right)}{A_{rec} G_{rec}} \tag{11.1}$$

Where $C_{p,htf}$ is the specific heat capacity of the fluid, and G_{rec} is the intensity of solar radiation on the receiver.

The collector efficiency can also be expressed as a function of the inlet temperature:

$$\eta_{rec} = \frac{F_R\, A_{ap}\left[S - \frac{A_{rec}}{A_{ap}} U_L (T_{in} - T_a)\right]}{A_{rec} G_{rec}} \tag{11.2}$$

Solar Receivers for Thermal Power Generation. https://doi.org/10.1016/B978-0-323-85271-5.00011-2

If direct measurement of G_{rec} is not feasible, it can be estimated from measurements of the direct normal irradiance on the aperture of the concentrator. In this case, the irradiance on the solar receiver can be estimated from:

$$G_{rec} = \rho\, A_{ap} G_{ap} / A_{rec} \tag{11.3}$$

In general, the absorbed solar energy is $S = \eta_{opt} G_{ap}$. The optical efficiency (η_{opt}) depends on the concentrator geometry, concentrator optics, receiver geometry, and receiver optics. For the design with $\eta_{opt} = \rho(\alpha\tau\gamma)_n K_\theta$, the collector efficiency can be expressed as:

$$\eta_{rec} = \frac{F_R \left[S - \frac{1}{C_r} U_L (T_{in} - T_a) \right]}{\rho\, G_{ap}} \tag{11.4}$$

Where $C_r = A_{ap}/A_{rec}$ is the concentration ratio of the system.

It is also possible to express the collector efficiency as a function of the average temperature of the fluid ($T_{av} = 0.5(T_{in} + T_{out})$).

Example 11.1

A concentrating solar collector is tested outdoors under steady-state conditions with an inlet temperature of 350 K and ambient temperature of 290 K. The intensity of solar radiation on the concentrator is 850 W m^{-2}, and the overall heat loss is 10 $Wm^{-2}K^{-1}$. The system has a concentration ratio of 1000, an optical efficiency of 0.80 and a thermal efficiency of 0.72. The concentrator has a reflectance of 0.95. Calculate heat removal factor.

Solution

From Eq. (11.4), the heat removal factor (F_R) is:

$$F_R = \frac{\eta_{rec} \;\; \rho\, G_{ap}}{\left[S - \frac{1}{C_r} U_L (T_{in} - T_a) \right]}$$

$$S = \eta_{opt}\, G_{ap}$$

Therefore,

$$F_R = \frac{0.72 * 0.95 * 850}{\left[0.80 * 850 - \frac{1}{1000} * 10 * (350 - 290) \right]} = 0.86$$

$$F_R = \frac{1.5 * 0.95 * 0.72}{\left[0.85 * 4 * 10^6 - \frac{1}{2000} * 5 * (350 - 290) \right]}$$

It is necessary to apply standards and codes for credible test results. In this context, some standards exist for testing flat-plate and concentrating solar collectors under simulated or outdoor environmental conditions. Solar and other simulators enable flexibility in system testing because outdoor weather conditions are generally unstable. These standards are premised on the collector theory presented and discussed in Chapter 10. In some cases, receivers are tested as integral components of a CSP system. For example, an outdoor testbed called RESOL was developed to measure the optical efficiency of parabolic trough concentrator (PTC) receiver tubes at the Plataforma Solar de Almería (PSA) in Spain (López-Martín & Valenzuela, 2018). The testbed utilizes natural sunlight to heat the system, thereby providing a measurement of the optical efficiency of a whole receiver tube. Nevertheless, a review of previous work reveals the existence of a gap in standards for testing of full CSP systems (Madhlopa, 2018). Selected important projects on design and testing of solar receivers are presented in this chapter.

11.2 Measurement of variables for performance evaluation of solar receivers

The efficiency of a solar receiver is simply the ratio of useful energy to the input energy (solar radiation). Solar radiation is the most important environmental parameter, and procedures for its measurement have been reported in Section 2.6. Operational variables must also be measured to facilitate the evaluation of the receiver performance. These variables fall into the following categories: temperature, flow rate, pressure, thermophysical properties of heat transfer fluid, and quality of steam (Janotte et al., 2017).

11.2.1 Temperature

Part of the absorbed solar radiation raises the temperature of the heat transfer fluid flowing through a receiver. So, it is necessary to measure the inlet and outlet temperatures of the fluid and other components. Temperature measurements are taken by using a thermometer that may operate on the principles of resistance, thermal expansion of a liquid-in-glass, thermoelectric effect, and radiation (Nicholas & White, 1994).

Resistance thermometers have a very high level of resolution (up to 1×10^{-3} K) with a wide range of operation (14–1233 K). Nevertheless, they require an external voltage or current signal that creates errors associated with resistance-measuring sensors. Pt100 thermometers have a platinum coil with a resistance of 100 Ω at 273.15 K, and their specific resistance rises linearly with the temperature of the body. In industrial applications, resistance thermometers are the most widely exploited variety of thermometers. Liquid-in-glass thermometers operate on the principle of thermal expansion. They can be applied in many fields within the range of 83–873 K, with a resolution of up to

1×10^{-3}. However, they are limited to taking a few measurements because the readings are taken manually. Thermocouple thermometers operate on the principle of thermoelectric effect, are compatible with most data loggers (Delta-T Devices, 2000), and can be applied in many areas. Errors in thermocouples may be caused by thermal, inhomogeneity, isothermal, reference junction, interference, wire resistance, and linearization factors. The operational principle of radiation thermometers is that the electromagnetic radiation from an incandescent object depends on its temperature. The intensity of radiation increases with the temperature of the surface (Bergman et al., 2011). Consequently, radiation sensors can be used remotely without touching the surface of the test object. Solar receivers operate at elevated temperatures, and so, remote measurement of receiver temperature is more suitable. However, this method is usually less accurate than those techniques that use contact thermometers.

11.2.2 Flow rate

Mass flow rate of the heat transfer fluid is an input variable to the computation of enthalpy at various points in the flow circuit (within and outside the receiver). Thus, it is necessary to measure the flow rate of the fluid. Fluid flow measurement is the measurement of smoothly moving particles that fill and conform to the piping in an uninterrupted stream to establish the quantity flowing (LaNasa & Upp, 2014). Fluid flow meters are used to measure the flow rate, and they can be classified into the following major types as shown in Fig. 11.1 (Janotte et al., 2017): Coriolis, ultrasonic, vortex, orifice, and electromagnetic meters.

A Coriolis gas-flow meter is made up of one or more pipes with axially displaced section(s) that are induced to oscillate at a resonant frequency (Cai, 2013). Both the upstream and downstream parts of the displaced section oscillate in phase with each other when the fluid within the displaced section is at rest. The frequency of oscillation may vary between 50 Hz and 1 kHz, and it is affected by the total mass of the system (Henry et al., 2013). Only Coriolis flow meters can directly measure mass the flow rate and be operated under a maximum temperature of 673 K. This temperature limit is a drawback for taking flow measurements of the heat transfer fluid in parabolic dish and solar tower technologies that generally operate above this temperature limit.

There are two types of ultrasonic flow meters: (a) transit-time effect and (b) Doppler effect. In the transit-time ultrasonic flow meter, a pair of transducers is fitted with their centrelines inclined at angle θ to the axis of a fluid channel. The time difference ($\Delta t = t_{up}$-t_{do}) between the signal transmitted from upstream to downstream and the signal from downstream to upstream is proportional to the rate of fluid flow in the channel. The upstream (t_{up}) and downstream (t_{do}) times of signal propagation are given by (Zhou et al., 2018):

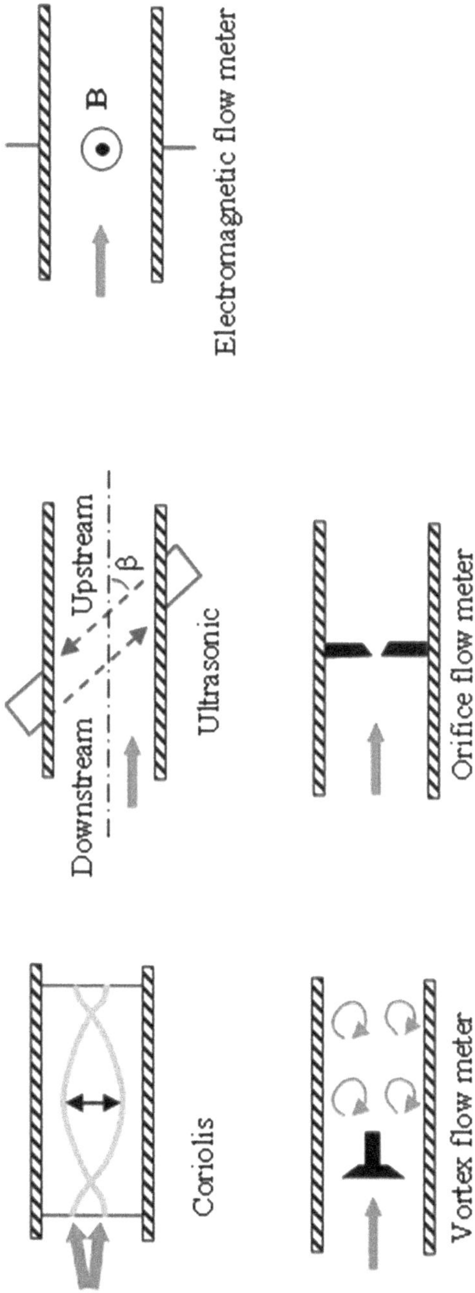

FIGURE 11.1 Principles of fluid flow measurement. *Adapted from Janotte et al. (2017).*

$$t_{up} = \frac{L}{V_u - V_{ag}\cos\theta} \tag{11.5}$$

$$t_{do} = \frac{L}{V_u + V_{ag}\cos\theta} \tag{11.6}$$

Where V_u is the speed of ultrasound, L is the path length between the two transducers, V_{ag} is the average flow velocity along the acoustic path, and θ is he angle of inclination of the center line of the transducers to the axis of the channel. The time-difference ultrasonic flow meter over-reads the values of gas flow rate when the gas flowing in the channel is wet (Xu et al., 2017). Nevertheless, it is the commonest type of flow meter.

An ultrasonic signal is transmitted across the flow channel and deflected by the particles inside the flow stream, in the principle of Doppler shift. The flowing particles cause a shift in the frequency of the propagating signal (Doppler effect) that is proportional to the flow speed of the particles. Some advantages of the ultrasonic flow meter are (Zhou et al., 2018): (a) absence of moving parts and therefore, avoidance of wear and tear associated with moving components, (b) absence of induced pressure drop across the component where flow rate is being measured, (c) application in measurements involving large diameter of pipes with a high degree of accuracy, and (d) possibility of bidirectional flow measurement.

Electromagnetic flowmeters operate based on the principle of Faraday's law that relates the electric potential induced in the measurement zone with a liquid velocity vector and a magnetic flux. An important requirement is that the liquid be able to conduct electricity. When the liquid flows through a channel, it induces a voltage U between two electrodes fitted on the walls of the channel. The induced voltage is defined by (Watral et al., 2015):

$$U = \int \mathbf{v} \text{ x } (\mathbf{B} \text{ x}\mathbf{J})dv \tag{11.7}$$

Where **B** is the magnetic flux, **J** is the virtual current density vector, and **v** is the liquid velocity vector.

A lot of research into electromagnetic flow meters has been conducted for more than 180 years. These meters are good competitors of the ultrasonic flow meters in the measurement of flow rate in open and closed channels. Recent developments have included an aspect of low-energy electromagnetic flow meters.

The relationship between the frequency of vortex behind a bluff body and the average velocity of fluid flow forms the operational basis of vortex flow meters (Hans & Windorfer, 2003). Pressure sensors are fitted on the walls of the channel or bluff body to detect separating vortices. A bluff body (obstruction) is placed in a flowing stream to produce vortices downstream of the obstruction in this type of flow meters. These flow meters are very flexible for measuring flow rates of gases, steam, and liquids. In addition, they are reliable because of the absence of moving parts and exhibit a relatively high level of accuracy. Nevertheless, the presence of an obstruction in the flow stream may increase the pressure drop.

Orifice flow meters use pressure drop to measure the flow rate of the fluid. As the fluid flows through the constriction, its velocity rises, thereby creating a pressure differential across the orifice. This pressure differential is related to the flow rate of the fluid. Measurement of pressure upstream and downstream yields the required data for computation of the flow rate. The orifice may have one or more holes. Orifice flow meters with multiple holes have a higher coefficient of discharge (Singh & Tharakan, 2015) with lower associated pressure loss. The commonest instrument for flow measurements is the orifice meter because of its compact size especially for larger flow rates, absence of any moving part, and low cost.

Example 11.2

An ultrasonic flow meter is used to measure the flowrate of a gas. The average flow velocity along the acoustic path of ultrasound is 20 ms^{-1} while the speed of ultrasound is 1540 ms^{-1} and the pathlength between the transducers is 50 cm. If the angle of inclination of the centre line of the transducers to the axis of the channel is 60°, calculate the time difference between the upstream and downstream flows.

Solution

From Eqs. (11.5) and (11.6),

$$\Delta t = t_{up} - t_{do} = L\left(\frac{1}{V_u - V_{ag}\cos\theta} - \frac{1}{V_u + V_{ag}\cos\theta}\right)$$

$$= 0.5\left(\frac{1}{1540 - 20 * \cos 60^\circ} - \frac{1}{1540 + 20 * \cos 60^\circ}\right)$$

$$= 4.2 \times 10^{-6}\text{s}$$

11.2.3 Pressure

Pressure measurements are carried out to determine pressure changes across various components of the CSP system, including solar receivers. Data on pressure levels are also useful for computation of thermophysical properties and for checking sealed tanks or for measuring volumetric flow rate with orifices and nozzles (Janotte et al., 2017). Some thermophysical properties (such as the density, specific heat capacity, and thermal conductivity of the heat transfer or working fluid) are sensitive to pressure.

Different approaches to pressure measurement are available. However, it is important to select a suitable method depending on the requisites of measurement pertaining to the pressure itself, operating temperature, type, and state of the heat transfer fluid (or working fluid), and associated uncertainty. A fluid at an elevated temperature may necessitate the thermal separation of the hot fluid and pressure sensor. For differential pressure measurements, short and straight lines of pressure are preferred. It is important to ascertain the vertical alignment to avert accumulation of gas in systems that operate with liquid heat transfer fluid.

11.2.4 Thermophysical properties of heat-transfer media

Solar receivers use different fluids (such as air, molten salts, water, nanofluids, suspended solid particles, and thermal oil) as heat transfer media. Properties of these fluids are required for various thermodynamic computations. Data on air and water properties is readily available for a wide range of applications because the two fluids have been studied and exploited for a long time. However, there is limited experience with the thermophysical properties thermal oils, molten salts, and advanced heat transfer fluids (such as nanofluids and suspended solid particles). Advanced heat transfer media are still being developed with the aim of improving the thermal efficiency of receivers and the overall performance of CSP systems.

The use of a fluid in a closed flow circuit over a prolonged term results in the degradation of the fluid. This drawback is observed in molten salts, thermal oils, and other fluids, and it is mainly enhanced by high operational temperatures. The system and its operational modes (such as transient mode of operation) may also influence the thermophysical properties of the fluids. Consequently, it is necessary to check the quality of heat transfer fluids on a regular basis to ensure that test results are accurate. Some of the important thermophysical properties that require monitoring are density, viscosity, specific heat capacity, and thermal conductivity.

11.2.5 Quality of steam

In some CSP designs, steam is directly generated by receivers (see Chapter 5). The quality of steam is measured by the mass fraction of steam in a saturated steam/water mixture. The liquid part wets the internal surface of the tubes through which steam flows and facilitates desirable heat transfer across the tube. In turn, this helps to regulate the temperature of the tubes within acceptable limits. Otherwise, overheating of tubes can cause receiver and overall system failure. An extremely high quality of steam can lead to accumulation of solids in the flow tubes, thereby reducing the rate of heat transfer and augmenting the temperature of the tubes (Xie et al., 2013). In contrast, a low fraction of steam reduces the efficiency the system. The steam fraction at the vaporizer outlet usually falls in the range 0.6–0.8 in a CSP system based on the Rankine cycle operating in the recirculation mode (Aurousseau et al., 2016). A high quality of steam is preferred to reduce the recirculation pumping power, but the steam quality should be adequately low to provide a safety margin from the tube dry-out.

11.3 Selected standard methods

Testing of solar collection systems involves more than just measurement of solar radiation, and therefore, other details are provided by relevant standards. Many standards for testing flat-plate and concentrating solar collectors exist at national and international levels. Some national standards are adapted from international standards. In this section, selected common standard methods are briefly described.

11.3.1 ISO 9806: 2017 solar energy–solar thermal collectors–test methods

The ISO 9806 standard is comprehensive and defines procedures for testing fluid heating solar collectors for performance, reliability, durability, and safety (ISO, 2017). This standard contains methods for performing tests outdoors under natural solar irradiance and natural and simulated wind and for conducting tests indoors under simulated solar irradiance and wind. Outdoor tests can be conducted as steady-state or as all-day measurements (under varying weather conditions). In addition, the standard covers a range of applications that include flat-plate and tracking-concentrating collectors for thermal power generation and process heat. These methods are applicable to laboratory and in-situ testing.

Processing of test data is performed according to stipulated procedures. One of the critical parameters for evaluation of collector performance is the thermal efficiency. It is computed according to the specific standard. Based on the inlet fluid temperature, the thermal efficiency of the collector equation is defined by Eq. (10.30).

In some concepts of thermal energy storage, a solar receiver is integrated with some storage to ensure a uniform heat output during short-term discontinuities of solar radiation because of the thermal buffering effect (Senthil & Cheralathan, 2019). However, this standard is not applicable to solar collectors with a thermal storage unit that is an integral part to such an extent that the collection process cannot be distinguished from the storage process for purposes of making the collector thermal performance measurements.

11.3.2 ASTM E905-87(2021) standard test method for determining thermal performance of tracking concentrating solar collectors

This standard test method includes the determination of thermal performance of tracking concentrating solar collectors that heat fluids for use in thermal systems (ASTM, 2021). It is applicable to one- or two-axis tracking reflecting concentrating collectors in which the heat transfer fluid flows into the collector through a single inlet and exits the collector through a single outlet, and to those collectors where a single inlet and outlet can be effectively provided, such as into parallel inlets and outlets of multiple collector modules. The method is intended for those collectors in which the effect of diffuse irradiance on performance is insignificant and whose performance can be characterized in terms of direct irradiance. The collector may be tested either as a thermal collection subsystem where the effects of tracking errors have been essentially eliminated from the thermal performance or as a system with the manufacturer-supplied tracking mechanism. In addition, only collectors with a minimum geometric concentration ratio of seven can be tested by using this method, and tests can only be conducted outdoor, under clear sky, quasi-steady state conditions.

This test method is not planned for and may not be applicable to phase-change or thermosyphon collectors, to any collector under operating conditions that involve phase-change, to fixed mirror-tracking receiver collectors, or to central receivers. It should be noted that phase change takes place in some receivers. For instance, direct steam generation involves water and steam flows within the receiver (see Chapter 8). So, such receivers cannot be tested using this standard.

11.3.3 ANSI/ASHRAE standard 93–2003 methods of testing to determine the thermal performance of solar collectors

This standard provides test methods for determining the thermal performance of solar collectors that use single-phase fluids and have insignificant internal energy storage (ASHRAE, 2003). It is applicable to concentrating and non-concentrating solar collectors in which a fluid enters the collector through a single inlet and exits the collector through a single outlet. Tests can be conducted outdoors under natural solar irradiance and indoors under simulated solar irradiance. The standard allows the determination of steady-state and quasi-steady-state thermal performances of collectors.

Nevertheless, this standard is not applicable to collectors: (a) with integrated thermal storage units, (b) that are not glazed, and (c) in which the heat transfer fluid changes phase. Unglazed flat-plate collectors can be tested in accordance with ASHRAE Standard 96–1980 (RA 1989), while collectors in which the heat transfer fluid undergoes phase change can be tested in accordance with ASHRAE Standard 109–1986 (RA 2003).

11.4 Progress in the development of solar receivers

Solar receivers must operate at high temperatures to achieve high performance of a CSP system. In theory, it has been shown that point focus receivers can attain temperatures of up to 1773 K (Avila-Marin, 2011). Nevertheless, worldwide, most of the CSP plants are based on the conventional Rankine cycle, with efficiencies of up to 42% (Petrakopoulou et al., 2017). It has been theoretically shown that a wet-cooled steam cycle can achieve an efficiency of 50% at operational temperature and pressure of 1023 K and 30 MPa, respectively (Dunham & Iverson, 2014). Moreover, point focus receivers (solar towers and parabolic dish concentrators) can supply heat at temperatures above 1023 K. For instance, the parabolic dish concentrator can achieve temperatures of up to 1773 K (Zhang et al., 2013). However, material considerations limit the maximum temperature of the steam engine to 900 K based on the current level of state of the art. The flux limit depends on the properties of the heat transfer fluid utilized, the fluid flow rate, and the thermal properties of the receiver materials. High heat flux on walls of fluid conduits induces high temperature gradients through the tube walls, resulting in elevated thermal stresses and strains (see Chapter 9). Consequently, research efforts are directed toward the development of high-temperature receivers that can drive advanced thermodynamic cycles to achieve high CSP efficiency.

11.4.1 Selected project developments

Based on geometric design, conventional high-temperature receivers for solar collection in three dimensions are classified into: (a) external receivers, (b) cavity receivers, and (c) volumetric receivers (Dutta, 2017). These receivers are used for solar collection in three dimensions, and they utilize liquids or gases as heat transfer fluids. Some of the recent major innovations in heat transfer media are the supercritical cycles, liquid metals, nanofluids, and suspended solid particles. Details of conventional and advanced heat transfer media are reported in Chapter 3.

11.4.1.1 External receivers

In 1982, the Department of Energy in United States of America (USA), along with an industry consortium began operating Solar One, a 10 MW demonstration project located at Daggett, California. An external receiver made of tubular panels was utilized to directly generate steam (with a design outlet temperature of 789 K) for a Rankine cycle. The receiver was mounted on a tower of height 40 m. The peak incident flux varied with panel number, and it was in the range of 100–300 kW m^{-2}. Receiver tube leaks were observed and found to be the main cause of the plant's unscheduled outages (Radosevich, 1988). The leaks were associated with cracks at two distinct locations on the panels. This pilot plant provided valuable lessons for consideration during the design of future solar tower plants.

After the experience from Solar One, the US Department of Energy and with the industry consortium, commenced operating Solar Two plant, an upgrade of the Solar One project. The Solar Two plant used molten salt as a heat transfer fluid and had a capacity of 10 MW at the same location. It was commissioned in 1995 and operated until 1999. This project demonstrated the ability of the solar tower technology to capture and store thermal energy efficiently and to generate electricity even when there is insufficient or no solar radiation. The project also fostered commercial interest in solar towers.

Another important milestone in the development of external receivers is the 20 MW Gemasolar project. This was the first commercial molten-salt solar tower plant commissioned in 2011 in southern Spain. The Gemasolar CSP plant utilizes an external receiver on top of a 140 m tower to heat molten salt from 563 to 838 K and to indirectly heat water and produce steam for a Rankine cycle.

Recently, in 2015, the Crescent Dunes Solar Energy Project in Tonopah, Nevada (USA) was commissioned. The Crescent Dunes plant (with a nominal capacity of 110 MW) also employs a molten-salt external receiver to indirectly generate steam for a Rankine cycle. The receiver is cylindrical with a height of 30 m, and it is mounted on top of a 195 m tower. It is irradiated by 10,347 heliostats, each with an area of 116 m^2.

11.4.1.2 Cavity receivers

Planta Solar 10 (PS 10) was the first commercial solar tower CSP plant producing grid-connected electricity to became operational in southern Spain in 2007. Later, in 2009, the Planta Solar 20 (PS 20) plant was commissioned to produce electricity. The PS 10 and PS 20 power plants use direct-steam cavity receivers to produce saturated steam at 523−573 K and 4.5 MPa for a power cycle employing wet cooling with net turbine capacities of 11MW_e and 20 MW_e, respectively. The cavity receivers are fitted on towers of heights 115 and 165 m for the PS 10 and PS 20 power plants, respectively. They accommodate tubular panels that heat the water/steam.

Application of superalloys in solar receivers has been demonstrated in some projects, one of which is the SOLUGAS project. The aim of this project was to prove the performance of an operational solar hybrid gas turbine system at a precommercial scale (Korzynietz et al., 2016). So, a full solar hybrid gas turbine system, with heliostat field and tower, was built near Seville in Spain. A cavity receiver made of tubular panels was designed for compressed air at a gauge pressure of 0.9 MPa and an outlet temperature of 1073 K. Both the absorber tubes and the collector were made of a nickel-based superalloy (Inconel 617), while the distributor was made of stainless steel (EN 1.4571). Consequently, these elements of the receiver could withstand the anticipated operational pressures and temperatures. Although solar gas turbines were tested in previous work, SOLUGAS is the first demonstration project at megawatt scale where compressed air can be preheated by a 3.2 MW_{th} receiver up to 1073 K and returned to the combustion chamber of a 4.6 MW Mercury 50 gas turbine.

11.4.1.3 Volumetric receivers

The earliest designs of volumetric receivers used wire mesh and metallic foams as absorber elements. In these designs, metal foam or the wire mesh was the absorber element that could be packed in modular cups. These designs were open to the environment, and ambient air would be used as the heat transfer fluid, and they appeared as billboards. Some of such designs were MK-1 that consisted of a mesh of thin wires of stainless steel (Fricker, 1986). This design was able to attain temperatures of up to 1123 K without any damage. Sultzer 2 had a similar design of wire mesh absorber, but it also included a metal sheet with perforations behind the absorber so that proper mass flow was provided to the absorber depending on the flux intensity (Becker & Sanchez, 1989). Nevertheless, the design developed hot spots and some distortions because of inadequate support. A proposal was made to build a full-scale 30 MW solar power plant in Jordan by the PHOEBUS Consortium (Grasse, 1991) but the plant could not be constructed because of some nontechnical factors. Even though attempts for commercializing these designs were underway, they suffered from the following limitations (Dutta, 2017): (a) inability to withstand high radiation fluxes, with an average sustainable flux of about 200 kWm^{-2}, (b) limited gas temperatures, and (c) vulnerability to hot spots. Open designs of volumetric receivers with ceramic foams such as the selective receiver have also been developed. In this receiver design, siliconized silicon carbide (SiSiC) with 40% porosity was utilized. The SiSiC was formed into square ducts of few millimeter width. The aperture of this design would be covered by quartz glass to reduce heat losses from the receiver. It was found that the design could sustain flux of up to 1 MW m^{-2} and attain temperatures of up to 1273 K.

Research attention has also been given to concepts of pressurized volumetric receivers. This added feature can enable the system to sustain much higher levels of flux and to directly heat the working fluid of advanced thermodynamic cycles (such as the Brayton cycle and supercritical carbon dioxide cycle), resulting in a higher overall efficiency of the CSP system. An example of this design is the pressure-loaded volumetric ceramic receiver (PLVCR) (Pritzkow, 1991). This receiver was designed to sustain a flux of 2 MW m^{-2} and to work at a temperature of 1273 K and a pressure of up to 1.0 MPa. The receiver efficiency was 71% for an outlet temperature of 1323 K. Another notable design of the pressured volumetric receivers is the directly irradiated annular pressurized receiver (DIAPR) (Karni et al., 1997). The receiver comprises many small fins termed as "porcupine" that are the absorbers of the concentrated solar radiation. The design has a window made of quartz glass that separates the absorber from the ambient environment. This design was able to sustain solar flux of up to 10 MWm^{-2} and achieve outlet temperature of about 1573 K at a pressure of 1−3 MPa and efficiency of 79%. Based on these studies, the design was taken up for upscaling, and a demonstration power plant was commissioned in 2009 in Israel. This is the world's first operational solar gas turbine power plant built at Kibbutz Samar in southern Israel by AORA Solar. The power plant utilizes 30 heliostat mirrors to concentrate solar energy onto the receiver and generates 100 kW_e and 170 kW_{th}, and it is capable of powering 70 households through the Israel's national grid system. The receiver heats up compressed air to about 1273 K before injecting it into a 100 kW gas turbine. AORA's second demonstration plant is situated in Almeria, Spain.

One of the CSP projects in which metallic and ceramic materials were used is the SOLGATE project. The goal of the project was to develop a solar receiver cluster capable of providing pressurized air at 1273 K. A pressurized solar receiver system was made up of three 400 kW_{th} modules each with a secondary concentrator. The modules were connected in series and formed a hexagonal aperture. The first module in the series was a low-temperature (LT) metal tubular module that was used in the outer low flux region for outlet temperatures of about 873 K. In the middle position was a pressurized medium-temperature (MT) volumetric receiver equipped with a metal wire mesh absorber. This module was utilized in the high flux region for a hot gas exit temperature of about 1073 K. The high-temperature (HT) module was a volumetric receiver with a porous ceramic absorber made of silicon carbide. The HT module was designed for an exit temperature of 1273 K, hence the need for using a ceramic porous material. Testing of the system was divided into two stages. The objective of the first stage was to demonstrate the ability of the gas turbine as well as the receiver to achieve design operating conditions of 1073 K with volumetric receivers equipped with metal wire mesh for both MT and HT. In the second stage, the receiver outlet air temperature was raised to 1273 K by replacing the metal wire mesh absorber in the HT receiver with a ceramic absorber. It should be noted that chipping and cracking of the quartz glass can occur because of structural changes in the recrystallization phase during temperature changes, thereby damaging the receiver window. Consequently, overheating the window above 1073 K must be avoided. To prevent the possible danger of recrystallization of the quartz glass, the outer window of this receiver module was cooled by using an active method of cooling. The SOLGATE plant commenced operating in 2001 to demonstrate the performance of a solar-hybrid gas turbine system driven by a solar tower (Avila-Marin, 2011; SOLGATE, 2005). During the first stage of testing, an outlet temperature of 1073 K was achieved, and the system delivered 230 kW_e to the grid without major problems. In the second stage, the temperature rose to 1233 K with about 770 W m^{-2} of direct normal irradiation and 70% efficiency.

11.4.1.4 Advanced liquid-based solar receivers

Nanofluids and liquid metals can be used as heat transfer fluids in HT applications. Many studies have examined the potential of using nanofluids as heat transfer media in solar receivers (see Chapter 5). At present, this medium is under active research. Some liquid metals are also promising candidates of heat transfer fluids.

Exploitation of liquid sodium has been demonstrated. In 2017, the Jemalong Solar Thermal Station (1.1 MW) was commissioned as a pilot project at Jemalong in Australia. This plant uses a 2 m^2 flat tubular receiver panel mounted on a tower, and liquid sodium is used as a primary heat transfer fluid in a steam Rankine cycle—based power plant. The solar field aperture has an area of 15,000 m^2 and tower height of 30 m. The liquid sodium enters and exits the receiver at temperatures of 543 and 833 K, respectively. In addition, the same fluid functions as a storage medium with a storage capacity of 3 h.

11.4.1.5 Solid-based solar receivers

A rare CSP plant with eight solar storage receivers made up of graphite blocks was commissioned in 2011 at Lake Cargelligo (New South Wales) in Australia. Each graphite block was mounted on a tower with a face-down aperture that is irradiated by surrounding heliostats. The block heats steam from 473 to 773 K that drives a 3 MWe steam-Rankine cycle (Ho, 2017).

Particle-based receivers are also under active research, but there are not yet commercial systems. The Solar Thermo Electric Magaldi (STEM) demonstration project using a modular 2 MW_t fluidized-bed particle receiver was commissioned in 2016 at Sicily (Ho, 2017). In this system, a ground-based fluidized-bed receiver is irradiated with concentrated solar radiation by using beam-down optics from a field of heliostats. The receiver can heat steam up to 793 K with a storage capacity of 6 h.

11.4.2 Challenges and opportunities in the development of solar receivers

Many new solar receiver designs and concepts of heat transfer have been developed over the past few decades. High receiver outlet temperatures are required to drive advanced thermodynamic cycles and to achieve high overall performance of CSP systems. Conventional receivers are globally exploited in commercial CSP plants. However, these plants operate at temperatures below 900 K, and the outlet temperatures of conventional receivers are also below this limit. Advanced concepts of solar receivers have higher temperatures, but they have not yet attained commercial stage due to some challenges encountered in the developmental process. These challenges can be classified into three categories: (a) maturity of technology, (b) financial and policy constraints, and (c) technical factors. Major technical factors include solar radiation, materials and supply chain, and receiver and expertise.

11.4.2.1 Technological maturity

Every technology is developed through stages from conception of the idea to commercial stages. In this regard, technological maturity is an important indicator of the closeness of a technology to commercial application (Selman & Chen, 2012). Research and development managers use this factor in decision-making about the advancement of a technology. Thus, research is a critical tool for upgrading technological maturity. It can contribute to the reduction of the energy cost of a generation technology, thereby augmenting its rate of diffusion into the market.

There are many studies on the development of solar receivers for power production. Line-focus tubular receivers are the most matured receivers. They are exploited on the parabolic trough concentrator (PTC) and linear Fresnel reflector (LFR) and use thermal oil, molten salt, water/steam, air, and organic liquids and liquid metals. At present, most of the CSP power plants are based on the parabolic trough concentrator, with line-focus tubular receivers (Fig. 11.2). Thermal oil is the most widely exploited heat transfer medium. However, it can only be used at temperatures of up to 673 K and has low thermal conductivity (Wang et al., 2013). On the other hand, advanced receivers (such as those based on supercritical fluids, nanofluids, liquid metals, and solid particles) are not yet commercially available. This indicates that they have a low level of maturity, and therefore, they require more research and demonstration to accelerate their rate of diffusion.

FIGURE 11.2 Parabolic trough concentrator with a tubular receiver for thermal desalination developed at Chancellor College, University of Malawi.

11.4.2.2 Financial and policy instruments

Project development requires financial resources for various expenses. Investment is a common pathway for funding the activities of a project, and investors aim at making some profit. Consequently, care is taken when choosing opportunities for investment. Uncertainty and risk factors are considered during decision-making about investing in a given project. In this vein, the existence of many similar previous successful projects helps in reducing the magnitude of uncertainty and risk. It is therefore difficult for emerging receivers to displace conventional receivers that have proven technical and financial performance.

Many receivers for high-temperature application are just emerging on the market, and so they need suitable financial and policy instruments to support their transition to commercial stage. Lack of financial motivation through the provision of grant facilities could be a barrier to the commercialization of receivers. Thus, suitable policy instruments are required to promote the development of innovative receivers. As shown Section 11.4.1, some testing and demonstration facilities have been established through collaboration between governments and industry. These demonstration facilities enable acquisition of experience, thereby reducing uncertainty and risk.

11.4.2.3 Solar and land resources

Solar radiation is the most important environmental factor that affects the performance of solar receivers. Characteristics and availability of solar radiation have been presented and discussed in Chapter 2. Solar receivers exploit only direct normal irradiance (DNI) to generate heat of high grade. The optical efficiency of receivers is high in locations with elevated levels of DNI (see Chapter 10). In this context, DNI data are required in experimental and modeling studies on the performance of receivers. Nevertheless, only a limited number of sites measure direct radiation on the ground. Satellite solar radiation data can be used, but they are not as accurate as the data obtained from ground-based sensors (Ali, 1990). In view of this observation, the challenge of accurate solar resource data may be a barrier to the development of solar receivers.

The aperture area of a concentrator is larger than that of a receiver. Thus, the solar field generally requires adequate land surface. For example, heliostat aperture area has been found to vary linearly with equivalent power capacity of the existing solar tower plants (Srilakshmi et al., 2015). This implies the number of heliostats and the size of required land increase with the capacity of the plant. Solar towers (8−12 m^2/MWh/year) and parabolic dish concentrators (8−12 m^2/MWh/year) have higher demand for land use than parabolic trough concentrators (6−8 m^2/MWh/year) and linear Fresnel reflectors (4−6 m^2/MWh/year) (Islam et al., 2018). Wasteland can be used for the development of solar fields, but this requires data on the condition of the terrain to estimate the potential available land where CSP technology can be installed. This kind of data are scarce in many countries, which may pose a challenge to the development of solar receivers and widespread deployment the CSP technology.

11.4.2.4 Materials and supply chain

Solar receivers operate at elevated temperatures. Consequently, structural and heat-transfer materials must withstand such temperatures and meet most of the other required characteristics to achieve high receiver performance. Important characteristics of structural materials and heat transfer media are: (a) operational temperature range and thermal stability, (b) thermophysical properties, (c) flow pressure, (d) operational aspects, and (e) availability and affordability. Details of these characteristics are reported in Chapters 5 and 9. Unfortunately, no single naturally occurring material possesses all the required characteristics, which poses a challenge of finding a feasible material. Each receiver design has some advantages and challenges (Table 11.1). Similarly, there are some challenges of using some materials as heat transfer media in different receiver designs. For example, air is readily available, nontoxic, and cheap, but it has low thermal conductivity. A summary of challenges associated with heat transfer media is presented in Tables 11.2–11.4 (Ho, 2016, 2017).

TABLE 11.1 Advantages and challenges of selected materials for solar absorber elements.

Structural material	Melting point (K)	Advantages	Challenges
Copper (cu)	1358	High thermal conductivity, corrosion-resistant, ductile	Low yield strength, low resistance to heat
Stainless steel (Grade 316)	1673	High resistance to rust, resistant to heat, high yield strength, cheaper than superalloys	Lower yield strength than superalloys
Inconel 617 (superalloy)	1605–50	High yield strength, resistant to heat, resistant to corrosion	More costly than stainless steel
Inconel 625 (superalloy)	1563–1623	High yield strength, resistant to heat, resistant to corrosion	More costly than stainless steel
Silicon carbide(SiC)	>3000	High strength, high thermal conductivity, low thermal expansion, high solar absorptance, and chemical inertness at elevated temperatures	Brittleness can cause failure from mechanical shock

TABLE 11.2 Receiver designs based on gas heat transfer media.

Receiver design	Outlet temperature (K)	Advantages	Challenges
Volumetric air receivers	>1073	Use of ambient air is cheap, potential for volumetric heating to reduce heat losses near aperture	Flow instabilities, high temperatures and radiative heat losses near the aperture due to limited light penetration, direct thermal storage is not possible
Tubular gas receivers	973–1073	Ability to utilize features and geometry (such as cavity) to augment solar radiation trapping	Low efficiency of heat transfer from tube walls to gas, fatigue in tube walls, and high cost of materials to withstand high-pressure, high-temperature flow, direct thermal storage is not possible
Microchannel gas receivers	973	Higher surface area and heat transfer to gas, ability to raise microchannel modules for higher capacities	Potential of elevating pressure drop, potential manufacturing complexity and costs, direct thermal storage is not possible

Molten salt is one of the most widely exploited heat transfer fluids, and it can also be used as a storage medium (see Chapter 5). Unfortunately, there is a limited number of suppliers of this product worldwide: SQM (Chile), Haifa Chemicals (Israel), and Durferrit Salts and Auxiliary Products (Germany) (Srilakshmi et al., 2015). The small number of suppliers can contribute to raising the cost of the product, and consequently, the cost of energy production. Research is underway to produce advanced molten salts with wider operational temperatures.

TABLE 11.3 Receiver designs based on liquid heat transfer media.

Receiver design	Outlet temperature (K)	Advantages	Challenges
Liquid sodium receiver	Up to 1073	High thermal conductivity	Highly flammable in air, reacts violently with water
Lead-bismuth receiver	1273	Has very high boiling point, higher heat capacity than liquid sodium	Has lower thermal conductivity than liquid sodium, very corrosive, costs 10–20 times more than solar salt
Halide salt receiver	Up to 1073	Can achieve high temperatures, relatively low cost	Highly corrosive when exposed air or water, requires regulated purification and premelting in a vacuum
Carbonate salt receiver	Up to 1073	Can attain high temperatures, does not require regulated purification and premelting procedures	High cost

Another critical element of technological advancement is the supply chain. A product is produced at a point of origin and distributed to points of consumption (Wang et al., 2015). Demand is found at points of consumption. Sufficient demand is required to sustain a market. In case of solar receivers, the supply chain is not well-established, and there is lack of local demand for these components to economically justify their production within a country. At present, CSP plants are operational in a limited number of countries, and the capacity share of this technology is still very low in many countries. However, the outlook for CSP technology is bright because the levelized cost energy is declining, and the CSP share of the total capacity is increasing with time. Thus, there are potential opportunities for investment in the production and distribution of receiver materials, manufacturing and distribution of receivers, and development and operation of CSP power plants. These economic activities also entail creation of job opportunities, thereby reducing unemployment.

TABLE 11.4 Receiver designs based on suspended solid particles.

Receiver design	Outlet temperature (K)	Advantages	Challenges
Free-falling direct absorption particle receiver	>973	Sustains very large levels of solar flux, can be scaled up to large capacities and particle mass flow rates	Elevated convective heat losses from entrained airflow, loses particles through open aperture
Obstructed-flow direct absorption particle receiver	>1073	Low particle velocity that raises residence time and heating, low entrainment of air flow and convective losses, low particle loss	Potential overheating of the obstruction structures with direct irradiance, extra costs of materials, complexity of system installation
Centrifugal direct absorption particle receiver	1173	Regulation of particle residence time using the rotational speed of receiver	Large rotating receiver
Enclosed falling particle receiver (gravity-driven flow)	>973	Absence of particle loss	Low heat-transfer efficiency between irradiated walls and particles, elevated outer-wall surface temperature and potential for hot spots, limitation on solar flux
Enclosed fluidized particle receiver	1023	High particle-side heat transfer coefficient, absence of particle loss	High parasitics of fluidization, potential heat loss from fluidization gas

11.4.2.5 Solar receiver and expertise

A solar receiver is the linchpin between the solar field and power block, and so it is a critical component of the CSP technology. For high-temperature application, the solar tower and parabolic dish concentrator are suitable. There are currently some commercial power plants based on the solar tower system that are in operation with no operational power plant based on the parabolic dish concentrator around the world, which shows that the solar tower technology has attained maturity.

At present, a few companies are into the business of manufacturing receivers, and a small number of companies are specialized in the design, construction, and implementation of solar tower technologies on a global scale (Srilakshmi et al., 2015). This implies that expertise in receiver technology is not readily available in many countries, which shows that there are opportunities for skills development and employment. Constraints to the development of suitable receivers could therefore contribute to the retardation in the deployment of CSP systems of high performance.

Nomenclature

A area (m^2)
B magnetic flux density (T)
G irradiance ($W\ m^{-2}$)
J electric current density ($A\ m^{-2}$)
T temperature (K)
v velocity ($m\ s^{-1}$)
$\dot{m}$ mass flow rate ($kg\ s^{-1}$)

Greek symbols

η efficiency (dimensionless)
θ angle (°)
ρ reflectance (dimensionless)

References

Ali, B. F. (1990). Theoretical study of main factors affecting the heliostat field design of tower power plant. *Energy Conversion and Management, 30*(2), 101–106. https://doi.org/10.1016/0196-8904(90)90019-U

ASHRAE. (2003). *ANSI/ASHRAE standard 93-2003: Methods of testing to determine the thermal performance of solar collectors*. Atlanta: American Society of Heating, Refrigeration and Air-conditioning Engineers (ASHRAE).

ASTM. (2021). *ASTM E905-87(2021) standard test method for determining thermal performance of tracking concentrating solar collectors*. Conshohocken: American Society for Testing and Materials.

Aurousseau, A., Vuillerme, V., & Bezian, J. (2016). Control systems for direct steam generation in linear concentrating solar power plants - a review. *Renewable and Sustainable Energy Reviews, 56*, 611–630. https://doi.org/10.1016/j.rser.2015.11.083

Avila-Marin, A. L. (2011). Volumetric receivers in solar thermal power plants with central receiver system technology: A review. *Solar Energy, 85*(5), 891–910. https://doi.org/10.1016/j.solener.2011.02.002

Becker, M., & Sanchez, M. (1989). *Wire pack volumetric receiver tests performed at the Plataforma Solar de Almeria, Spain in 1987 and 1988, SSPS Technical Report, No. 2/89.*

Bergman, T. L., Lavine, A. S., Incropera, F. P., & DeWitt, D. P. (2011). *Fundamentals of heat and mass transfer* (7th ed.). New Jersey: John Wiley & Sons.

Cai, Y. (2013). *Measuring gas flow with MEMS sensors*. Electronic Products. http://www.electronicproducts.com.

Delta-T Devices. (2000). *User manual for DL2e data logger*. Cambridge: Delta-T Devices.

Dunham, M. T., & Iverson, B. D. (2014). High-efficiency thermodynamic power cycles for concentrated solar power systems. *Renewable and Sustainable Energy Reviews, 30*, 758–770. https://doi.org/10.1016/j.rser.2013.11.010

Dutta, P. (2017). High temperature solar receiver and thermal storage systems. *Applied Thermal Engineering, 124*, 624–632. https://doi.org/10.1016/j.applthermaleng.2017.06.028

Ebert, M., Benitez, D., Röger, M., Korzynietz, R., & Brioso, J. A. (2016). Efficiency determination of tubular solar receivers in central receiver systems. *Solar Energy, 139*, 179–189. https://doi.org/10.1016/j.solener.2016.08.047

Fricker, H. W. (1986). Tests with a small volumetric wire receiver. In *Proceedings of the third international Workshop on solar thermal central receiver systems*. Konstanz, Berlin, Heidelberg, New York: Springer-Verlag.

Grasse, W. (1991). PHOEBUS international 30 MW_e solar tower plant. *Solar Energy Materials, 24*, 82–94. https://doi.org/10.1016/0165-1633(91)90050-U

Grove, D. M., & Davis, T. P. (1992). *Engineering, quality and experimental design*. Essex: Longman Scientific & Technical.

Hans, V., & Windorfer, H. (2003). Comparison of pressure and ultrasound measurements in vortex flow meters. *Measurements, 33*, 121–133.

Henry, M., Tombs, M., Zamora, M., & Zhou, F. (2013). Coriolis mass flow metering for three-phase flow: A case study. *Flow Measurement and Instrumentation, 30*, 112–122. https://doi.org/10.1016/j.flowmeasinst.2013.01.003

Ho, C. K. (2016). A review of high-temperature particle receivers for concentrating solar power. *Applied Thermal Engineering, 109*, 958–969. https://doi.org/10.1016/j.applthermaleng.2016.04.103

Ho, C. K. (2017). Advances in central receivers for concentrating solar applications. *Solar Energy, 152*, 38–56. https://doi.org/10.1016/j.solener.2017.03.048

Islam, M. T., Huda, N., Abdullah, A. B., & Saidur, R. (2018). A comprehensive review of state-of-the-art concentrating solar power (CSP) technologies: Current status and research trends. *Renewable and Sustainable Energy Reviews, 91*, 987–1018. https://doi.org/10.1016/j.rser.2018.04.097

ISO. (2017). *ISO Standard 9806:2017 Solar energy- solar thermal collectors - test methods*. Geneva: International Organization of Standardization (ISO).

Janotte, N., Wilbert, S., Sallaberrt, F., Schroedter-Homscheidt, M., & Ramirez, L. (2017). Principles of CSP performance assessment. In P. Heller (Ed.), *The Performance of concentrated solar power (CSP) systems: modelling, measurement and assessment* (pp. 31–64). Woodhead Publishing.

Karni, J., Kribus, A., Doron, P., Rubin, R., Fiterman, A., & Sagie, D. (1997). The DIAPR: A high pressure high-temperature solar receiver. *Transactions of ASME Journal of Solar Energy Engineering, 119*, 74–78. https://doi.org/10.1115/1.2871853

Korzynietz, R., Brioso, J. A., del Río, A., Quero, M., Gallas, M., Uhlig, R., Ebert, M., Buck, R., & Teraji, D. (2016). Solugas - comprehensive analysis of the solar hybrid Brayton plant. *Solar Energy, 135*, 578–589. https://doi.org/10.1016/j.solener.2016.06.020

LaNasa, P. J., & Upp, E. L. (2014). Introduction. *Fluid Flow Measurement: A Practical Guide to Accurate Flow Measurement* (pp. 1–17). Elsevier Inc.

López-Martín, R., & Valenzuela, L. (2018). Optical efficiency measurement of solar receiver tubes: A testbed and case studies. *Case Studies in Thermal Engineering, 12*, 414–422. https://doi.org/10.1016/j.csite.2018.06.002

Madhlopa, A. (2018). *Principles of solar gas turbines for electricity generation*. Springer International Publishing AG. https://doi.org/10.1007/978-3-319-68388-1_7

Nicholas, J., & White, D. (1994). *Traceable temperatures: An introduction to temperature measurement and calibration*. New York: John Wiley & Sons.

Petrakopoulou, F., Sánchez-Delgado, S., Marugán-Cruz, C., & Santana, D. (2017). Improving the efficiency of gas turbine systems with volumetric solar receivers. *Energy Conversion and Management, 149*, 579–592. https://doi.org/10.1016/j.enconman.2017.07.058

Pritzkow, W. E. C. (1991). Pressure loaded volumetric ceramic receiver. *Solar Energy Materials, 24*, 498–507. https://doi.org/10.1016/0165-1633(91)90086-Z

Radosevich, L. G. (1988). *Final report on the power production phase of the 10 MWe solar thermal central receiver pilot plant, SAND87-8022*. Albuquerque, NM: Sandia National Laboratories.

Selman, J. R., & Chen, C. (2012). Scientific and technical maturity of molten carbonate technology. *International Journal of Hydrogen Energy, 37*, 19280–19288. https://doi.org/10.1016/j.ijhydene.2012.06.016

Senthil, R., & Cheralathan, M. (2019). Enhancement of the thermal energy storage capacity of a parabolic dish concentrated solar receiver using phase change materials. *Journal of Energy Storage, 25*, 100841. https://doi.org/10.1016/j.est.2019.100841

Singh, V., & Tharakan, T. (2015). Numerical simulations for multi-hole orifice flow meter. *Flow Measurement and Instrumentation, 45*, 375–383. https://doi.org/10.1016/j.flowmeasinst.2015.08.004

SOLGATE. (2005). *Solar hybrid gas turbine electric power system: Final Publishable Report*. Luxembourg: European Communities.

Srilakshmi, G., Venkatesh, V., Thirumalai, N., & Suresh, N. (2015). Challenges and opportunities for solar tower technology in India. *Renewable and Sustainable Energy Reviews, 45*, 698–709. https://doi.org/10.1016/j.rser.2015.02.016

Wang, Y., Wallace, S., Shen, B., & Choi, T. (2015). Service supply chain management: A review of operational models. *European Journal of Operational Research, 247*, 685–698. https://doi.org/10.1016/j.ejor.2015.05.053

Wang, C., Yang, J., & Ding, Y. (2013). Phase transfer based synthesis and thermophysical properties of Au/Therminol VP-1 nanofluids. *Progress in Natural Science: Materials International, 23*(3), 338–342. https://doi.org/10.1016/j.pnsc.2013.04.006

Watral, Z., Jakubowski, J., & Michalski, A. (2015). Electromagnetic flow meters for open channels: Current state and development prospects. *Flow Measurement and Instruments, 42*, 16–25. https://doi.org/10.1016/j.flowmeasinst.2015.01.003

Xie, L., Zhao, Y., Aziz, D., Jin, X., Geng, L., Goberdhansingh, E., Qi, F., & Huang, B. (2013). Soft sensors for online steam quality measurements of OTSGs. *Journal of Process Control, 23*, 990–1000. https://doi.org/10.1016/j.jprocont.2013.05.006

Xu, Y., Yu, P., Zhu, Z., Yuan, C, & Zhang, T. (2017). Over-reading modeling of the ultrasonic flow meter in wet gas measurement. *Measurement, 98*, 17–24. https://doi.org/10.1016/j.measurement.2016.11.007

Zhang, H. L., Baeyens, J., Degreve, J., & Cacéres, G. (2013). Concentrated solar power plants: Review and design methodology. *Renewable and Sustainable Energy Reviews, 22*, 466–481. https://doi.org/10.1016/j.rser.2013.01.032

Zhou, H., Ji, T., Wang, R., Ge, X., Tang, X., & Tang, S. (2018). Multipath ultrasonic gas flowmeter based on multiple reference waves. *Ultrasonics, 82*, 145–152. https://doi.org/10.1016/j.ultras.2017.07.010

Appendix A

Units of measurement

Basic units

Length	metre (m)
Mass	kilogram (kg)
Temperature	Kelvin (K)
Time	second (s)

Conversion of units

Length

1 ft = 0.3048 m
1 in = 25.4 mm
1 mile = 1.609 km

Area

1 ft^2 = 0.09290304 m^2
1 $mile^2$ = 2.590 km^2
1 hectare = 10^4 m^2

Volume

1 litre L = 10^{-3} m^3
1 ft^3 = 28.32 L
1 U.K. gal = 4.546 L
1 U.S. gal = 3.785 L

Mass

1 lb = 0.45359237 kg
1 oz = 28.35 g

Volume flow rate

1 cubic foot per minute (cfm) = 0.4719 L s^{-1}
1 lb h^{-1} = 0.0001256 kg s^{-1}

Conversion of units

Pressure

1 atmosphere atm = 101.325 kPa
1 bar = 10^5 Pa
1 mm H_2O = 9.80665 Pa
1 mm Hg = 133.3 Pa
1 Pascal Pa = 1 N m^{-2}
1 psi = 6.894 kPa

Energy

1 Joule J = 1 Nm
1 kWh = 3.6 MJ
1 Btu = 1.055 kJ
1 kcal = 4.1868 kJ

Power

1 Btu/h = 0.2931 W
1 hp = 0.7457 kW
1 W ft^{-2} = 10.76 W m^{-2}
1 Btu ft^{-2} h^{-1} = 3.155 W m^{-2}

Viscosity

1 cP (centipoise) = 10^{-3} Pa s
1 lbf h ft^{-2} = 0.1724 MPa s

Speed

1 ft min^{-1} = 0.00508 m s^{-1}
1 mile h^{-1} = 0.4770 m s^{-1}

Force

1 Newton (N) = 1 kg m s^{-2}

Force
1 Newton (N) = 1 kg m s^{-2}

Important prefixes

Pico	P	10^{-12}
Nano	N	10^{-9}
Micro	μ	10^{-6}
Milli	M	10^{-3}
Kilo	K	10^{3}
Mega	M	10^{6}
Giga	G	10^{9}
Terra	T	10^{12}

Temperature conversion
Fahrenheit (F) to Celsius (C):
$C = \frac{5}{9}(F - 32)$
Celsius to Kelvin (K):
$K = C + 273.15$

Density
1 g $cm^{-3} = 10^{3}$ kg m^{-3}
1 lb $ft^{-3} = 16.02$ kg m^{-3}

Appendix B

Selected constants

Avogadro constant, $N_A = 6.02252 * 10^{23}$ molecules per gmole

Planck constant, $h = 6.62607 * 10^{-34}\ m^2\ kg\ s^{-1}$

Boltzmann constant = $1.38054 * 10^{-23}\ J\ K^{-1}$

Speed of light = $2.9979 * 10^8\ m\ s^{-1}$

Universal gas constant $R = 8.3143\ kJ\ kmole^{-1}\ K^{-1}$

Appendix C

Properties of selected materials

The density (ρ), specific heat capacity at constant pressure (C_p), dynamic viscosity (μ), kinematic viscosity (ν), thermal conductivity (k), thermal diffusivity (ξ), and Prandtl number (Pr) of selected materials are presented in this section (Tables C.1, C.2 and C.3).

TABLE C.1 Properties of air at a temperature (T) and atmospheric pressure.

T (K)	ρ (kg m^{-3})	C_p (J kg^{-1})	μ (Pa s)	k (W $m^{-1}K^{-1}$)	Pr
250	1.3947	1006	0.00001596	0.0223	0.720
300	1.1614	1007	0.00001846	0.0263	0.707
350	0.9950	1009	0.00002082	0.0300	0.700
400	0.8711	1014	0.00002301	0.0338	0.690
450	0.7740	1021	0.00002507	0.0373	0.686
500	0.6964	1030	0.00002701	0.0407	0.684
550	0.6329	1040	0.00002884	0.0439	0.683
600	0.5804	1051	0.00003058	0.0469	0.685
650	0.5356	1063	0.00003225	0.0497	0.690
700	0.4975	1075	0.00003388	0.0524	0.695
750	0.4643	1087	0.00003546	0.0549	0.702
800	0.4354	1099	0.00003698	0.0573	0.709
850	0.4097	1110	0.00003843	0.0596	0.716
900	0.3868	1121	0.00003981	0.0620	0.720
950	0.3666	1131	0.00004113	0.0643	0.723

Continued

TABLE C.1 Properties of air at a temperature (T) and atmospheric pressure.—cont'd

T (K)	ρ (kg m^{-3})	C_p (J kg^{-1})	μ (Pa s)	k (W $m^{-1}K^{-1}$)	Pr
1000	0.3482	1141	0.00004244	0.0667	0.726
1100	0.3166	1159	0.00004490	0.0715	0.728
1200	0.2902	1175	0.00004730	0.0763	0.728
1300	0.2679	1189	0.00004960	0.0820	0.719
1400	0.2488	1207	0.00005300	0.0910	0.703
1500	0.2322	1230	0.00005570	0.1000	0.685
1600	0.2177	1248	0.00005840	0.1060	0.688
1700	0.2049	1267	0.00006110	0.1130	0.685
1800	0.1935	1286	0.00006370	0.1200	0.683
1900	0.1833	1307	0.00006630	0.1280	0.677
2000	0.1741	1337	0.00006890	0.1370	0.672
2100	0.1658	1372	0.00007150	0.1470	0.667
2200	0.1582	1417	0.00007400	0.1600	0.655
2300	0.1513	1478	0.00007660	0.1750	0.647
2400	0.1448	1558	0.00007920	0.1960	0.630
2500	0.1389	1665	0.00008180	0.2220	0.613
3000	0.1135	2726	0.00009550	0.4860	0.536

TABLE C.2 Properties of carbon dioxide at a temperature (T) and atmospheric pressure.

T (K)	ρ (kg m^{-3})	C_p (kJ kg^{-1})	μ (Pa s)	K (W $m^{-1}K^{-1}$)	Pr
280	1.9022	0.830	0.0000140	0.01520	0.765
300	1.773	0.851	0.0000149	0.01655	0.766
320	1.6609	0.872	0.0000156	0.01805	0.754
340	1.5618	0.891	0.0000165	0.01970	0.746
360	1.4743	0.908	0.0000173	0.02120	0.741
380	1.3961	0.926	0.0000181	0.02275	0.737
400	1.3257	0.942	0.0000190	0.02430	0.737
450	1.1782	0.981	0.0000210	0.02830	0.728

TABLE C.2 Properties of carbon dioxide at a temperature (T) and atmospheric pressure.—cont'd

T (K)	ρ (kg m^{-3})	C_p (kJ kg^{-1})	μ (Pa s)	K (W $m^{-1}K^{-1}$)	Pr
500	1.0594	1.020	0.0000231	0.03250	0.725
550	0.9625	1.050	0.0000251	0.03660	0.721
600	0.8826	1.080	0.0000270	0.04070	0.717
650	0.8143	1.100	0.0000288	0.04450	0.712
700	0.7564	1.130	0.0000305	0.04810	0.717
750	0.7057	1.150	0.0000321	0.05170	0.714
800	0.6614	1.170	0.0000337	0.05510	0.716

TABLE C.3 Properties of water vapour at a temperature (T) and atmospheric pressure.

T (K)	ρ (kg m^{-3})	C_p (kJ kg^{-1})	μ (Ns m^{-2})	k (W $m^{-1}K^{-1}$)	Pr
380	0.5863	2.060	0.00001271	0.0246	1.060
400	0.5542	2.014	0.00001344	0.0261	1.040
450	0.4902	1.980	0.00001525	0.0299	1.010
500	0.4405	1.985	0.00001704	0.0339	0.998
550	0.4005	1.997	0.00001884	0.0379	0.993
600	0.3652	2.026	0.00002067	0.0422	0.993
650	0.3380	2.056	0.00002247	0.0464	0.996
700	0.3140	2.085	0.00002426	0.0505	1.000
750	0.2931	2.119	0.00002604	0.0549	1.000
800	0.2739	2.152	0.00002786	0.0592	1.010
850	0.2579	2.186	0.00002969	0.0637	1.020

Index

Note: 'Page numbers followed by "f" indicate figures those followed by "t" indicate tables.'

D

E

R

S

T

U

V

W

Y

Z

FSC
www.fsc.org
MIX
Papier | Fördert gute Waldnutzung
FSC® C083411

Kolibri 360 GmbH
Ferdinand-Jühlke-Straße 7
99095 Erfurt, Deutschland
produktsicherheit@kolibri360.de